Aircraft Ownership

A Legal and Tax Guide

Raymond C. Speciale, Esq., C.P.A.

McGraw-Hill

New York   Chicago   San Francisco   Lisbon   London   Madrid
Mexico City   Milan   New Delhi   San Juan   Seoul
Singapore   Sydney   Toronto

The *McGraw·Hill* Companies

Cataloging-in-Publication Data is on file with the Library of Congress

Copyright © 2003 by The McGraw-Hill Companies, Inc. All rights reserved.
Printed in the United States of America. Except as permitted under the
United States Copyright Act of 1976, no part of this publication may be reproduced or distributed in any form or by any means, or stored in a data base or
retrieval system, without the prior written permission of the publisher.

34567890 BKM BKM 0987

P/N 140765-0
PART OF
ISBN 0-07-140764-2

*The sponsoring editor for this book was Larry S. Hager, the editing supervisor
was Stephen M. Smith, and the production supervisor was Pamela A. Pelton.
It was set in Century Schoolbook by Wayne Palmer of McGraw-Hill Professional's Hightstown, N.J., composition unit.*

McGraw-Hill books are available at special quantity discounts to use as premiums and sales promotions, or for use in corporate training programs. For
more information, please write to the Director of Special Sales, McGraw-Hill
Professional, Two Penn Plaza, New York, NY 10121-2298. Or contact your
local bookstore.

ISBN-13: 978-0-07-140764-9

ISBN-10: 0-07-140764-2

To Xiaoping, Anna Maria, and our angel boy Nicholas

Contents

Preface

Over the last 15 years, I have found that general aviation aircraft owners are a great pleasure to work with. They are quick to understand legal and tax problems related to their aircraft and they typically react well when advice is given—even if it involves some bitter legal medicine. I'm not sure why, but another thing I've observed is that most aircraft owners possess a healthy sense of humor and enormous stores of patience. It could be that these two qualities are essential ingredients for anyone interested in entering (and remaining in) the world of general aviation aircraft ownership.

I decided to write this book after giving a seminar on aviation tax matters in Fort Lauderdale, Florida, during an Aircraft Owners and Pilots Association Expo in 2001. A number of aircraft owners came to me after the presentation and suggested that I expand on my presentation with a series of articles or a book. With this encouragement and a review of my files it became clear to me that certain legal and tax issues keep coming up time and time again for light aircraft owners. It seemed a worthwhile undertaking to condense my experiences, observations, and advice into a book that could serve as a reference guide for aircraft owners.

This book was not necessarily designed to be read cover to cover. Each part and chapter tackles discrete legal and tax issues related to general aviation aircraft ownership. When I set out to write the book, it was my hope that aircraft owners would refer to it from time to time when they had legal or tax questions related to their aircraft. However, my biggest fear is that the book might be misused as a substitute for seeking personal counsel from a legal or tax professional. In the end, I'm confident that the good sense of most aircraft owners will prevail and they will use this book as a tool to identify and better understand the legal and tax issues they face. With a fundamental understanding of these issues, they will be able to make more-educated decisions with the help of counsel familiar with their particular situation.

The first part of the book deals largely with issues you'll encounter when you decide to buy or sell an aircraft. Chapter 1 contains a detailed overview of the necessary ingredients for a buy-sell agreement. This is also a good time to consider the type of legal entity that you want to own your aircraft. A discussion of the pros and cons of each possible form of aircraft ownership is included in Chap. 2. The review of corporations, sole proprietorships, co-ownerships, partnerships, and limited liability companies should provide you with the information you'll need to make an informed decision.

Once you start thinking about becoming an aircraft owner, you will quickly learn that you are entering a highly regulated world. Part 2 of the book deals with the legal and regulatory requirements for properly recording your ownership of an aircraft and registering your aircraft. Chapter 3 focuses on the requirements for recording an interest in an aircraft. Registration of an aircraft for U.S. citizens is covered in Chap. 4. If you are not a U.S. citizen, you should refer to Chap. 5 for some ideas that might allow your aircraft to qualify for U.S. registration.

As an aircraft owner, you'll also learn that one of your more pressing responsibilities is the proper storage and maintenance of your aircraft. Part 3 of this book involves a review of storage- and maintenance-related issues. Chapter 6 takes a detailed look at hangar and tie-down leases using real-life illustrations to sort through some rather complex legal issues you will surely face as an aircraft owner. Maintenance of your aircraft is another issue you won't be able to avoid. Chapter 7 provides a neat breakdown of your regulatory responsibilities for aircraft maintenance and Chap. 8 addresses the sometimes-thorny question of aircraft airworthiness—just when is your aircraft considered airworthy?

In today's litigious world, aircraft owners are more concerned than ever about liability. The fourth part of the book addresses liability-related legal issues. First, the basics of aircraft insurance are addressed in Chap. 9 with a checklist of issues and considerations you should be aware of when you are looking for aircraft insurance. Chapter 10 includes a detailed discussion related to the often-misunderstood subject of liability releases. You may also be interested in knowing that you have exposure to liability through a myriad of federal and state laws and regulations. Chapter 11 takes a look at some of the more substantial liability exposure created by these laws and regulations.

With the cost of aircraft ownership on the rise, many are tempted to turn to effective cost-sharing methods. Whereas spreading your costs through leasing or multiple ownership arrangements may sound tempting, it is important that you understand the legal hazards that might bite you when you least expect a problem. The fifth part of the book addresses aircraft cost-sharing methods, with Chap. 12 focusing on leases and Chap. 13 focusing on multiple ownership arrangements through co-ownerships, partnerships, corporations, and/or limited liability companies.

No one likes to talk about taxes, but you can't avoid the subject once you purchase an aircraft. Part 6 of this book focuses on the most common tax questions posed by aircraft owners. If you use your aircraft for business purposes, Chap. 14 is a must for you. This chapter provides a detailed analysis of significant court cases and compliance issues related to reporting and defending your aircraft-related tax deductions. If you are looking to dispose of an aircraft you use substantially for business, you might want to look at Chap. 15 to see if a tax-free exchange of aircraft is for you. For many aircraft owners, leasing an aircraft to a flight school or FBO is a great way to defray costs of aircraft ownership—but what impact can this have on your taxes? Read Chap. 16 for some answers. Finally, Chap. 17 wraps things up with a discussion of state and local taxes including sales and use taxes, registration fees, and personal property taxes related to aircraft.

The CD that accompanies this book includes a number of common legal documents and agreements you might run across as an aircraft owner. The documents in the CD come from real-life examples of transactions related to aircraft. However, the CD should be treated with caution—these types of legal documents require the watchful eye of a professional looking out for your best interest. The CD is best used as a discussion tool for you and your legal and/or tax counsel. The CD was not intended to enable or encourage you to become a do-it-yourself lawyer.

Raymond C. Speciale, Esq., C.P.A.

Acknowledgments

After a year of juggling teaching duties, a law practice, the arrival of my baby daughter, and writing a book, I owe a great deal of thanks to several people for their patience and support. I am most grateful to my wife Xiaoping for putting up with me during this process and for her invaluable technical assistance in helping me piece together the manuscript. It is a great gift to have such a capable and loving partner.

I also want to thank all of my colleagues and friends at the Law Offices of Yodice Associates for their insights and assistance throughout the past year. I am particularly indebted to Ron Golden, Esq., Diane Fox, C.P.A., and Dean Torgerson for letting me bounce around ideas and questions whenever I darkened their doorsteps.

The folks at the Aircraft Owners and Pilots Association were also very helpful in giving me feedback and encouragement. I want to specially thank Woody Cahall, Rodney Martz, and Rob Hackman for their assistance in the early stages of my preparation for this book.

Last but not least, I want to thank my colleagues Drs. Kirk Davidson, John Hook, and Bill Forgang at Mount Saint Mary's College for encouraging me to write a book. Seeing the satisfaction they derive from laboring over their books gave me the motivation to give it a try.

Buying or Selling an Aircraft

Buying and Selling an Aircraft

The purchase or sale of an aircraft is a big step. Many aircraft purchasers and sellers don't realize the legal ramifications of an aircraft buy/sell transaction. This can be a dangerous mistake. Careful planning and professional counsel are necessary to traverse the legal and technical issues you are bound to encounter when you are buying or selling an airplane.

This chapter is intended to walk you through the process. First, we'll review the legal issues related to aircraft sales agreements, warranties, transfer of title, and transfer of risk of loss. Next you'll get a guided tour through the major items you should address in a written agreement to buy or sell an airplane. The last section of this chapter will provide a review of legal issues you should consider if you are planning on financing an aircraft purchase.

Laws Governing Aircraft Sales Transactions

In order to achieve uniformity in the laws dealing with sales and other commercial transactions, virtually every state in the United States has adopted a uniform code of commercial law. This code is known as the Uniform Commercial Code (generally referred to as the UCC). It has been in existence since the late 1930s.

The UCC has a special section, Article 2, that is devoted entirely to sales transactions. For purposes of Article 2, a sales transaction is defined as a sale of "goods." The UCC defines goods as tangible and movable personal property.

Because an aircraft is tangible (you can touch and feel it) and movable (we certainly hope so), it qualifies as "goods" under the UCC. That means any purchase or sale of an aircraft is subject to the requirements of the UCC.

All too often, aircraft purchasers and sellers make the mistake of thinking that the UCC applies only when professional sellers or buyers are involved in the transaction. That's simply not true. The UCC pertains to all sales of goods, regardless of whether professional buyers and sellers (the UCC calls them "merchants") are involved.

So how does all of this affect you and your decision to buy or sell an aircraft? Actually, it will have a significant effect on how you structure and follow through on your aircraft buy/sell transaction. The discussion below focuses on some of the major impacts of the UCC on your aircraft buy/sell deal.

The need for a written agreement

Many aircraft buyers and sellers have been stung by a cavalier attitude toward their aircraft sales transaction—you don't want to join their ranks. Often, they don't even consider reducing their deal to a written agreement. That can be a big mistake.

Section 2-201 of the UCC requires that any sales transaction for the sale of goods that exceeds $500 or more must be in writing to be enforceable. Therefore, if you do not prepare some kind of writing evidencing an agreement, the courts may not be able to enforce the verbal terms of your deal.

Regrettably, aircraft owners have often been convinced that a bill of sale will serve as a written "contract" for the purchase or sale of their airplane. This is usually not true. The bill of sale is an important document. However, it does nothing more or less than transfer title to an aircraft. It is not meant to be, nor should it be looked upon, as a substitute for a written agreement designed to spell out the intentions of the parties to an aircraft purchase or sale.

How formal and detailed must your agreement be? The UCC is fairly lenient on this issue. All that the UCC requires is that you prepare some form of writing (1) sufficient to indicate that you have entered into a contract with someone else, (2) signed by the party against whom you are seeking enforcement, and (3) including at least some kind of a statement that you are buying or selling an airplane.

However, good common sense and case law dictate that it would be a lot smarter to set out the details of your deal in a well-drafted aircraft sales agreement. The simple act of working through a written agreement forces you to consider the risks of the transaction. Just as important, the detailed agreement allows both you and the other party an opportunity to formally record your intentions and avoid misunderstandings or miscommunications later. If your airplane purchase or sale runs into problems, a verbal agreement or a weak written agreement may only inflame an already dicey situation.

If your agreement was only verbal, it may not even be enforceable under the UCC. Even if it is enforceable because your behavior or the behavior of the other party indicates that an agreement existed, it will now be a battle of your word against the other party's word. That's not a good place to be when you are talking about the sale of an airplane that might have involved tens or hundreds of thousands of dollars.

If you are working with a weak written agreement, you also have hazards to deal with. The UCC states that if a written agreement is in place, your oral statements may not be used to contradict anything in the writing. Therefore, if you have a written agreement that is not complete or sufficiently detailed, and you and the other party verbally agreed to modifications or additions later, evidence of those modifications or additions may not be considered if a dispute arises.

All in all, the best advice is for you to get a professionally drafted buy/sell agreement. Your costs of preventing problems will be minimal compared to the costs you might incur if you have to untangle problems later. You can take a look at the ingredients for a basic airplane buy/sell agreement a bit later in this chapter. The CD that accompanies this book also contains a specimen agreement you and your lawyer might find helpful as a starting point for drafting your own agreement.

Warranties

The UCC also provides guidance on warranties. Creating or disclaiming warranties can be one of the most important elements of your aircraft buy/sell deal.

First, let's take a look at how the law defines a warranty. Generally speaking, a warranty is considered to be a promise or guarantee that the goods you are selling or buying will meet certain standards. Therefore, if the airplane you are selling fails to meet those standards, then the buyer may have the ability to recover damages against you.

If you are involved in an aircraft buy/sell transaction, you have to make some decisions regarding warranties. If you are the buyer, are you going to demand warranties as part of your deal? You may certainly be interested in extensive warranties—especially if you are purchasing a brand new airplane, or an airplane that is still covered by manufacturer's warranties. However, even if you are purchasing a used aircraft, you might expect certain warranties to apply. For instance, the seller may be representing that her aircraft had a complete engine overhaul a specified number of flight hours before your intended purchase. You may want to require that the seller warrant that important representation before you make your purchase.

If you are the seller, you most likely want to avoid being held liable for any warranties. Again, warranties may be an unavoidable part of doing business if you are a manufacturer selling a brand new airplane. However, if you are selling an aircraft you purchased used or an airplane you have used for many years, you may have a justifiable concern if you are asked to make warranties concerning issues that may predate your purchase of the aircraft. You may not even want to warrant the condition of the aircraft in any manner. After all, you have placed your faith in the maintenance logbook of the aircraft and the certificated mechanics that worked on your aircraft. Do you want to stick your neck out based on someone else's work?

Before you start making demands or decisions regarding warranties on the aircraft you are buying or selling, you should be aware of the different types of warranties recognized by the law in the UCC. There are essentially three types of warranties recognized by the UCC—express, implied, and warranties of title.

Express warranties. Express warranties are warranties that come from the actions or words of the aircraft seller. If you are an aircraft seller, you may be surprised to find that your actions or words can create a warranty. This is true even if you do not specifically use the words "warranty" or "guarantee" in discussions with your buyer (this should help you more clearly see the need for a written agreement).

UCC Section 2-313 tells us that express warranties can be created in three different ways. First, you can create an express warranty by making an affirmative statement of fact or tacitly guaranteeing that your airplane has certain features or capabilities. A second method of creating a warranty is to use certain descriptive words in reference to your aircraft. One final way for you to create a warranty is to use a sample or model intended to represent the airplane you may be selling. Under any circumstance, if you are the buyer and you wish to hold a seller to an express warranty, you will have to show that you relied on the warranty in making the decision to purchase your aircraft.

Affirmative statements of fact. If you are an aircraft seller, affirmative statements of fact are probably the most likely source of express warranties. Certainly, an express, written warranty by an aircraft manufacturer qualifies as an affirmative statement of fact. However, you can make affirmative statements of fact that create warranties in more subtle ways. For instance, a verbal statement that your aircraft has been painted within the past year may create an express warranty.

In even more subtle ways, your aircraft logbook may be making express warranties on your behalf—and without your knowledge. See how this happens to an unsuspecting seller in Case 1-1.

Case 1-1

Edward Miles, Richard W. Keenan and Kenneth L. "Dusty" Burrow, Appellants, v. John F. Kavanaugh, Appellee
Court of Appeals of Florida
350 So. 2d 1090 (1977)

OPINION BY: Hubbart, J.

OPINION: This is an action for breach of express warranty and misrepresentation in the sale of an airplane. Judgment was rendered for the plaintiff-buyer and the seller-defendant appeals. Party defendants responsible for repairing the airplane prior to the sale also appeal. We affirm.

In March, 1973, the plaintiff [John Kavanaugh] answered a newspaper ad placed by the defendant [Richard Keenan] advertising the sale of a used 1956 Cessna 172 private airplane. The plaintiff and defendant Keenan met on several occasions to examine the airplane and to discuss the sale. The defendant Keenan stated that the engine in the airplane had recently been completely overhauled during which time a number of new mechanical parts had been placed in the engine. The defendant Keenan gave the plaintiff an engine and propeller logbook detailing the mechanical repair and flight history of the airplane which the plaintiff carefully inspected.

The logbook reflected that on May 16, 1972, the engine had been given a major overhaul in which new mechanical parts were placed in the engine all in conformity with the manufacturer's engine overhaul manual. The repair work had been done by the defendant [Kenneth L. "Dusty" Burrow] whose work was certified in the logbook by the defendant F.A.A. inspector [Edward Miles]. Based on the accuracy of this information, the plaintiff purchased the airplane from the defendant Keenan. The plaintiff specifically testified that he would not have purchased the airplane had he not been able to inspect and rely upon the information contained in the logbook.

The plaintiff flew the airplane without incident for several months. Thereafter, he experienced a harrowing engine malfunction while the airplane was in flight. On December 5, 1973, he took off from a narrow airstrip in the Everglades approximately fifty miles out of Miami. After takeoff, the engine began to lose power, shake violently and emit a loud clanking sound. The plaintiff was barely able to land on the Everglades airstrip without crashing.

Subsequent thereto, the plaintiff had to arrange at considerable expense for the airplane to be transported in parts to an aircraft repair shop and there completely reoverhauled. It was there discovered that the prior overhaul had not included new parts as represented and that the prior overhaul had been performed in a completely defective manner. All parties to this appeal agree that the logbook contained inaccurate, misleading and false information about the prior repair history of the airplane.

The plaintiff paid approximately $350 to transport the airplane from the Everglades for repairs and $5,700 for the re-overhaul job. In addition, the plaintiff estimated his loss of use of the airplane during this repair period to be $600.

The plaintiff sued the defendant Keenan and the defendants Burrow and Miles for breach of express warranty and misrepresentation. After a non-jury trial, the court awarded a judgment in favor of the plaintiff against all defendants in the amount of $5,800. The defendant Keenan appeals questioning his liability on the sale of the airplane as well as the amount of damages awarded. The defendants Burrow and Miles appeal solely on the damages issue.

I

The first issue presented by this appeal is whether a private party, who sells his used airplane to a buyer and to induce the sale shows the buyer an engine and propeller logbook setting forth the repair history of the airplane, expressly warrants the accuracy of the information contained in the logbook within the meaning of Florida's Uniform Commercial Code, Section 672.313, Florida Statutes (1975). We hold that the seller expressly so warrants the accuracy of the information contained in the logbook where it forms part of the basis of the bargain between the parties.

The controlling law in this case is set forth at Section 672.313, Florida Statutes (1975), as follows:

"672.313 Express warranties by affirmation, promise, description, sample.

(1) Express warranties by the seller are created as follows:

(a) Any affirmation of fact or promise made by the seller to the buyer which relates to the goods and becomes part of the basis of the bargain creates an express warranty that the goods shall conform to the affirmation or promise.

(b) *Any description of the goods which is made part of the basis of the bargain creates an express warranty that the goods shall conform to the description.* [Emphasis added.]

(c) Any sample or model which is made part of the basis of the bargain creates an express warranty that the whole of the goods shall conform to the sample or model.

(2) *It is not necessary to the creation of an express warranty that the seller use formal words such as 'warrant' or 'guarantee' or that he have a specific intention to make a warranty,* but an affirmation merely of the value of the goods or a statement

purporting to be merely the seller's opinion or commendation of the goods does not create a warranty." [Emphasis added.]

The official comments of the above provision of Florida's Uniform Commercial Code is instructive on the issue presented in this case and state in part as follows:

"(1)(b) makes specific some of the principles set forth above when a description of the goods is given by the seller.

A description need not be by words. Technical specifications, blueprints and the like can afford more exact description than mere language and if made part of the basis of the bargain goods must conform with them." [Emphasis added.]

In the instant case, the defendant Keenan gave the plaintiff-buyer the engine and propeller logbook which, much like a blueprint, set out in some detail the prior repair and flight history of the airplane. The accuracy of the information contained in the logbook formed the basis of the bargain as the plaintiff relied upon the accuracy of such information and would not have purchased the airplane if he had not been permitted to see the logbook. The logbook thus constituted a description of the goods purchased by the plaintiff and an express warranty of the accuracy of such description.

The defendant Keenan argues that he never in so many words warranted the accuracy of the information contained in the logbook and was in fact ignorant of the admittedly false information on the prior repair history of the airplane. The simple answer to that argument is that an express warranty need not be by words, but can be by conduct as well, such as, the showing of a blueprint or other description of the goods sold to the buyer. Moreover, fraud is not an essential ingredient of an action for breach of express warranty and indeed it is not even necessary that the seller have a specific intention to make an express warranty. It is sufficient that the warranty was made which formed part of the basis of the bargain. We find such an express warranty in this case through Keenan's showing of the logbook to the plaintiff without which this sale would never have been made. For breach of such warranty, the defendant Keenan is liable to the plaintiff. *Downs v. Shouse,* 18 Ariz. App. 225, 501 P.2d 401 (1972).

II

The second issue presented by this appeal is whether the measure of damages in a breach of warranty action in the sale of a defective airplane may include the expense of transporting the airplane for repairs, the expense of overhauling the airplane, and damages due to a loss of use of the airplane during repairs. We hold that such expenses and losses are recoverable when proximately caused by the breach of warranty.

The law is clear under Florida's Uniform Commercial Code that the measure of damages in a breach of warranty action where the goods have been accepted by the buyer include any consequential damage proximately caused by the breach of warranty. Sections 672.714(3), 672.715(2), Florida Statutes (1975); *Council Bros., Inc. v. Ray Burner Co.,* 473 F.2d 400 (5th Cir. 1973). In the instant case, the plaintiff's expenses in transporting the airplane for repairs and overhauling it to conform to the express warranty as contained in the logbook plus the loss of use of the airplane during the repairs were all proximately caused by the breach of express warranty. Consequently, all such expenses were recoverable in this case.

The defendants do not contest the above principles of law but contend instead that the overhaul bill was excessive and unnecessary. Without going into detail as to each part replaced and the technical mechanical nature of the overhaul, we are satisfied that the bill was reasonably related to restoring the airplane to its original warranted condition and see no valid reason for upsetting the award made. *Downs v. Shouse,* 18 Ariz. App. 225, 501 P.2d 401 (1972). The judgment appealed from is in all respects affirmed.

The moral of this case is that aircraft sellers must be sensitive to the fact that they may be making warranties whenever they present logbooks for inspection by buyers. Whether they are willing to be held liable for those warranties is a question that should be addressed in a well-drafted aircraft buy/sell agreement.

If you are buying a new airplane, the seller will usually make an express, written warranty. Often the warranty will be a "limited" warranty. Usually the warranty is considered limited because it will extend for a set amount of time after your purchase (usually a few years). Your new aircraft warranty will often be limited to defects in material and workmanship under normal use. Every warranty is different, so you are obliged to read the warranty on your new aircraft carefully. One big question you may want answered is whether the warranty covers airworthiness directives (ADs) (see Chap. 7) issued within the warranty period. You may also want to know exactly what parts are covered, and those that are specifically not covered by the warranty. Many manufacturers' warranties have an extensive list of components and parts not covered by the warranty.

Description of the airplane. Another way you can (knowingly or unknowingly) create a warranty is to use descriptive language in the process of selling your aircraft. Generally speaking, the law says that if you use certain "trade terms" to describe your aircraft, you may have created an express warranty. One possible example of such a warranty would be a description of an aircraft as being "airworthy" or "aerobatic." Here again, a well-drafted buy/sell agreement will go a long way toward clarifying your intentions and the intentions of the other party.

Use of "samples or models." The law also states that if you use a sample or model, you may create a warranty. Practically speaking, this type of warranty may only apply if you are an aircraft dealer or a purchaser buying an aircraft from a dealer. The most obvious example of the creation of this type of warranty would include a demonstration flight in an aircraft that is meant to replicate the flight characteristics of the aircraft to be sold. If the aircraft you purchased or sold does not measure up to the aircraft used as a demonstrator, it may trigger a claim that express warranties were breached.

Reliance by purchaser. UCC section 2-313 indicates that an express warranty is only created if the buyer reasonably relied upon the warranty in deciding to make the purchase of an aircraft. Sometimes courts have said that in order for an express warranty to be created, the warranty must have become the "basis of the bargain."

The first circumstance where this would come into play is the situation where the seller makes certain statements about an airplane and the situation indicates that the parties intended for the statements to become a part of the aircraft sales agreement. This would be apparent in a circumstance where the seller indicates in a written sales agreement that the aircraft she is selling is IFR capable. If the buyer indicated that he intends to do substantial IFR flying, the representation that the aircraft is IFR capable has obviously become part of the "basis of the bargain" and an express warranty has been created.

A more difficult situation may exist if the statement was not made in the written sales agreement, but sales literature posted by the seller in pamphlets or website advertisements indicates that the aircraft being sold is IFR capable.

The UCC and cases interpreting the UCC now seem to indicate that an express warranty has been created unless the seller can prove otherwise.

What happens when aircraft sellers make promises about the performance of the aircraft that they sold after the sale has taken place? Suppose the buyer contacts the seller after purchasing an aircraft and requests assurances that the aircraft can achieve a certain true air speed under specified conditions. Does the seller's promise that the aircraft can perform accordingly become part of the "basis of the bargain"? Under current interpretation of the law, the seller's post-sale promises can become the basis of the bargain, creating an express warranty. The courts have looked at this kind of postsale statement as a modification of the agreement, even if the buyer did not pay anything extra for the promise.

Another difficult situation may arise when sellers make statements during negotiations that fail to appear in a written sales agreement between the parties. In this scenario, a buyer who later claims breach of contract may have a difficult time pursuing the claim because the courts often view the written agreement as the final statement of the parties regarding the transaction.

As usual, the best advice is to make sure you clearly spell out what your intentions are in writing. Both aircraft buyers and sellers need to ensure that their intentions and understandings are adequately detailed with little chance left for misunderstanding.

Implied warranties.   Many aircraft sellers are unaware that even if they don't create express warranties, they may create implied warranties by merely selling their aircraft. The UCC identifies two types of implied warranties that are relevant to aircraft sales. The first type is called an implied warranty of merchantibility. The second type of warranty is known as an implied warranty of fitness for a particular purpose.

Implied warranty of merchantibility.   The implied warranty of merchantibility applies only if the seller is considered to be an aircraft merchant. Under the law, a merchant is considered to be someone who regularly "deals in goods of the kind" sold. Therefore, if you are an occasional private seller, you are not likely to be subject to the implied warranty of merchantibility. On the other hand, if you are an aircraft dealer (even part-time) you will probably be subject to this warranty. Under certain interpretations of the law, you may even qualify as a merchant if you merely buy and sell aircraft on a frequent basis.

In order for the aircraft you buy or sell to be merchantable, it must be "fit for the ordinary purpose for which such goods are used." Inasmuch as we're talking about airplanes, this may essentially boil down to a question of whether the aircraft is airworthy. This implied warranty does not guarantee that the aircraft you buy or sell is of high quality. It only implies that the airplane is "average."

However, this can become a tricky question when it comes to airplanes. The FAA essentially defines an airworthy aircraft as an aircraft that conforms to its initial type certificate. Whether any aircraft is airworthy under this definition is difficult to say. (For a fuller discussion of this issue please refer to Chap. 8.) This is all the more reason to carefully consider how you wish to deal with this issue in an aircraft buy/sell agreement.

Implied warranty of fitness for a particular purpose. This requirement applies to all sellers of aircraft and is often referred to as a warranty of fitness. UCC Section 2-315 states: "Where the seller at the time of contracting has reason to know any particular purpose for which the goods are required and that the buyer is relying on the seller's skill or judgment to furnish suitable goods, there is . . . an implied warranty that the goods shall be fit for such purpose."

Although there is some confusion in the law as to whether this warranty applies only to use outside the ordinary purpose for an aircraft, it may be safer to assume that warranty applies to all uses of an aircraft. It's obvious that any seller of an aircraft would be put on notice that the buyer is purchasing the airplane for flight. Therefore, it could be argued that any seller of the aircraft could be held liable on an implied warranty of fitness.

Warranties of title. In aircraft sales, a warranty of good title held by the seller automatically exists. The UCC creates two types of automatic warranties of title.

The first type of title warranty is the seller's warranty that he/she is passing good title to the buyer and that he/she has the authority to make the transfer of title. Obviously, this warranty is breached if the seller is a thief and stole the airplane from another person. Fortunately, this does not become an issue in most aircraft purchases.

The second type of title warranty is the seller's warranty that he/she is passing title to an aircraft that is free from any security interest or other lien or encumbrance that the buyer is not aware of at the time of the sales agreement. This type of title warranty would be breached if the seller sold an aircraft with liens or other encumbrances without notifying the buyer about the liens or encumbrances.

Disclaimer of warranties. As indicated above, the sale of an aircraft can trigger three types of warranties: express warranties, implied warranties, and title warranties. In many cases, the aircraft seller may not desire the exposure to liability that some or all of these warranties may bring. The UCC will permit the seller to disclaim or eliminate these warranties if the seller follows certain rules. To a certain extent, the law allows disclaimers in order to permit the parties involved in a buy/sell transaction to design their own deal with freedom. This freedom also brings a responsibility to the parties to ensure that the deal they design is a deal they can live with.

The law will permit a seller to disclaim express warranties. However, as you might imagine, this may become a very clumsy sort of situation. After all, the seller has created an express warranty and now wishes to take special pains to disclaim the warranty. Because of this, the law makes it rather difficult for a seller to disclaim an express warranty. Specifically, the UCC states that if a disclaimer of an express warranty contains language that is inconsistent with the express warranty, it is not valid as a disclaimer. This makes it extremely difficult for a seller to properly disclaim express warranties. It is simple enough to draw a lesson from this discussion—if you don't want to be bound by a warranty, don't make it in the first place. Another lesson learned the hard way is presented in Case 1-2 involving a seller who argues that he disclaimed an express warranty created by his aircraft maintenance logbook.

Case 1-2

Limited Flying Club, Inc., an Iowa Corporation, James E. Vining, Vernon H. Witt, and George C. Clausen, Appellees, v. Gerald O. Wood and Eugene O. Wood, d/b/a Wood Aviation, Appellants
United States Court of Appeals, Eighth Circuit
632 F.2d 51 (1980)

OPINION BY: Heaney, J.

OPINION: Limited Flying Club, Inc., and its three members brought this diversity action alleging fraudulent misrepresentation and breach of express and implied warranties in connection with the sale of a used airplane by Gerald and Eugene Wood. The case was tried to a magistrate, sitting without a jury, who dismissed the warranty counts but awarded compensatory and punitive damages for fraud. The district court adopted the magistrate's findings. We reverse and remand for consideration of damages in connection with the claim of breach of express warranty.

We summarize the facts in the light most favorable to the magistrate's findings. In the spring of 1973, Eugene Wood purchased a 1965 Mooney Mark IV airplane in Tucson, Arizona. Prior to the purchase, the plane was involved in two forced "wheels up" landings. In the most recent, the plane's surface and structure were extensively damaged. The airplane was towed to a warehouse hangar operated by Wood at Ryan Field, an airport near Tucson. Eugene Wood then arranged for his son, Gerald, and George Mickelson, a mechanic licensed by the Federal Aviation Administration, to repair the aircraft. Gerald and Mickelson inspected the damage to the airplane, planned the repairs and ordered the necessary parts. In May, 1973, Gerald attended aviation school and in June, he passed the FAA-required examination and received his Airframe and Powerplant (A & P) license authorizing him to make major aircraft repairs. Gerald and Mickelson removed the skins from the wings and belly of the plane and made repairs and replaced parts in many areas. Certain repairs and alterations were major, including the installation of an engine; the repair of a structural rim in the right wing; the repair and replacement of wing skins and inspection covers; the replacement of belly fairings, a former assembly, a bulkhead, fairings, and a panel assembly; the replacement of belly skins, fuselage bottom skins and bulkheads; the replacement of landing gear linkage; the replacement of elevation linkage; and the installation of replacement retroacting links. These repairs were itemized in the airplane's logbook. Form 337, which is required by the FAA to be filed for each major repair, was filed only for the repair of the structural rim in the right wing.

Mickelson, who held an Inspection Authorization (I.A.) license, approved and certified the airplane as airworthy in July, 1974. The airplane was again certified as airworthy in August, 1975, by David Ateah, who also held an I.A. license. Ateah was employed by Eugene Wood to inspect the airplane and was paid $35 for the two and one-half to three-hour inspection. Eugene flew the plane frequently after its return to service in 1974, taking his family with him on some trips and flying for distances of up to 1,200 miles.

In late December of 1975, appellee James Vining was visiting Ryan Field and noticed the Mooney in Wood's hangar. He spoke with an unidentified person who told him the Mooney would be for sale. That person told Vining that the plane had previously been damaged in a belly landing. A few days later, Vining returned to the airfield and met Gerald Wood, who gave him the impression that the airplane was for sale and told Vining he should contact his father. Vining came back a few days later and met Eugene, who told him the plane might be for sale for approximately $13,000. Eugene described the Mooney as a "nice little airplane" and "a good little airplane."

In early January, Vining returned to his home in Clinton, Iowa, and agreed with Vernon Witt and George Clausen to jointly purchase Wood's Mooney. Vining made a number of telephone calls to Wood, attempting to arrange the purchase. Eventually, the parties agreed that Vining would travel to Tucson to take possession of the airplane.

Vining and a pilot, Leo Cozzolino, arrived in Tucson on June 12, 1976. They visually inspected the airplane and Eugene showed them the logbook, reviewed its entries with them and discussed its two previous belly landings. Eugene suggested that Vining have the airplane inspected and certified before returning to Iowa; however, Vining was anxious to return home and stated his preference to have the inspection done there. Eugene then flew Cozzolino in the Mooney to Tucson International Airport, some fifteen to twenty miles away, to pick up a radio which was to be installed. Cozzolino flew the plane back to Ryan Field and the sale was completed; Vining paid Eugene $14,200 for the airplane and Eugene delivered a Bill of Sale. At Eugene's request, Vining signed a typewritten document that stated as follows:

June 12, 1976: After inspection and trial flight, which have met with my approval, of Mooney N 7875 V, I have agreed to accept the aircraft on an "as is"–"where is" basis, for the amount previously agreed upon.

Cozzolino flew the airplane and Vining back to Clinton, Iowa, that day, with intermediate stops in Albuquerque, New Mexico, and Hutchinson, Kansas. During the next six weeks, the airplane was flown sixteen to eighteen hours and no problems arose.

The airplane was taken to Straley Flying Service in Clinton, Iowa, in August, 1976, for its annual inspection. The plane was grounded upon discovery of a number of major defects.

The plane was then flown by special ferry permit to Niederhauser Airways in Waterloo, the authorized Mooney dealer for the State of Iowa, where it was inspected and the following defects found:

1. Tunnel cover bent and ripped loose; 2. Wing skins improperly riveted and not fit flush (distorted—not predrilled and aligned); 3. Flap hinge ground out; 4. Right wing skins improperly installed; 5. Center panel damaged; 6. Bottom side leading edge bent–also improper rivets, dents filled with putty and filler both main and center panel (illegal); 7. 1/4 to 3/8 slope in stabilizer; 8. Compression bend in tubing aft of firewall; 9. Illegal spliced stringers; 10. Damaged belly panel; 11. Defective truss illegally repaired at Station 33 (Exhibit 21); 12. Illegally repaired nose gear truss.

The plane was then flown, again by special ferry permit, to Kerrville, Texas, where it was inspected by Charles Dugosh, an expert in the construction of Mooney airplanes. He found many defects in the fuselage bottom, the wings, the fuselage, the nose gear truss and the stabilizer. Both Dugosh and Richard Carley, the mechanic who inspected the airplane in Iowa, testified that many of these defects would be observable on a normal annual inspection.

At some point after the defects were discovered, Vining telephoned Eugene and told him of the problems with the airplane. Eugene offered to buy the plane back. Vining testified that Eugene offered $10,000, and Eugene testified that he offered another club member $13,000. Vining rejected this offer and had the plane repaired at a cost of $12,534.33.

The magistrate found that the plaintiffs had proven fraud by a preponderance of the evidence and awarded compensatory damages of $12,534.33 (the cost of repairs) and punitive damages of $15,000. The district court held that the magistrate had applied an incorrect burden of proof, and remanded the case to the magistrate to determine whether each element of the case had been proven by clear, satisfactory and convincing evidence. The magistrate subsequently found that this required burden of proof had been met and affirmed the damage award. The district court adopted the magistrate's memorandum as supplemented.

* * *

II. Express Warranty

The magistrate concluded without discussion that no express warranty was present in this case. We disagree. The magistrate found and the evidence demonstrates that Eugene Wood showed Vining and Cozzolino the airplane logbook, including the certificates of airworthiness, and went over it with them. Vining testified that he relied on the logbook and certificates of airworthiness and it is clear that they became part of the basis of the bargain between the parties.

The law governing the creation of an express warranty is set forth in Iowa Code § 554.2313, which is identical to the Uniform Commercial Code provision:

(1) Express warranties by the seller are created as follows: (a) Any affirmation of fact or promise made by the seller to the buyer which relates to the goods and becomes part of the basis of the bargain creates an express warranty that the goods shall conform to the affirmation or promise. (b) Any description of the goods which is made part of the basis of the bargain creates an express warranty that the goods shall conform to the description. (2) It is not necessary to the creation of an express warranty that the seller use formal words such as "warrant" or "guarantee" or that he have a specific intention to make a warranty, but an affirmation merely of the value of the goods or a statement purporting to be merely the seller's opinion or commendation of the goods does not create a warranty.

U.C.C. official comment 5 to that section provides that "(a) description need not be by words. Technical specifications, blueprints and the like can afford more exact description than mere language and if made part of the basis of the bargain goods must conform with them."

In this case, the seller provided the buyer with the logbook which set forth the repair and inspection history of the airplane. Vining and his pilot examined those entries and relied on the certifications of the airplane as airworthy. Those certifications consequently formed part of the basis of the bargain as a description of the goods, similar to a description that might be provided by a blueprint. Under these circumstances, Eugene expressly warranted the accuracy of that description—the airworthiness of the plane—and is liable for damages arising from the breach of that warranty. Accord, *Miles v. Kavanaugh,* 350 So.2d 1090 (Fla.App.1977).

We reversed the magistrate's finding of fraudulent misrepresentation because the evidence did not demonstrate that Eugene represented the airplane's condition in reckless disregard of the truth or falsity of his representations. To create an express warranty of the plane's airworthiness, however, there is no requirement that Eugene have actual knowledge of the airplane's airworthiness or lack of airworthiness. His representation of the plane as airworthy, based on the description in the logbook, created the warranty.

There remains, however, the question of the effect of the "as is"–"where is" disclaimer. It is fairly clear that such a provision operates to disclaim implied warranties. Iowa

Code § 554.2316(3)(a). The effect of such a disclaimer on express warranties, however, is less clear. The Iowa Code provides as follows:

Words or conduct relevant to the creation of an express warranty and words or conduct tending to negate or limit warranty shall be construed wherever reasonable as consistent with each other; but subject to the provisions of this Article on parol or extrinsic evidence (Section 554.2202) negation or limitation is inoperative to the extent that such construction is unreasonable.

Iowa Code § 554.2316(1).

Although the disclaimer does not explicitly address the affirmations contained in the logbook, it states in general terms that the buyer accepts the airplane "as is." Because these words tend to limit or negate the seller's warranties, the Code requires that the provision be construed as consistent with the warranty, if such a construction is reasonable. That construction is reasonable under these facts. Eugene presented the logbook to Vining and Cozzolino for their inspection and went over its entries with them prior to the time Vining signed the "as is" disclaimer. Eugene delivered the logbook with the airplane to Vining after the disclaimer was signed. The disclaimer itself made no reference to the description of the airplane contained in the logbook. The "as is" clause, then, can fairly be read to disclaim all implied warranties, leaving the written express warranties of the logbook, including the warranty of airworthiness, intact.

The Code's provision on parol evidence does not preclude consideration of the express warranty. That section states:

Terms with respect to which the confirmatory memoranda of the parties agree or which are otherwise set forth in a writing intended by the parties as a final expression of their agreement with respect to such terms as are included therein may not be contradicted by evidence of any prior agreement or of a contemporaneous oral agreement but may be explained or supplemented(b) by evidence of consistent additional terms unless the court finds the writing to have been intended also as a complete and exclusive statement of the terms of the agreement.

Iowa Code § 554.2202.

The "as is"–"where is" clause was certainly not a "complete and exclusive statement of the terms of the agreement" between Eugene and Vining. The description of the airplane as set forth in the logbook is, as we have indicated, a consistent additional term and may be introduced to explain the actual agreement between the parties.

We turn, finally, to the question of damages. We are unable to determine on the basis of this record the amount of damages recoverable for the breach of express warranty. The measure of damages for breach of warranty is set out in the Iowa Code as follows:

(2) The measure of damages for breach of warranty is the difference at the time and place of acceptance between the value of the goods accepted and the value they would have had if they had been as warranted, unless special circumstances show proximate damages of a different amount. (3) In a proper case any incidental and consequential damages under the next section may also be recovered.

Iowa Code § 554.2714(2), (3).

It remains for the trier of fact to determine the amount of damages under this rule. We note, however, that he should determine whether Eugene offered to buy the airplane back upon learning of its defects and if so, for what price. If he determines that such an offer was made, he should determine whether the damages caused by the breach could have been mitigated by acceptance of that offer.

For aircraft buyers and sellers, dealing with implied warranties is often a trickier part of the deal. Although the seller wants to avoid warranties, the buyer may insist on certain quality assurances. There are two applicable methods for a seller to disclaim warranties in an aircraft sale under the UCC. The first involves the use of certain disclaimer language. The second involves a buyer's inspection of the aircraft.

The most common way for a seller to avoid an implied warranty is to state in clear and unambiguous language that the airplane is being sold "as is" or "with all faults." This language will serve to disclaim warranties of merchantibility and warranties of fitness. It is also important that the disclaimer language be conspicuous if it is contained in a writing (a written disclaimer is highly recommended). Using bold type or larger letters for the disclaimer language will certainly help make the language more conspicuous to the reader.

Another way that implied warranties can be disclaimed is through the actions of the buyer. Specifically, when a buyer either fully examines an airplane or refuses to inspect an airplane, he/she may have eliminated his/her ability to claim a warranty theory of recovery at a later date. This disclaimer will be applicable to any reasonably apparent defects from the prepurchase inspection. It will therefore not apply to hidden or latent defects that would not be discoverable in a routine prepurchase inspection.

As far as title warranties are concerned, it would be very rare for a buyer to be willing to waive this kind of warranty. The warranty of title is automatic and as a buyer you would be taking an undue risk to waive the right to receive an aircraft that your seller actually owns. Therefore, it is unlikely that you will see a waiver of the warranty of title in an aircraft buy/sell transaction. One exception to this general rule may be judicial sales or sales by a sheriff, executor, or foreclosing lien holder.

Transfer of title and risk of loss. The UCC also provides for when title to an aircraft is actually transferred. Along with this comes guidance from the UCC on when you bear the risk for any loss related to the aircraft you are preparing to buy or sell. All of these rules tie in to the very important question of when you obtain an insurable interest in the aircraft you intend to buy or sell. These issues are discussed immediately below.

Transfer of title. The passing of aircraft title from the seller to the buyer can be accomplished in a number of ways. The law allows you to agree to title passage in any manner and on any conditions agreed to by you and the other party to your airplane buy/sell transaction.

If the parties do not expressly agree on when title changes hands, the transfer of title in aircraft purchases will generally occur when the seller delivers the aircraft to the buyer. If the parties don't expect delivery of the aircraft, aircraft title will usually change hands when the bill of sale is transferred from the seller to the buyer.

Risk of loss. Because of the movable nature of aircraft, it is imperative that the parties determine in advance who bears the risk of loss should the aircraft be damaged or destroyed while it is the subject of a sales contract. As you might guess, the risk of loss in such a case initially rests with the seller. However, it will ultimately shift to the buyer. The question of when that shift occurs is an important one in an aircraft sales transaction.

The fact that the parties may have insurance coverage does not diminish the importance of this question. The real issue may then be whose insurance company will have to pay for damages to the aircraft. It is also quite possible that insurance coverage will be inadequate to cover for the entire loss. This makes the issue all the more important because any losses may be coming directly out of the pockets of the aircraft seller or buyer.

The UCC has certain rules in place that apply to typical aircraft sales transactions. The first rule is the most important—the parties have all legal authority to make any agreement with respect to the time that risk of loss changes hands. You should take advantage of this rule. Shape the transaction and the agreement to meet the needs of the parties and specifically state when risk of loss will pass from the seller to the buyer.

If you don't specify when risk of loss passes from the seller to the buyer, you are probably subjecting yourself to the vagaries of whether the seller is a merchant or a nonmerchant. If the seller is a merchant, risk of loss does not pass from the seller to the buyer until the buyer actually takes delivery of the aircraft. If the seller is not a merchant, the risk of loss will pass when the seller "tenders" or offers delivery or pickup to the buyer.

Insurable interest.   Obviously, it is in your best interest as a buyer or seller to ensure that you have adequate insurance coverage for the aircraft that is the subject of a sales agreement. However, it is important to note that an insurance policy is only valid if the party seeking protection has an insurable interest in the airplane. The answer to this question may require some additional study. However, the UCC has provisions addressing this issue that are relevant to aircraft sales transactions.

For the seller the rules are pretty simple. As long as a seller has title to an aircraft, he or she has an insurable interest. However, not so obviously, the seller of an aircraft may retain an insurable interest in the aircraft even after title passes to the buyer. This is the case if the seller retains a lien or mortgage to secure payment on the airplane.

On the other hand, the aircraft buyer has an insurable interest once the aircraft has been specifically identified as the subject of a sales contract. This essentially means that a seller and buyer can both have an insurable interest in the same aircraft at the same time.

An application of the risk of loss and insurable interest rules can be found in Case 1-3.

Case 1-3

James Bowman, Appellee, v. American Home Assurance Company, a corporation, Appellant
Supreme Court of Nebraska
213 N.W.2d 446 (1973)

OPINION BY: White, J.

OPINION: This is an action on a contract of insurance. The insured, James Bowman, seeks to have his insurer pay for damage done to an airplane covered under a policy issued by the insurer, American Home Assurance Company. The insurer argues that the insured did not have title as defined under the Uniform Commercial Code at the time of the loss and therefore asserts that the insured had no insurable interest at the time of the loss and should be denied recovery. The insured had the verdict and judgment at trial. The insurer appeals. We affirm.

James Bowman, the appellee-insured, and Keith Moeller were engaged in a partnership doing business as Bowman Hydro-Vat in Fremont, Nebraska. The partnership purchased a twin-engine Cessna in 1969. Upon purchasing the aircraft, Bowman applied for and obtained a policy of insurance issued by the appellant-insurer, American Home Assurance Company, insuring the aircraft during the period from December 23, 1969, through December 23, 1970. The named insured under the policy was James Bowman.

In December of 1970, Bowman and James Hemmer entered into negotiations for the sale of the plane to Hemmer. On December 12, 1970, Bowman and Hemmer agreed upon a price of $18,500 for the purchase of the aircraft and Hemmer paid $15,000 down. The remaining $3,500 was to be paid later in cash or through its equivalent in aircraft instrument instruction which Hemmer was to give to Bowman. Hemmer requested a bill of sale signed by both partners to protect himself, and to comply with the Federal Aviation Administration requirements for the transfer of an aircraft. Bowman testified that it was agreed that he was to remain the owner of the aircraft until "we were able to fill out the necessary paperwork." Hemmer, the buyer, testified numerous times that he was to be the owner when he received the bill of sale. This was to allow the buyer to comply with the Federal Aviation Administration requirements and make arrangements for insurance prior to the time he was to become the owner. Bowman also testified that he told Hemmer that he would leave his insurance in effect until it expired on December 23, 1970, only 11 days later. Bowman retained possession of the plane.

On December 15, 3 days later, Bowman contacted Hemmer and asked Hemmer if he wanted to go with him on a business trip to Kansas. The purpose of the flight was for Bowman to transact some business in Kansas. Upon their return to Fremont, Hemmer asked Bowman for permission to use the plane on the following Friday and Saturday. Bowman and Hemmer specifically examined Bowman's insurance policy to ascertain whether it would provide coverage while Hemmer flew the plane. Bowman then gave Hemmer permission and Hemmer took the plane to Columbus. On December 16, Hemmer flew the plane from Columbus to Fremont to obtain the bill of sale, but it had not been signed so he returned to Columbus without it.

On December 18, Hemmer flew the plane to Mitchell, South Dakota, pursuant to the permission granted by Bowman. In attempting to take off from Mitchell, the tip of a wing caught in a snow bank causing extensive damage to the aircraft.

Hemmer testified that the Federal Aviation Administration regulations require that a registration certificate be in an aircraft before title to the plane can be transferred to a new owner, and that once the bill of sale is received, it is attached to a new registration application and sent to the Federal Aviation Administration. A pink copy of the new registration is placed in the aircraft to serve as a temporary registration. This paperwork had not been completed at the time of the loss because the bill of sale had not yet been received. The registration certificate in the aircraft at the time of the accident showed James Bowman as the owner.

The signature of Bowman's partner was obtained and the bill of sale was mailed to Hemmer on December 18, the day of the accident. Hemmer received the bill of sale on December 20. The bill of sale was in blank form and had not been filled out at the time Hemmer received it. It was understood that Hemmer was to fill out the necessary information on the bill of sale. Bowman filed an accident report after the accident and indicated he was the owner.

The controversy between the parties centers around two provisions of the Uniform Commercial Code. Section 2-501, U.C.C., provides in part:

"(2) The seller retains an insurable interest in goods so long as title to or any security interest in the goods remains in him * * *."

The case was tried to the jury on the theory that the seller retained an insurable interest in the goods as long as title to the goods remained with the seller. Both parties agree this was the appropriate standard for submission of the case to the jury. Section 2-401, U.C.C., details the concept of passage of title:

"(2) *Unless otherwise explicitly agreed* title passes to the buyer at the time and place at which the seller completes his performance with reference to the physical delivery of the goods, despite any reservation of a security interest and even though a document of title is to be delivered at a different time or place * * *." (Emphasis supplied.)

As section 2-501, U.C.C., provides, the seller has an insurable interest until title passes to the buyer. Under section 2-401, U.C.C., title passes to the buyer (1) at the time and place where the seller completes his performance with reference to the physical delivery of the goods or (2) at any other time explicitly agreed to by the parties. As dictated by the Uniform Commercial Code, the trial court submitted two factual questions to the jury. First, whether the seller had completed physical delivery of the goods. Second, whether there was an explicit agreement between the buyer and the seller as to the time when title was to pass. A jury finding in favor of the insured upon either of these factual issues supports the verdict. Where reasonable minds might draw different inferences or conclusions from the evidence, it is within the province of the jury to decide the issues of fact and the jury verdict will not be set aside unless it is clearly wrong. *Mustard v. St. Paul Fire & Marine Ins. Co.,* 183 Neb. 15, 157 N.W.2d 865 (1968).

There was substantial evidence from which the jury could have inferred that the seller had not completed physical delivery of the goods. The evidence shows that the buyer was only given limited use of the plane to make the trip to South Dakota. The buyer even asked the seller for permission to use the plane for this one trip. The seller had only granted a limited possession of the plane to the buyer, even though the buyer had possession of the plane for 3 days prior to the accident.

The jury could have also inferred from the evidence that the buyer and seller had an explicit agreement that title was to pass upon the completion of the "necessary paperwork." The code itself provides no definition of the term "explicit" as used in section 2-401, U. C. C. In *Harney v. Spellman,* 113 Ill. App. 2d 463, 251 N. E. 2d 265, 6 U. C. C. Rptng. Serv. 1185 (1969), the court defined "explicit" in reference to section 2-401, U. C. C., as follows: "The term 'explicit' means that which is so clearly stated or distinctly set forth that there is no doubt as to its meaning." See, also, *Binkley Co. v. Teledyne Mid-America Corp.,* 333 F. Supp. 1183 (E. D. Mo., 1971), affirmed 460 F.2d 276 (8th Cir., 1972), for a discussion of the term explicit as used in another section of the code.

Bowman testified that it was agreed that he was to remain the owner of the aircraft until "we were able to fill out the necessary paperwork." The only other testimony by Bowman on this subject was the testimony of Bowman on cross-examination that he thought a sale had occurred on December 12, 1970. On redirect however, Bowman testified that he meant an agreement to sell the plane was entered into on December 12, 1970. The buyer testified numerous times that there was an agreement that he was to be the owner when he received the bill of sale. This was to allow the buyer to complete the Federal Aviation Administration requirements and make arrangements for insurance prior to the time he was to become the owner. From all this testimony the jury could infer that there was an explicit agreement. The credibility of the witnesses and weight to be given to their testimony are for the triers of fact. *First Nat. Bank of Omaha v. First Cadco Corp.,* 189 Neb. 734, 205 N.W.2d 115 (1973). On the record as a whole, we cannot conclude that there was not sufficient evidence to support a jury finding that there was an explicit agreement as to the passage of title of the aircraft.

The evidence showed that the bill of sale sent to the buyer from the sellers on the day of the accident was not signed by both sellers until the day of the accident. The bill of sale was not received by the buyer until several days after the accident, and even at this time it remained in blank form. The parties knew that the buyer would fill in and complete the bill of sale after he received it. Thus, it is clear from the evidence that "completion of the necessary paperwork" involved more than the mere signing of the bill of sale by both sellers. It at least included receipt of the blank bill of sale by the buyer, but it could also have included the action of the buyer in filling out the necessary Federal Aviation Administration papers and completing the bill of sale. None of the above steps had been completed at the time of the accident, and therefore the time of the completion of the necessary paperwork had not occurred at the time of the accident.

In summary, the jury could reasonably have found from substantial evidence that either (1) the seller had not completed physical delivery of the goods under section 2-401, U.C.C., or (2) there was an explicit agreement for title to pass upon completion of the necessary paperwork which had not occurred at the time of the accident. Under either of these findings title had not passed to the buyer under section 2-401, U.C.C., and therefore under section 2-501, U.C.C., the sellers retained an insurable interest. For these reasons we affirm the judgment of the District Court.

Affirmed.

You should also view this case as another example of how important it is to carefully spell your intentions out in a written buy/sell agreement. Although the aircraft owner prevailed in this case, it is would have been a lot cheaper and easier to simply avoid the kind of confusion experienced in this case.

Key Ingredients for an Aircraft Buy/Sell Agreement

By now you should have a sense that a well-drafted agreement is one of your best protections against unpleasant surprises in your aircraft buy/sell transaction. In the following paragraphs, we will review some of the ingredients necessary to draft an aircraft sales agreement that will clarify the buyer's and seller's intentions.

Remember, however, that each transaction has its own unique context and circumstances. The items presented are for discussion purposes between you and your legal counsel. Some of the suggestions may be appropriate for your deal, and some may not. In the end, the effort you expend on drafting the agreement may also help you clarify your expectations from the transaction— whether you are the aircraft buyer or seller.

Identification of the parties

This may seem a bit obvious, but you should be certain that the written agreement properly identifies each of the parties. You will want to ensure that the agreement includes the correct name and address of the seller. The place where people often make mistakes is in the use of the seller's individual name when in fact, the aircraft is actually owned by a corporation, limited liability company (LLC), or other entity. The person or entity owning the aircraft should be identified as the seller.

If you are the buyer, you also want to carefully decide who will own the aircraft. Do circumstances warrant that you own the aircraft in your individual name or the name of a corporation, LLC, partnership, or co-ownership? If you are a buyer who is looking to answer these questions, you should refer to Chap. 2 for a closer look at entity selection and the advantages, disadvantages, and the legal and tax ramifications of your selection.

Whether the agreement involves individuals or legal entities, your agreement should identify the mailing address and physical location of each party. Additionally, if your agreement involves a legal entity such as a limited liability company or a corporation, you should identify the state where the entity is authorized or chartered.

Recitals

Sometimes, it helps to set the stage for a contract if an introduction is included. Lawyers like to use "recitals" at the beginning of a contract for this purpose. Recitals are nothing more than an introduction to the background of the agreement. If the contract is simple, there may not be a great need for recitals as a lead-in to the agreement. However, if your transaction involves a trade-in, an anticipated tax-free exchange, or any other more complex issues, recitals may be appropriate. In many cases, your lawyer may refer to the recitals as the "Whereas" clauses. You will find a basic set of recitals for a simple aircraft sales transaction in the specimen aircraft buy/sell agreement found in the accompanying CD.

Description of the aircraft

It is always advisable for you to describe the aircraft you are buying or selling in some detail. Most well-drafted agreements for the purchase and sale of an aircraft will identify the aircraft with the following information:

- Manufacturer
- Model

- Serial number

- FAA registration number (N-number)

In this section of the agreement, it might also be appropriate to list certain accessories and/or equipment that might be included with the aircraft. If you are the seller, you might limit the transferred items to a specific list you have prepared. If you are the buyer, you may want to include as much as possible. Nowadays, that may be made a bit easier as a result of Internet marketing of aircraft. Some buyers will simply copy the features and accessories pages from the aircraft seller's advertisements and Internet features list. The Internet page(s) detailing the features of the aircraft can then be used as an exhibit to the agreement with language in the contract stating that the exhibit will be "incorporated by reference into the agreement." Essentially, this means that the representations made in the exhibits are now part of the seller's promises. These pages can make for a very complete list of accessories and may be helpful to both buyer and seller so they will know what they are getting or giving in this important transaction.

Price

This is the time to make sure that everyone knows the price of the aircraft. In most cases, this will be simple enough. However, in some new aircraft sales contracts there may be provisions allowing for the escalation of the price by the seller under certain circumstances, including the passage of a time before the deal can be closed. In order to protect yourself as a buyer in these situations, you must ensure that you will be notified of any price escalation in writing and that upon receipt of such notice, you have the absolute right to cancel the contract.

If the pricing includes a deposit, the contract should make it very clear whether the deposit is refundable or not. If it is refundable, the conditions and timing of the refund should be clearly spelled out. In some cases, it may be possible that the deposit will be held for extended periods of time before a sale is consummated. If this is the case and you are an aircraft buyer, it may be a good idea for you to consider interest accruing on the deposit if it is refundable.

Closing

Closing the transaction is a big step in your aircraft sale or purchase. The basic mechanics of a closing involve a transfer of aircraft title to the buyer with a transfer of funds to the aircraft seller. This sounds simple enough. However, many transactions involve additional complexities. For instance, will the transaction be financed? If so, how much will be financed by a lender? Is the seller financing any part of the transaction? Does the aircraft for sale have any liens that must be released? Are the parties going to meet at a closing table or will the aircraft be sold via long distance? Were deposits made for the aircraft sale? Are the deposits refundable or nonrefundable? How will payment be made? Will it be via wire transfer? Will a certified check or a cashier's check be required?

Because of the multitude of variations in any given aircraft buy/sell transaction, you should at least consider the possibility of using an escrow agent to assist in the smooth transfer of title and funds. An escrow agent acts as a neutral third party. The escrow agent is usually a bank or a special aircraft escrow service located at or near the FAA Aircraft Registry in Oklahoma City.

Here's how an escrow transaction works. In the initial stages, the buyer and/or her finance company deposit necessary funds with an escrow agent. Usually the funds are wire transferred to the escrow agent (if the transaction is financed, your bank will probably send the escrow agent any necessary documents to establish its future lien). At this stage, it may also be advisable for the buyer to submit her FAA application for registration to the escrow agent. The escrow agent is then responsible for holding these funds, documents, and the buyer's application for FAA registration. In some cases the funds may be held in an interest-bearing account.

At approximately the same time, the seller deposits the necessary aircraft title documents with the escrow agent. Usually this involves a deposit of a properly executed FAA bill of sale transferring title to the airplane from the seller to the buyer. If the seller has financed the aircraft and liens exist on the aircraft, the escrow agent will call on the seller's bank to submit any necessary documents to release its liens once any money owed to the seller's bank is paid through the escrow.

Once the buyer's money and the seller's title transfer documents arrive in the escrow agent's possession, the transaction is ready to be completed. To complete the transaction, the escrow agent will usually perform a last-minute title search to make sure that clear title can be passed to the buyer. If everything checks out okay, the escrow agent will wire transfer the buyer's funds (including the buyer's deposit and funds financed by the buyer's bank) to the seller and/or the current lienholders on the aircraft. Simultaneously, the escrow agent will file the aircraft bill of sale with the FAA along with the buyer's registration application. At this time, any lien releases necessary to pass free and clear title to the aircraft to the buyer are usually filed with the FAA. See Figs. 1-1 and 1-2 illustrating a typical escrow transaction.

It is highly advisable to consider the use of an escrow agent for your aircraft sale or purchase transaction. The peace of mind it can give you will be well worth the relatively small price of the service.

Contingencies

It is almost inevitable that your aircraft purchase or sale will have certain contingencies. A contractual contingency means that if something happens (or doesn't happen), your deal may be terminated. One of the most common contingencies for an aircraft transaction is the need for the buyer to be satisfied with the condition of the aircraft after a thorough prepurchase inspection. Often, the contract can give the buyer a certain amount of time to perform the inspection and make a decision on the airplane. Your contract should also indicate who is responsible for the costs of the prepurchase inspection. Generally, it will be the buyer's responsibility to pay for the inspection. But that is not always the case.

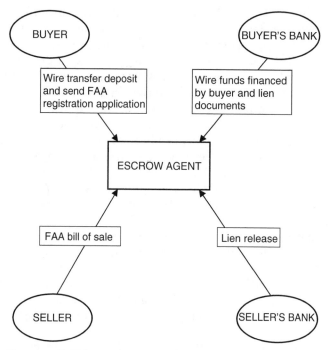

Figure 1-1   Initial steps in an aircraft escrow transaction.

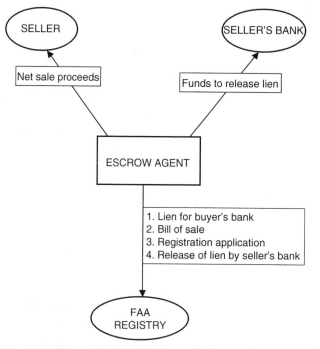

Figure 1-2   Closing the aircraft escrow transaction.

Another common contingency in an aircraft buy/sell agreement is the financing contingency. If you are the purchaser, you may not be able to (or desire to) pay for the aircraft you are purchasing in cash. It will be important for you to have the ability to terminate the transaction if you are unable to get a loan. It is equally important that the contract provide for an appropriate amount of time to allow for the loan to be approved and to have the necessary funds in place.

Less common contingencies might require that a buyer procure a hangar or tie-down spot at a particular airport for the transaction to proceed. From time to time, noncitizen aircraft purchasers may want to condition the purchase on their ability to get an aircraft registered with their own country's registry or the U.S. aircraft registry. A full discussion of this situation is found in Chap. 5.

Delivery of aircraft

In most aircraft purchase and sale transactions, the buyer and seller may be in different locations. It is very important for your sales contract to specify when, where, and how the aircraft will be delivered from the seller to the buyer.

In some cases, the place of delivery may have sales and or use tax implications. If the place of delivery is an interim location for the aircraft, you will want to ensure that the aircraft's temporary stay does not inadvertently trigger sales or use tax (see Chap. 17 for a fuller discussion of this issue).

Transfer of title/risk of loss

As indicated earlier in this chapter, it is very important to ensure that your contract clearly states when title to the aircraft transfers and when risk of loss transfers from the seller to the buyer. This should be fully spelled out so the parties will be able to properly plan for their insurance needs.

Tax issues

From the seller's perspective, it may be important to note that any sales tax, use tax, or other taxes are solely the responsibility of the aircraft purchaser. However, there may be occasions where the seller is responsible for the collection of tax on the aircraft sales transaction. Any questions regarding responsibility for taxes should be addressed in the agreement.

Title warranties

If you are the buyer, it is going to be very important for you to get an unconditional guarantee that the aircraft you are purchasing from your seller really belongs to the seller. However, this guarantee may not be worth much to you once you have paid the seller and the seller is nowhere to be found when problems arise. This is why you will need to do a thorough title search (see Chap. 3).

Prepurchase inspection

Of course, if you are an aircraft buyer, it is always good practice to perform a thorough prepurchase inspection of the aircraft you are proposing to purchase. From your perspective, the entire agreement should be contingent on the findings of the prepurchase inspection. This is also a good time to think about the contract you may enter into with the company you are hiring to do your prepurchase inspection. Does the agreement spell out what you want done for the inspection? Are there any protections for you in case the inspection misses something it should have caught? If you are purchasing an aircraft with no implied or express warranties, your prepurchase inspection becomes all the more important. Treat this part of your transaction with great care.

Aircraft condition and warranties

As indicated earlier, this may be one of the most significant issues in your contract. If you are the seller, you will want to push for an "as is" sale. If you are the buyer, you may be seeking certain warranties regarding the condition of your aircraft. The resolution of some of these issues may depend on whether the seller is a broker, manufacturer, or private individual or entity. Remember that entries in aircraft maintenance logbooks may also produce express warranties. Do you want to make such warranties if you are the aircraft seller? In any case, you should think all of this through and reduce it to writing in your agreement.

Liquidated damages provisions

In certain circumstances, it might be appropriate for your aircraft sales contract to contain a liquidated damages provision. A liquidated damages provision is a contractual promise by one or both parties to the agreement to pay a certain amount in damages in case of a breach.

A liquidated damages provision will only be enforceable if it is a reasonable estimate of the damages that will be incurred if one of the parties breaches the aircraft sales contract. If the amount of damages more closely resembles a penalty or punishment (because it is unreasonable), then your liquidated damages provision may not be enforceable.

As an example, let's say that you are the aircraft seller and you have just contracted with a buyer for the sale. You may be able to estimate that if the buyer breaches the contract, you will be liable for additional insurance payments, hangar or tie-down fees, demonstration flights, advertising costs, and other reasonable expenses related to renewed efforts to sell the aircraft after the buyer's breach. If these expenses can be reasonably estimated, you may wish to specify a liquidated damage amount in your contract.

If you are a buyer, you may also have an interest in a liquidated damage provision. For instance, if the seller backs out of the sale after an agreement has been made, you may be able to estimate the costs of your prepurchase inspection, travel costs, legal fees, and other costs that will be incurred if your aircraft purchase falls through.

Not every contract needs a provision for liquidated damages. However, this type of planning may eliminate costly disputes over how much compensation you should pay or be paid if your aircraft buy/sell deal goes sour.

Agreement modifications

It is always advisable for you to include a provision that states that any modifications to your written agreement must be made in writing and signed by both parties. It is not uncommon for certain elements of a deal to change after a contract is signed.

For example, after an agreement has been signed, your prepurchase inspection may uncover a radio that is not properly functioning. Based on this discovery, you and the seller may agree to a reduction of price for the aircraft.

If you have an agreement in writing for a certain sales price, you should modify the original sales price by drafting a modification to the agreement, signed by both parties, indicating the newly reduced sales price. If you fail to do this, the law may make it difficult for you to later argue that the sales price was anything different than indicated in your signed aircraft sales agreement.

Indemnification and legal fees

What happens if things go bad with your aircraft transaction? You may want to expressly provide that if you are damaged as a result of the other party's breach of your aircraft buy/sell agreement, the other party will be responsible for making you whole. This may include attorney fees and other costs you may have incurred in pursuing your claim for breach of contract and damages. As a general rule, this type of provision may be more beneficial to buyers than sellers. Typically, the buyer performs his end of the deal by simply paying the amount due under the contract. However, more often, the seller has more duties in terms of delivery, providing an airworthy aircraft, living up to warranties, and other potential duties. Think it through with your counsel before you put an indemnification clause in your agreement—it could be a double-edged sword.

Dispute resolution and governing law

It is most likely that your aircraft sales transaction will go off without a hitch. However, if it does not, the costs to remedy the situation could be high. You should ensure that your agreement addresses the possibility of disputes.

Because buyers and sellers often live in different states, selection of the state law to govern disputes is critical. You may also help clarify things if you specifically identify the state court or federal court that you wish to have jurisdiction over your dispute. If a dispute arises, it may cost more to determine where a trial will occur than the costs for the trial itself. If you do some planning when you draft your contract, you may save yourself substantial legal fees later on.

Sometimes you may also want to agree that any disputes will be submitted to arbitration. There are various arbitration services throughout the United States that may be able to help you come to a resolution of your dispute more cheaply than the court system. You can specify in this section whether you want your arbitration to be binding or nonbinding. This is an important consideration that will require your consultation with qualified legal counsel.

Signatures and dates

Finally, each party will need to sign and date the aircraft sales agreement. It would be best if both parties to the agreement sign originals that each of them can keep. That means that two copies of the agreement will require original signatures.

Sometimes, time is short and you may need to use fax signatures. You may also want to agree that each party can sign one copy of the agreement. This is called signature by counterparts. If either of these situations applies to you, you should indicate in the agreement that fax signatures or signatures by counterparts are acceptable.

Legal Aspects of Aircraft Financing

One of the bigger issues you may have to deal with when you are planning an aircraft purchase is financing the transaction. The airplane you purchase may carry a higher price tag than your home. Therefore, there may be a need for you to borrow some or all of the money needed to pay the aircraft purchase price. This section will first discuss some of the key legal issues you will deal with in the application process. Next, you'll get an overview of the typical legal documents that you will encounter in just about any aircraft financing deal.

The application process

If you want to get a loan to pay for an airplane, be prepared to collect the same kind of information that you would have to gather if you were purchasing real estate. Lenders want to ensure that you will be a good credit risk. If you don't respond fully to the lender's questions, you may not get the loan you need for your purchase.

One thing you may want to do before you get started with the application process is to evaluate your "credit score." For years, potential creditors have had access to a credit score number assigned to each person who applied for credit. Until recent changes in the law, you usually could not access your own credit score. Now you can.

Your score will fall between 300 and 900 with 900 being the highest possible credit score. Your score is determined on the basis of the following criteria:

- Payment history (about 35 percent of your score)
- Amounts owed (about 30 percent of your score)
- Length of credit (about 15 percent of your score)
- Pattern of credit use (about 10 percent of your score)
- Types of credit in use (about 10 percent of your score)

As a general matter, if your score is 650 or above, you have a good to excellent credit history and you should not encounter any problems with your credit check. A score of 600 to 650 indicates average to good credit histories and should make it feasible to obtain a loan at going rates. If your score is below 600, you might expect inquiries and potential problems obtaining a loan.

Beyond your credit history, lenders will also want to know about your current and potential earnings picture. Essentially, your prospective lender wants to make sure you generate enough income or have enough assets to make your loan a safe bet. To get this information, the prospective lender will probably request a personal financial statement and/or copies of past filed income tax returns.

A personal financial statement will request information regarding your employment history and annual income. The lender may verify this information with a phone call to your employer with your signature on a waiver allowing the information to be released by your employer. The financial statement usually includes information concerning your current bank accounts along with a detailed balance sheet which requires your detailed disclosure of all your assets and liabilities.

Your lender may also request copies of prior years' tax returns for you and/or your business. This may be especially important to the lender if you are self-employed. If you lost your copies of previously filed tax returns, you can recover them by filing a request with the Internal Revenue Service on an IRS Form 4506. If you simply want an IRS transcript of your income tax information for a particular year (name, social security number, exemptions, filing status, self-employment tax, adjusted gross income, taxable income, and tax shown on return), there is no charge for the service. If you want a copy of a tax return, the IRS will charge you a fee (currently $23.00 per tax period). Sometimes your lender will ask you to sign an IRS Form 4506 in case it wants to get copies of the return directly from the IRS.

Loan documents

There are usually three loan documents you will probably get during the course of your airplane purchase. These three documents consist of a promissory note, security agreement, and perhaps a personal guarantee.

Promissory note. The primary loan document you will be asked to sign is called a promissory note. When you sign the promissory note, you are making a promise to pay the holder of the note the amount owed plus any interest as stated in your note.

You should keep in mind that your note might be held by a succession of different holders who you will have to pay. This is possible because in most cases, your note will be a negotiable instrument. That essentially means that it can be transferred from one holder to another.

The promissory note will indicate the amount that you owe to the holder, plus a stated interest rate as negotiated between you and your lender. The rates and the terms of the loan can vary widely in various buy/sell agreements. When you review your note, it is important to note the due dates of principal and interest payments. It is also important for you to carefully review the provisions defining default. Defaulting on a loan is serious business. This is not time for you to be engaging in a financially risky transaction.

Security agreement.   The promissory note is the lender's legally enforceable promise that you will repay the loan. However, this promise may not mean much to the lender if you do not have any assets to repay the loan.

In view of this risk, the lender will want to get a security interest in your aircraft. Essentially, this means that if you default on your promise to repay the loan plus interest, the lender may repossess your aircraft. If you lender is a bank and you keep funds at the bank or are required to keep funds at the bank (this is often called a compensating balance), the lender will have the right to take those funds.

If your lender repossesses your aircraft, you generally have a right to try to redeem the aircraft before it is sold. This will often mean a full payment of the loan balance or a substantial piece of the loan balance. However, you may fail in your attempt to redeem your aircraft because someone else purchased the aircraft when the lender put the plane up for sale. If someone else purchases the aircraft, the difference between what they paid and your loan balance is called the deficiency. That is the amount that you may now be required to pay to the lender to bring closure to the loan.

It is also important to note that in most cases, any change in titie to your aircraft may trigger the need to pay off your loan in full. This type of loan provision is often called a "due on sale" clause. That essentially means that you cannot transfer title to any other person or entity as long as a loan balance exists without authorization from your lender.

Guarantees.   If you decide to form a separate entity to own your aircraft (see Chap. 2), your lender may require you and/or your fellow members/shareholders/partners to sign a guarantee making you personally responsible for the repayment of the promissory note. The lender will do this because in most cases, the entity you have formed may have no assets other than an airplane and most likely has no earnings or credit history. Therefore, the lender wants to be able to turn to you as a potential safety net for payment of the aircraft-financing loan.

Essentially, your guarantee will state that if the aircraft-owning entity defaults on its obligations to the lender, the lender will turn to you for payment. In order to collect, the lender may pursue your personal assets and/or garnish your salary or wages. A personal guarantee is something to take very seriously. The law will enforce them against you. Your personal credit and assets will be in jeopardy if you default on your obligations under a guarantee.

2

Entity Choices for Aircraft Ownership

One of the more important decisions you'll have to make as an aircraft owner is the choice of legal entity or form for aircraft ownership. There are several different choices. Careful consideration of your circumstances and the intended uses of your aircraft will go a long way to helping you make the best decision.

Here's a list of the specific forms of ownership and types of legal entities you might consider using for aircraft ownership:

- Individual ownership
- Partnership
- Co-ownership
- Limited liability company or limited liability partnership
- Corporation

Each of these forms of aircraft ownership has its positives and negatives. This chapter will provide you with an overview of the various choices in an effort to guide you toward the best fit for your circumstances.

If you are considering entering into an ownership arrangement with multiple owners, you should review this chapter for a legal overview of each entity and then turn to Chap. 13 for more detailed guidance on how to operate an aircraft within the entity structure you are selecting. If you are planning to use your aircraft for business purposes, you may also want to review this chapter and turn to Part 6 of this book for a discussion of tax issues.

Individual Ownership

Owning an aircraft as an individual is perhaps the simplest form of aircraft ownership. If your only intention is to use your aircraft for personal purposes and you don't intend to permit anyone else to operate your aircraft, this may be a form of ownership for you to consider.

Formation and liability

As an aircraft purchaser, all you will need is a bill of sale transferring legal title to the aircraft to your name. The registration certificate you receive from the FAA will carry your name and designated address.

Along with the simplicity of this arrangement, there are also some disadvantages. If you are like most aircraft owners and potential aircraft owners, you may have concerns about possible legal liability if an aircraft you own is involved in an incident or accident injuring persons or property.

As an individual owner of an aircraft, you may be exposed to personal liability for any injury to property or persons caused by your aircraft. It is because of this that many aircraft owners seek a different form of aircraft ownership. They don't want to be personally responsible for such potentially large sums of money.

However, if you are the only person who will be pilot in command of your aircraft, you will always be exposed to liability if your aircraft does injury or harm to persons or property. Setting up a separate legal entity will not shield you from this exposure. Therefore, if you will be the only person owning and piloting an aircraft, individual ownership may be a viable choice for you.

Of course, that does not mean selecting individual ownership of an aircraft means that you are surrendering your ability to protect yourself against damages caused by accidents. To give yourself some peace of mind, you will have to carefully select the right insurance and insurance coverage. For a more detailed discussion of insurance and liability issues, see Part 4.

Taxation

As far as tax ramifications are concerned, there will be no income tax issues if you are merely using an individually owned aircraft solely for personal use. If you are self-employed and you use your aircraft to further your business interests, you will probably be reporting your business income and expenses on an IRS Form 1040 Schedule C. The Schedule C is nothing more than an income statement prepared for the IRS. Any income from your business (including income or expenses related to your aircraft) will be taxed at your personal income tax rate. If you are an employee using your aircraft to further your employer's business, your aircraft expenses will most likely be accounted for on an IRS Form 2106 as an employee business expense. Details on the deductibility and proper reporting of aircraft business expenses can be found in Part 6.

Partnership

If you are sharing the ownership of your aircraft with one or more other owners, then you might consider a partnership. However, you must keep in mind that a partnership will only be appropriate for aircraft ownership purposes if your situation meets certain legal tests as outlined below.

Formation and liability

A partnership is defined as an association of two or more persons to carry on a business for profit as co-owners. Therefore, one or more persons volunteering to be a part of the partnership may create a legal partnership. You are not required to evidence a partnership with any formal or written agreement.

Another important point to note is that the law does not allow for recognition of a partnership simply because there are two or more persons involved. As indicated above, there must be a joining of the parties to own a business for the purpose of making profit. Without a profit motive, you are really entering into an association that is more appropriately classified as a co-ownership.

For certain purposes a partnership is considered a legal entity separate and apart from its owners. For instance, a partnership can purchase and dispose of an aircraft in the partnership name.

However, for purposes of liability protection, a partnership is still treated as a mere aggregate of all the partners. This means that in an aircraft partnership, you may be held personally liable for the negligence of another partner in the operation of the aircraft.

If you are going to enter into a partnership for the purposes of aircraft ownership, it is very important to have a solid partnership agreement. A more detailed discussion of aircraft partnership agreements is found in Chap. 13.

Taxation

As far as tax issues go, a partnership is considered to be a "pass-through" entity with an individual partner's share of profit or loss being passed on to the individual partner. At that point, the individual partner's profits will be taxed at his/her individual income tax rates.

For tax reporting purposes, a partnership files a tax return on an IRS Form 1065. Any profit or loss from the partnership is then distributed proportionately to the individual partners and taxed at the individual partners' personal income tax rates. A report form called an IRS 1065 K-1 is distributed to each individual partner indicating his/her share of the partnership's profits or losses.

Co-Ownership

Another possible arrangement for aircraft ownership is the co-ownership. Many people tend to confuse a partnership and a co-ownership. As indicated above, a partnership involves two or more persons owning an aircraft as part of a business with intent to make a profit. A co-ownership is simply two or more persons joining together to own an aircraft. A co-ownership arrangement does not require any profit-making intentions on the part of the aircraft owners.

Just as in a partnership, individual co-owners are personally liable for any damages caused by an aircraft owned by the co-owners. Each co-owner is potentially liable to the full extent of any damages caused by the aircraft. Again, the only way to properly protect against this exposure is to obtain adequate insurance coverage.

Because a co-ownership by definition does not exist as a business enterprise, it is not, as a general matter, subject to taxation. Therefore, in most cases there is no need for an income tax report to be filed by an aircraft co-ownership.

However, if you are thinking about co-ownership, you still have the need for an ownership agreement and aircraft operating agreement. This will ensure that each owner has a clear understanding of his/her rights and duties as a part owner of an aircraft. A fuller discussion of aircraft co-ownerships and related agreements is found in Chap. 13.

Limited Liability Company

Because of the increasing desire of aircraft owners to decrease exposure to liability, the limited liability company (LLC) option has become much more attractive in recent years. An LLC is a form of an unincorporated business association. Before 1990 only a handful of states recognized LLCs. At present date, all states have laws recognizing LLCs. Although the legal framework for LLCs varies from state to state, there are many common threads in the LLC laws.

LLCs may be attractive to you as an aircraft owner because they are noncorporate business organizations that provide limited liability to all of the LLCs owners (LLC owners are usually referred to as members). At the same time, you or someone else may act as a designated manager of the LLC.

Formation

If you want to form an LLC, you must comply with state law and rules for establishing an LLC. Some states will require your LLC to have at least two members. However, a growing number of states now permit one-person LLCs. Once formed, your LLC is a separate legal entity that is separate and distinct from its members. As a member, you are generally not personally liable for the LLC's debts and obligations. Your LLC may own property (including an aircraft) and contract in its own name and carry on any lawful business (with some exceptions in certain states).

The paperwork you'll be required to file for the formation of an LLC varies from state to state. However, in almost every case, you must prepare and file articles of organization with the appropriate state agency. A few states may limit the life of an LLC to 30 years or less. Some states may also require a termination or dissolution date for the LLC.

Specimen LLC articles of organization have been provided in the CD that comes with this book. However, you should seek counsel familiar with local or state law to prepare your LLC articles of organization. Every state has its own special requirements for LLC articles of organization and you will want to ensure that you have drafted the document in accordance with applicable state law.

When you select the name of your LLC, you will most likely be required to use the letters "LLC" in the name of the company. A few states allow the use of the designation "limited company" or the abbreviation LC.

To get started, you and any other members of the LLC may contribute cash, property, services, or promissory notes, among other possible contributions. Members can be held liable for contributions they have promised to make to the LLC.

If your LLC has more than one member, it is very important that the members of the LLC prepare an operating agreement. The operating agreement is a governing document that details how the LLC will operate and the rights and duties of the LLC's members. Several states require that the operating agreement be in writing. A basic LLC operating agreement can be found in the accompanying CD. In most states, the LLC operating agreement should include, at a minimum, provisions related to:

1. The manner in which the business and affairs of the LLC will be managed and how the LLC will be operated

2. Managing the LLC

3. How the members will share the LLC's assets and earnings

4. The rights of members to assign all or partial interests in their LLC membership

5. Admission of new members

6. Procedures for the amendment of the LLC's operating agreement

7. An address for the LLC

8. An address for a designated resident agent for the LLC

Beyond the basic LLC operating agreement, it will be very important for you to draft an aircraft operating agreement that will spell out members' rights and duties related to the LLC's aircraft. A sample aircraft operating agreement can also be found in the CD in the back of the book.

An LLC with an aircraft as its major asset may find a new home for the aircraft outside of its state of origin. You will not have to create a new LLC for this situation. Instead, you should simply register your current LLC as a foreign limited liability company in your new state. There is often a fee for this registration. Sometimes there may be a substantial penalty if you started to engage in business as an LLC before you applied for registration. Before you've moved your aircraft to a different state, it is always a good idea to make sure that you are in compliance with your new state's laws. You should also consider any local or state income tax ramifications, including sales/use taxes and/or registration fees or personal property taxes (see Chap. 17).

Member's rights

As a member of an LLC you do not have a property interest in any particular aircraft or other property owned by the LLC. Instead, you get a personal property interest in the following two items: (1) a financial interest which is the right to any company distributions and (2) a management interest in the LLC.

Your basic financial interest in the LLC includes the right to share in profits and losses and the right to company distributions. If your LLC has a provision in its operating agreement for the sharing of profits, profits must be distributed accordingly. If no provision is contained in your operating agreement, then the percentage must be distributed on the basis of the value of the members' contributions.

Much like profits and losses, as a member of an LLC you will have a right to share in any distributions of the LLC's assets. If your LLC's operating agreement fails to specifically address the proportionate distributions to each member, the distributions will generally be made on the basis of the value of your contributions to the LLC.

In most states, a member of an LLC has the right to withdraw from the LLC and receive payment for his/her membership share. In order to secure this right, you generally must give appropriate notice as indicated by the law or in your LLC's operating agreement.

Almost every state LLC statute provides that as a member, you have the right to participate in the management of the LLC. Every LLC statute allows an LLC to be managed by one or more managers who may be, but do not have to be, members. In a member-managed LLC, the members have actual and apparent authority to bind the LLC. In a manager-managed LLC, the managers have the authority, and the members generally do not have the ability to exercise real or apparent authority to bind the LLC.

Several LLC statutes specify the voting rights of members, subject to a contrary agreement in the LLC's operating agreement. Typically, your LLC's members will be given the authority to vote on proposals that will:

- Adopt or amend the LLC's operating agreement.
- Admit a new member.
- Sell all or substantially all of the assets of the LLC.
- Merge an LLC with another LLC.

As a general rule, LLC members will have the right to bring an action on behalf of an LLC to recover a judgment in its favor if the managers or members with authority to bring the action have refused to do so. Members may also be permitted to assign their interest in an LLC. This does not mean that the person assigned the interest becomes a member of the LLC. The assignment will merely permit the assignee to assume the financial interests of the assigning member in the LLC.

Members' duties

Just as in partnerships, the general duties of loyalty and care are also applicable to LLCs. Some states expressly impose these duties by law while others rely on common law to preserve these duties. To a large extent, the person who has the duties is determined by whether the LLC is member-managed or manager-managed.

If your LLC is manager-managed, the appointed manager generally will have a duty of due care. In several states, this duty will expose the manager to liability for grossly negligent, reckless, or willful misconduct. In other states, the LLC manager will be liable if it is deemed that his/her conduct does not meet a "prudent man" standard (this standard asks whether his/her actions comport with what a prudent man would do in a similar situation). Managers will generally also have a fiduciary duty to the LLC and other members of the LLC, whereas the members have no fiduciary duties. If an LLC is member-managed, the members all carry the same duty of loyalty and care that managers are exposed to in manager-managed LLCs.

Liability

Perhaps the most attractive feature of an LLC is the limited personal liability it offers its members and managers. Most LLC statutes expressly provide that no member shall be personally liable for any debt, obligation, or liability of the LLC solely by reason of being a member or acting as a manager of the LLC.

However, members may be liable to the LLC under certain circumstances. A member may be personally liable to the LLC for failure to make an agreed-upon contribution to the LLC. A member may also be liable to the LLC if he/she receives a return of contribution that violates the LLC's operating agreement. A handful of states require that members who receive returns of their capital contributions without violating the LLC's operating agreement or any statutory language may be liable to the LLC for a period of time if funds are needed to pay creditors. Of course, it is very important that members of an LLC keep their personal funds in accounts that are separate from the LLC's accounts. Mixing personal and LLC funds can create exposure to personal liability for LLC debts.

Dissolution

An LLC is generally dissolved automatically upon any of the following events:

- The dissociation of a member
- The expiration of the LLC in accordance with the LLC's articles of organization
- Any specified event that causes dissolution
- The unanimous written consent of all LLC members
- A judicial decree of dissolution

Dissociation of a member means that a member has terminated his/her membership in the LLC. This can be caused by the death, withdrawal, incompetence, expulsion, or bankruptcy of the member. Almost every state law permits the LLC to continue to do business through unanimous consent or a majority vote.

If an LLC is dissolved, the assets of the LLC will be distributed as follows:

- First, assets will be distributed to the LLC's creditors, including members and managers who may have loaned money or property to the LLC.

- Second, assets will be distributed to members and former members in satisfaction of liabilities for unpaid distributions, except as otherwise agreed.

- Third, assets will be distributed to members for return of their contributions, except as otherwise agreed.

- Finally, assets remaining will be distributed to members in the proportion they share LLC distributions, except as otherwise agreed.

Taxation

An LLC has the advantage of being very flexible for taxation purposes. It can either pass its profits or losses directly to its members as a "pass-through" entity, or it can be treated as a separate corporate entity.

A domestic single-member LLC is considered by default to be a sole proprietorship. This means that all of its income and expenses will be included in its member's personal income tax return. Specifically, these LLC income and expense items will be reported on an IRS Schedule C (Form 1040). In fact, a single-member LLC generally is not required to apply for a special tax identification number from the IRS. The IRS is willing to use the member's social security number as the LLC's tax identification number.

A multimember, domestic LLC will be treated by default as a partnership for tax purposes. That means that each member will report his/her share of profits or losses on a personal tax return. Specifically, the profits or losses will be reported on an IRS Partnership Tax Return (Form 1065). Each individual member will have his or her share of profits or losses passed through to an individual tax return through a Form 1065 K-1.

An LLC can also elect to be treated as an entity separate and apart from its owners. If it elects this treatment, it must file an IRS Form 8832. It will then be taxed as a separate entity and will use corporate income tax returns (IRS Form 1120).

The selection of the proper tax treatment of your LLC is an important step. It must be done carefully and with the guidance of a professional tax adviser.

Corporations

In the early days of the United States, corporations were formed only by special acts of state legislative bodies. There was a fear that they might permit large accumulations of money and other resources. Over time, those fears subsided and most states eventually passed laws permitting the formation of corporations without special legislative acts. Now, all states have corporation codes that are generally uniform from state to state.

Formation of corporations

The usual steps in the formation of a corporation are as follows:

1. Planning and structuring
2. Preparing articles of incorporation

3. Filing the articles of incorporation with the state

4. State approval of articles or issuance of certificate of incorporation

5. Organizational meeting of the corporation's directors

6. Drafting and approval of bylaws

Planning and structuring.   This is the start-up phase of the corporation where plans are developed for the future corporation. There are several important issues that must be considered at this stage.

One of the first items for consideration is the selection of the state in which you choose to incorporate and operate. The two do not have to be the same. You may decide that it is best to incorporate in a particular state because of tax or other considerations. However, for the sake of convenience (or tax savings), the corporation's aircraft will be operated from a base in another state. If you do this, you should ensure that you properly register your corporation in the state where the aircraft is located.

This is also the time to decide whether you want your corporation to be a stock or nonstock corporation. Sometimes it is advantageous for certain flying clubs and other co-ownerships to establish a nonstock corporation. The owners of such corporations are called members. Owners of stock corporations are issued shares of stock and are referred to as stockholders or shareholders.

In some states, you will also have the option of forming a close or closely held corporation. This option may work well for sole shareholder corporations or corporations with very few members. It will allow the corporation to be managed by the shareholders directly. This typically eliminates the need for a board of directors.

Articles of incorporation

An attorney licensed in your state should prepare your corporation's articles of incorporation. The document itself is relatively simple. However, the processes of filing the articles, paying fees, and requesting appropriate copies typically require the expertise of legal counsel. As a general matter, most articles of incorporation require the following:

1. A statement that the incorporator is not a minor (usually this means 18 years of age or older).

2. A clear indication of the name of the corporation (in most states, the name must be followed by the words or abbreviations "corporation," "company," "limited," "incorporated," "Inc.," or "Ltd.").

3. The duration of the corporation's existence (usually perpetual).

4. The purpose for the corporation's existence.

5. The address of the corporation's principal office.

6. The address of the corporation's registered agent (the registered agent is a designated resident of the state of incorporation who will receive service of process in case a lawsuit is filed against your corporation).

7. If the corporation issues capital stock, the articles of incorporation should indicate how many shares are authorized to be issued (this number should be determined by your attorney because it may have tax ramifications).

8. Some states may require the names and addresses of each of the initial directors or incorporators.

9. In most cases the articles of incorporation must be signed by an incorporator (you should note that an incorporator may have no function other than signing the articles of incorporation—he/she does not have to be a shareholder, member, director, or officer of the corporation).

A "generic" set of articles of incorporation for a stock and a nonstock corporation can be found in the accompanying CD. Please note that articles of incorporation are very state-specific and the sample documents included in the CD are to be used only as a guide. Your attorney will know the rules for the state where you want to locate your aircraft corporation.

Once the articles are prepared, they must be mailed to the appropriate state office for processing and approval. One problem that may come up at this point is name availability. If the name you have chosen for your corporation is already in use by another corporation, you cannot use that name. One way you can avoid the problem is to "reserve" your name with the state. However, this may cost more than you are willing to pay. Your best bet may be to check on your chosen name's availability ahead of time (often a phone call to the state or a check of the state's records on the Internet will suffice). This may not guarantee that the name will still be available when you file, but there is a pretty good chance your chosen corporate name will still be there for you if you file shortly after you check. Experience indicates that most states will take about one to three weeks to process articles of incorporation that have been mailed.

If you are in a big hurry, you may want to check to see if your state has an expedited service for processing your articles of incorporation. This may allow for a 24- or 48-hour turnaround. Usually this service carries a price and requires fax or overnight mail, but it may be worth it to you if you need your corporation established quickly.

It is often prudent for you to request "certified" copies of your articles of incorporation or extra copies of your certificate of incorporation from your state incorporation office. It may cost you a few extra dollars, but it is a good idea to keep a certified copy with your corporation's records.

Bylaws

Your corporation's bylaws are the internal rules and regulations of the corporation. The bylaws should be prepared at or around the same time that your corporation's articles of incorporation are filed.

There is no need to file your bylaws with the state incorporation offices. However, every state requires that each corporation have bylaws drafted.

As a general matter, bylaws should include, at a minimum, the following:

1. The location of the corporation's principal offices.
2. A section regarding members or shareholders that details:
   a. Who will qualify as a member or shareholder.
   b. The usual day and time of annual meetings.
   c. Rules for special meetings.
   d. A description of where meetings will take place.
   e. Details for providing notice of meetings.
   f. How many members/shareholders must be present for voting to take place at meetings (this number is generally called a "quorum").
   g. Provisions for meetings by informal voting by speakerphone or written unanimous consent.
   h. For a nonstock or membership corporation, provisions detailing the terms of membership.
3. If the corporation has a board of directors or board of trustees, there should be a section that outlines:
   a. The general powers of the board.
   b. Number of directors allowed by the corporation.
   c. The term for any directors.
   d. A procedure for electing directors.
   e. Procedures for removing directors.
   f. Procedures for directors who resign.
   g. Meetings of the board.
   h. The possibility of informal actions by the board without the need for meetings.
4. All corporations will require officers; therefore, there should be provisions regarding the selection and duties of officers. This section can include the following:
   a. An identification of the officers necessary for the corporation (most states require at least a president, treasurer, and secretary).
   b. The length of term for each officer.
   c. Procedures for removing an officer.
   d. Procedures for the resignation of an officer.
   e. An outline of the duties and powers of each office held.
5. The bylaws should clearly indicate how the document might be revised or modified (i.e., Will it take a vote of members/shareholders or only a vote of directors?).
6. The bylaws should also provide for the indemnification of board members and/or officers.

A sample set of bylaws is provided in the CD that accompanies this book. As always, keep in mind that different states might have different requirements for bylaws. More important, each corporation will have its own special management issues that my need to be addressed by a set of bylaws. Therefore, you should seek the assistance of an attorney in your state to draft corporate bylaws.

Organizational minutes

Before your corporation can begin conducting business, it must hold an organizational meeting. At the organizational meeting, the corporation's initial board (or incorporators) will approve and adopt its bylaws, appoint directors or trustees, and appoint officers for the corporation. Usually the corporation will also use this opportunity to authorize the issuance of shares of stock (if the corporation is a stock corporation), approve any preincorporation contracts, and select a bank. If the corporation is going to file for tax-exempt status with the Internal Revenue Service, then this may be a good time to authorize the application for tax-exempt status.

In many cases, it may be difficult for the shareholders or members of a corporation to get together for a meeting. That's where the provisions for written informal actions will come in handy. With such provisions in place, this entire process can be dealt with in writing with no need for an actual meeting. Sample organizational minutes may be found in the accompanying CD.

Shareholder's agreement

If your corporation has more than one shareholder, a written agreement among the shareholders is highly recommended. A well-drafted shareholder agreement will detail the rights and responsibilities of each shareholder and the corporation. Typically, the parties to the shareholder's agreement will be the corporation, and each of the shareholders. Some of the common provisions that may be found in a shareholder's agreement include the following:

1. Identification of all parties and the number of shares held by each shareholder.

2. A provision outlining restrictions on the ability of a shareholder to dispose of his/her shares of stock.

3. Provisions permitting the corporation a "right of first refusal" on shares tendered for purchase at a price determined by formula or agreement.

4. Rules allowing the corporation to purchase shares of stock upon the death of a shareholder.

5. A formula or approach to periodically determining the value of the corporation's shares of stock.

6. Any necessary provisions for the corporation to purchase life insurance on shareholders (the idea is to get the insurance proceeds to the corporation to allow it to pay for the deceased shareholder's share of the aircraft).

7. Provisions for the corporation to finance the purchase of shares upon death or intended transfer of a shareholder's share.

A sample shareholder's agreement is found in the CD in the back of the book. As always, you must keep in mind that your situation may differ from the situation addressed in the sample agreement. The specimen agreement provided should be used solely for the benefit of you and your counsel in drafting a shareholder agreement that will suit the needs of your specific situation.

Ongoing records and bookkeeping requirements

In order for a corporation to retain its status as an entity separate from its owner(s), it must maintain certain formalities. First, there must usually be regular meetings of the board of directors and/or shareholders. These meetings do not have to be physical gatherings. If provided for in the bylaws, the meetings can be accomplished by telephone or using an informal written action of the directors and or shareholders. It may sometimes seem cumbersome or unnecessary to comply with these formalities. However, if you do not, you may be risking the possibility of being held personally liable if the corporation is sued for damages in the future.

Most states will also require that the corporation pay an annual fee to keep its corporate status alive. Some states call the fee a personal property tax. In other places it might be labeled a franchise tax. Regardless of what the tax is named, it is incumbent upon you to keep careful track that these requirements are complied with in order to ensure that the corporation does not lose its status as a corporation.

One thing you do not want to neglect is the opening of a corporate bank account. The mixing of your personal funds with the corporation's funds is a sure-fire way to put any protection you may have from personal liability at risk. Therefore, money from the corporation needs to be kept separate and apart from the owner's funds.

Liability protection

If you have kept up with the formalities of keeping your corporation legally alive, your corporation should effectively protect you from personal liability to creditors. This may be particularly important in a corporation that owns an aircraft that may be flown by many different pilots.

As a general rule, shareholders or members of a corporation may only be subject to liability that equals their investments or contributions to the corporation. This is obviously very significant because it eliminates the ability of the corporation's creditors to come after your personal assets.

Dissolution

As indicated earlier, one of the more important features of a corporation is its perpetual existence. However, this does not mean that your aircraft corporation must exist forever. There is always the possibility that circumstances will change and you (or other shareholders/members) will wish to terminate the corporation. There is also the possibility that unforeseen difficulties will arise in the management of your aircraft corporation and it will be involuntarily terminated.

Voluntary termination.   Most states will permit either the shareholders or (in limited circumstances) the incorporators to terminate a corporation. Generally speaking, a board of trustees or board of directors will not have the authority to terminate a corporation.

If your aircraft corporation has never been established, the incorporators can usually dissolve it. Usually this would happen when the corporation has not yet issued shares of stock or accepted memberships and not transacted any business. If this is the case, the incorporators can simply file articles of dissolution and the state will then recognize the dissolution, in many cases by issuing a certificate of dissolution.

If the corporation has been active and transacted business, it must be dissolved by an action of the shareholders. This process varies from state to state, but it usually starts with a resolution by the board of directors of the corporation to dissolve the company. The shareholders are then given an opportunity to meet and vote on the proposed dissolution. In most states a two-thirds majority is required to authorize the dissolution of a corporation. In a minority of states a simple majority vote will be necessary. If the appropriate number of shareholder votes authorizes the dissolution, articles of dissolution can be filed with the state with the state returning a certificate of dissolution.

Involuntary dissolution.   If problems arise in the management of the corporation, the corporation may be involuntarily dissolved. Most states provide that a shareholder can file a lawsuit requesting the court to dissolve a corporation. However, the courts will allow this remedy only in certain circumstances. An outline of those circumstances is presented below.

Oppression of minority shareholders.   In most states, oppression of minority shareholders is grounds for dissolution of a corporation. Oppression of minority shareholders involves unfair treatment by majority shareholders. An example might be a corporate flying club where a few majority shareholders typically make the aircraft unavailable to minority shareholders.

Deadlock.   Many states will provide for the dissolution of a corporation if it can be proved that the corporation's management can't function because it is deadlocked. This is not a common occurrence but it is possible with a closely held company. And many corporations holding aircraft as assets are indeed closely held with only a handful of shareholders.

Mismanagement.   Courts are usually reluctant to interfere with the management of a corporation. However, there may be cases where it is established that the corporation is so grossly mismanaged that it should be dissolved to protect shareholders and creditors.

Dissolution by the state.   The states provide the legal authority for corporations to exist. They also have the authority to terminate a corporation's legal existence.

TABLE 2-1   Comparison of Different Entity Choices

	Individual	Co-ownership	Partnership	Limited liability company	Corporation
Owner(s)	One only	More than one	More than one	One or more than one	One or more than one
Life of entity or organization	Limited by owner's choice or death	Limited by owners' choice or death	Limited by owners' choice or death	In most states unlimited; in some states 30–50 years	Usually indefinite
Personal liability for organization's liabilities	Owner is personally liable	Owners are personally liable	Partners are personally liable	Members are not personally liable	Stockholders/ members are not personally liable
Business entity tax treatment	All profits/losses directly affect owner's personal tax liability	Not applicable	Partner's share of profits/losses passes through to affect owner's personal tax liability	Members can elect to have profits/losses pass through to personal taxes or to have LLC treated as separate entity for tax purposes	Treated as separate entity for tax purposes unless shareholders elect subchapter S status and corporation qualifies for S treatment

States will generally terminate a corporation's existence for failure to file required reports, failure to pay taxes or other state fees, and failure to appoint or maintain a registered agent. If a corporation is terminated for any one of these reasons, it is usually quite easy to reinstate the corporation. Most states allow for a corporation to remedy the underlying problem and file articles of revival that will automatically reinstate the corporation.

Dissolution by creditors.   Many states now allow creditors to force the dissolution of a corporation. This would normally require the creditors to show that they have an unsatisfied claim or judgment and that the corporation is insolvent.

Summary

Choosing the right ownership entity for your aircraft will be an important decision. Sometimes the decision will be relatively simple. In other cases it may require that you seek professional legal and/or tax assistance. Table 2-1 is a summary of some of the key features for each form of aircraft ownership available.

Recording and Registration

3

Recording Aircraft Ownership and Liens

An aircraft can be a big investment for most people and/or businesses. In view of the size of the investment, it is not surprising that most prudent aircraft purchasers would want to know that the airplane they are buying really belongs to the person they are buying it from.

When it comes to buying a piece of real estate, the process is pretty simple. Before you purchase the real estate, you or your lawyer typically hire a title search company to review the local property records to determine whether there is a clear chain of title to the person selling you the real estate.

Aircraft create a somewhat different situation because aircraft are by their very nature movable. Therefore, unlike real estate, it becomes a much more difficult chore to track down a clear chain of title if you have to follow a trail of previous owners from one locality to the next. The fact that aircraft buyers and sellers are often located in different states makes the purchase and sale of aircraft a somewhat unique process.

This chapter will review many of the recording-related issues that are typically encountered by aircraft owners. It will start with an overview of the FAA's federal recording and registration system. Next, you'll walk through a typical example of how the recording system works and get some detailed advice on complying with the mechanics of recording rules. We'll also investigate some special issues that aircraft owners occasionally run across, including encumbrances against engines, propellers, and other aircraft parts; artisan's or mechanic's liens; tax liens; and gifting an interest in an aircraft or transferring it to a trust. This chapter will conclude with an in-depth look at aircraft title insurance policies and what they can and can't do for you.

The Federal Registry

In order to alleviate confusion and uncertainty in determining aircraft ownership issues, the federal government established a system for recording documents that transfer title or affect interests in aircraft ownership. The federal system was first established in 1938. In its current form, the law requires the FAA to maintain a system for recording documents affecting title to aircraft. It was really designed as a sort of federal clearinghouse for recording of aircraft-related title documents so that persons could have ready access to information about an aircraft's ownership history.

Specifically, the law states that any bills of sale, contracts of conditional sale, mortgages, assignments of mortgages, or other instruments affecting title to aircraft are subject to recording. Very important, the law states that no document or conveyance will be valid against any person (other than persons involved in the conveyance or a person who has actual notice of the conveyance) until it is recorded with the FAA.

What does this mean to you? It simply means that if you want to make sure that you own your aircraft, you had better record the bill of sale immediately after you purchase the airplane. Case 3-1 is a very important decision of the United States Supreme Court that demonstrates what can happen if you don't record with the FAA.

Case 3-1

Philko Aviation, Inc., v. Shacket et ux.
Supreme Court of the United States
462 U.S. 406 (1983)

OPINION BY: White, J.

OPINION: This case presents the question whether the Federal Aviation Act of 1958 (Act), 72 Stat. 737, as amended, 49 U.S.C. § 1301 *et seq.* (1976 ed. and Supp. V), prohibits all transfers of title to aircraft from having validity against innocent third parties unless the transfer has been evidenced by a written instrument, and the instrument has been recorded with the Federal Aviation Administration (FAA). We conclude that the Act does have such effect.

On April 19, 1978, at an airport in Illinois, a corporation operated by Roger Smith sold a new airplane to respondents. Respondents, the Shackets, paid the sale price in full and took possession of the aircraft, and they have been in possession ever since. Smith, however, did not give respondents the original bills of sale reflecting the chain of title to the plane. He instead gave them only photocopies and his assurance that he would "take care of the paperwork," which the Shackets understood to include the recordation of the original bills of sale with the FAA. Insofar as the present record reveals, the Shackets never attempted to record their title with the FAA.

Unfortunately for all, Smith did not keep his word but instead commenced a fraudulent scheme. Shortly after the sale to the Shackets, Smith purported to sell the same airplane to petitioner, Philko Aviation. According to Philko, Smith said that the plane was in Michigan having electronic equipment installed. Nevertheless, Philko and its financing bank were satisfied that all was in order, for they had examined the original bills of sale and had checked the aircraft's title against FAA records. At closing, Smith gave Philko the title documents, but, of course, he did not and could not have given Philko possession of the aircraft. Philko's bank subsequently recorded the title documents with the FAA.

After the fraud became apparent, the Shackets filed the present declaratory judgment action to determine title to the plane. Philko argued that it had title because the Shackets had never recorded their interest in the airplane with the FAA. Philko relied on § 503(c) of the Act, 72 Stat. 773, as amended, 49 U.S.C. § 1403(c), which provides that no conveyance or instrument affecting the title to any civil aircraft shall be valid against third parties not having actual notice of the sale, until such conveyance or other instrument is filed for recordation with the FAA. However, the District Court awarded summary judgment in favor of the Shackets, *Shacket v. Roger Smith Aircraft Sales, Inc.,* 497 F. Supp. 1262 (ND Ill. 1980), and the Court of Appeals affirmed, reasoning that § 503(c) did not pre-empt substantive state law regarding title transfers, and that, under the Illinois Uniform Commercial Code, Ill. Rev. Stat., ch. 26, P1-101 *et seq.* (1981), the Shackets had title but Philko did not. 681 F. 2d 506 (1982). We granted certiorari, 459 U.S. 1069 (1982), and we now reverse and remand for further proceedings.

Section 503(a)(1) of the Act, 49 U.S.C. § 1403(a)(1), directs the Secretary of Transportation to establish and maintain a system for the recording of any "conveyance which affects the title to, or any interest in, any civil aircraft of the United States."

Section 503(c), 49 U.S.C. § 1403(c), states:

"No conveyance or instrument the recording of which is provided for by [§ 503(a)(1)] shall be valid in respect of such aircraft . . . against any person other than the person by whom the conveyance or other instrument is made or given, his heir or devisee, or any person having actual notice thereof, until such conveyance or other instrument is filed for recordation in the office of the Secretary of Transportation."

The statutory definition of "conveyance" defines the term as "a bill of sale, contract of conditional sale, mortgage, assignment of mortgage, or other instrument affecting title to, or interest in, property." 49 U.S.C. § 1301(20) (1976 ed., Supp. V). If § 503(c) were to be interpreted literally in accordance with the statutory definition, that section would not require every transfer to be documented and recorded; it would only invalidate unrecorded title *instruments,* rather than unrecorded title *transfers.* Under this interpretation, a claimant might be able to prevail against an innocent third party by establishing his title without relying on an instrument. In the present case, for example, the Shackets could not prove their title on the basis of an unrecorded bill of sale or other writing purporting to evidence a transfer of title to them, even if state law did not require recordation of such instruments, but they might still prevail, since Illinois law does not require written evidence of a sale "with respect to goods for which payment has been made and accepted or which have been received and accepted." Ill. Rev. Stat., ch. 26, P2-201(3)(c) (1981).

We are convinced, however, that Congress did not intend § 503(c) to be interpreted in this manner. Rather, § 503(c) means that every aircraft transfer must be evidenced by an instrument, and every such instrument must be recorded, before the rights of innocent third parties can be affected. Furthermore, because of these federal requirements, state laws permitting undocumented or unrecorded transfers are pre-empted, for there is a direct conflict between § 503(c) and such state laws, and the federal law must prevail.

These conclusions are dictated by the legislative history. The House and House Conference Committee Reports, and the section-by-section analysis of one of the bill's drafters, all expressly declare that the federal statute "requires" the recordation of "every transfer . . . of any interest in a civil aircraft." The House Conference Report explains: "This section requires the recordation with the Authority of every transfer made after the effective date of the section, of any interest in a civil aircraft of the United States. The conveyance evidencing *each such transfer* is to be recorded with an index in a recording system to be established by the Authority." Thus, since Congress intended to require the recordation of a conveyance evidencing *each transfer* of an interest in aircraft, Congress must have intended to pre-empt any state law under which a transfer without a recordable conveyance would be valid against innocent transferees or lienholders who have recorded.

Any other construction would defeat the primary congressional purpose for the enactment of § 503(c), which was to create "a central clearing house for recordation of titles so that a person, wherever he may be, will know where he can find ready access to the claims against, or liens, or other legal interests in an aircraft." Hearings on H.R. 9738 before the House Committee on Interstate and Foreign Commerce, 75th Cong., 3d Sess., 407 (1938) (testimony of F. Fagg, Director of Air Commerce, Dept. of Commerce). Here, state law does not require any documentation whatsoever for a valid transfer of an aircraft to be effected. An oral sale is fully valid against third parties once the buyer takes possession of the plane. If the state law allowing this result were not pre-empted by § 503(c), then any buyer in possession would have absolutely no need or incentive to record his title with the FAA, and he could refuse to do so with impunity, and thereby prevent the "central clearing house" from providing "ready access" to information about his claim. This is not what Congress intended.

In the absence of the statutory definition of conveyance, our reading of § 503(c) would be by far the most natural one, because the term "conveyance" is first defined in the dictionary as "the action of conveying," *i.e.,* "the act by which title to property . . . is transferred." Webster's Third New International Dictionary 499 (P. Gove ed. 1976). Had Congress defined "conveyance" in accordance with this definition, then § 503(c) plainly would have required the recordation of every transfer. Congress' failure to adopt this definition is not dispositive, however, since the statutory definition is expressly not applicable if "the context otherwise requires." 49 U.S.C. § 1301 (1976 ed. and Supp. V). Even in the absence of such a caveat, we need not read the statutory definition mechanically into § 503(c), since to do so would render the recording system ineffective and thus would defeat the purpose of the legislation. A statutory definition should not be applied in such a manner. *Lawson v. Suwannee Fruit & S.S. Co.,* 336 U.S. 198, 201 (1949). Accordingly, we hold that state laws allowing undocumented or unrecorded transfers of interests in aircraft to affect innocent third parties are pre-empted by the federal Act.

In support of the judgment below, respondents rely on *In re Gary Aircraft Corp.,* 681 F. 2d 365 (CA5 1982), which rejected the contention that § 503 pre-empted all state laws dealing with priority of interests in aircraft. The Court of Appeals held that the first person to record his interest with the FAA is not assured of priority, which is determined by reference to state law. We are inclined to agree with this rationale, but it does not help the Shackets. Although state law determines priorities, all interests must be federally recorded before they can obtain whatever priority to which they are entitled under state law. As one commentator has explained: "The only situation in which priority appears to be determined by operation of the [federal] statute is where the security holder has failed to record his interest. Such failure invalidates the conveyance as to innocent third persons. But recordation itself merely validates; it does not grant priority." Scott, Liens in Aircraft: Priorities, 25 J. Air L. & Commerce 193, 203 (1958) (footnote omitted). Accord, Sigman, The Wild Blue Yonder: Interests in Aircraft under Our Federal System, 46 So. Cal. L. Rev. 316, 324-325 (1973) (although recordation does not establish priority, "failure to record . . . serves to subordinate"); Note, 36 Wash. & Lee L. Rev. 205, 212-213 (1979).

In view of the foregoing, we find that the courts below erred by granting the Shackets summary judgment on the basis that if an unrecorded transfer of an aircraft is valid under state law, it has validity as against innocent third parties. Of course, it is undisputed that the sale to the Shackets was valid and binding as between the parties. Hence, if Philko had actual notice of the transfer to the Shackets or if, under state law, Philko failed to acquire or perfect the interest that it purports to assert for reasons wholly unrelated to the sale to the Shackets, Philko would not have an enforceable interest, and the Shackets would retain possession of the aircraft. Furthermore, we do not think that the federal law imposes a standard with which it is impossible to comply. There may be situations in which the transferee has used reasonable diligence to file and cannot be faulted for the failure of the crucial documents to be of record. But because of the manner in which this case was disposed of on summary judgment, matters such as these were not considered, and these issues remain open on remand. The judgment of the Court of Appeals is reversed, and the case is remanded for further proceedings consistent with this opinion.

So ordered.

From a practical perspective, this case instructs that if you purchase an aircraft and fail to record your bill of sale, someone else may be able to buy the same airplane from your seller. If that happens, their right to the aircraft will supersede yours because they would have had no way to know about your purchase of the aircraft when they searched the federal registry. Sure, you might be able to successfully sue the seller for defrauding you, but will you ever collect on your judgment? That's a question that you'd rather not have to deal with—ever.

If nothing else, the *Philko* case also reinforces the need for escrow agents in certain aircraft buy/sell transactions. The next time you hear someone tell you, "Don't worry, I'll take care of the paperwork," think about the *Philko* case and consider the risks before you continue.

From a lawyer's perspective, this case sends a clear signal that as far as notice of aircraft ownership is concerned the federal law preempts all state laws. This means that any interest in an aircraft must be recorded with the federal registry to be valid against innocent third parties (parties without knowledge of previous transactions). Even more important, it signals that any state law that conflicts with the federal recording system will be overruled by the federal law.

This case also addresses another interesting issue for lawyers. If the federal law preempts state law when it comes to issues of notice, does it also preempt state law on questions of priority of ownership? As indicated in this case, the Court answers that question in the negative. As the Court indicated in *Philko*: "Although state law determines priorities, all interests must be federally recorded before they can obtain whatever priority to which they were entitled under state law." Therefore, state laws regarding ownership and priorities in ownership interest are still controlling when it comes to aircraft. However, a failure to record an interest in an aircraft at the federal registry will serve to invalidate any interest you or a creditor may have had under state law. In the end, the simple moral of the story is to promptly record any ownership or security interest you may have in an aircraft with the federal registry.

The *Compass* case (see Case 3-2) followed the *Philko* case. This case deals with the difficult question of who gets priority when a creditor obtains a judgment lien against an aircraft after the aircraft has been sold by the judgment debtor and the judgment creditor files the lien documents before the subsequent purchaser records its purchase.

Case 3-2

Compass Insurance Company, Appellee, v. H. P. Moore, and Moore Flying, Inc., Appellants
United States Court of Appeals for the Eighth Circuit
806 F.2d 796 (1986)

OPINION BY: Ross, Circuit Judge
OPINION: This appeal involves a priority dispute between appellee, Compass Insurance Company (Compass), who obtained a judgment lien on a Cessna aircraft, and appellants, H. P. Moore and Moore Flying, Inc. (Moore), who purchased the airplane. Moore bought the airplane before Compass obtained its judgment lien, but Compass registered its lien documents with the Federal Aviation Administration (FAA) before Moore recorded its purchase.

Compass's claim to the airplane arose as follows. In December 1981, Compass, a New York corporation, obtained a judgment in Dade County, Florida, against Hollywood Flying Service, Inc. (Hollywood) in the amount of $111,088.09. In March 1982, the Dade County, Florida Circuit Court granted Compass an "Order Declaring Judgment Lien" on the Cessna aircraft and another airplane. (A title search of the FAA register in February 1982 had revealed Hollywood as the record owner of the Cessna airplane.) The judgment lien represents Compass's only interest in the aircraft. Then Compass filed its judgment lien with the FAA, and on June 8, 1982, the FAA recorded Compass's interest in the airplane on the FAA register.

Meanwhile in August 1979, title to the aircraft had been transferred in a sale by Hollywood to Sam Vires, d/b/a Mid-South Aircraft Sales (Mid-South), a Tennessee company. In November 1979, Vires sold the airplane to Mid-South. In May 1981, Mid-South sold the aircraft to appellant Moore Flying, Inc., a Missouri corporation. H. P. Moore took possession of the airplane and moved it to Missouri in May 1981.

The 1979 sales from Hollywood to Vires and Vires to Mid-South and the May 1981 sale from Mid-South to Moore were not registered with the FAA until June 30, 1983, approximately one year after Compass had recorded its judgment lien with the FAA. However, by the time Compass obtained its Florida judgment lien on the aircraft in March 1982, the airplane had been purchased by Moore and had been in the possession of Moore in Missouri for ten months.

The Federal Aviation Act established a federal recording system for conveyances of interests in aircraft. Section 503(c) of the Act, 49 U.S.C. § 1403(c) (1982) states that:
> no conveyance or instrument the recording of which is provided for by [the Act] shall be valid in respect of such aircraft * * * against any person other than the person by whom the conveyance or other instrument is made or given, his heir or devisee, or any person having actual notice thereof, until such conveyance or other instrument is filed for recordation in the office of the Secretary of Transportation * * *.

Section 503(d), 49 U.S.C. § 1403(d) (1982) states that a conveyance or instrument recorded under the Act "shall from the time of its filing for recordation be valid as to all persons without further or other recordation * * *."

As this court discussed in *Armstrong v. State Bank of Towner (In re Gelking),* 754 F.2d 778, 780-81 (8th Cir.), *cert. denied,* 473 U.S. 906, 105 S. Ct. 3529, 87 L. Ed. 2d

653 (1985), sections 503(c) and (d) of the Act establish that perfection of an interest in aircraft occurs on the date the instruments creating the interest are filed for recordation. Moreover, "recordation is necessary * * * to reap the benefits of any priority [that] filing for recordation may have established." *Id., citing Philko Aviation, Inc. v. Shacket,* 462 U.S. 406, 76 L. Ed. 2d 678, 103 S. Ct. 2476 (1983).

The Supreme Court's decision in *Philko Aviation, Inc. v. Shacket* established three principles. First, an interest in aircraft which is never recorded with the FAA will have no effect against the rights of any third parties who lack actual notice of the interest. Second, all state laws permitting undocumented or unrecorded transfers to affect the interests of third parties are preempted by the federal Act. Third, as between competing interests which are recorded with the FAA, state law determines priorities.

Although recordation does not establish priority, "failure to record * * * serves to subordinate."
* * *
Failure [to record] invalidates the conveyance as to innocent third persons. But recordation itself merely validates; it does not grant priority.
* * *
Although state law determines priorities, all interests must be federally recorded before they can obtain whatever priority to which they are entitled under state law. *Id.* at 413 (citations omitted).

The district court entered summary judgment in favor of Compass. 621 F. Supp. 125. We reverse because the judgment lien obtained by Compass in March 1982 could not attach to property which at that time was no longer owned by or in the possession of the judgment debtor, Hollywood, and was no longer in the jurisdiction of the Florida Circuit Court. *See, e.g., Bergquist v. Anderson-Greenwood Aviation Corp. (In re Bellanca Aircraft Corp.),* 56 Bankr. 339 (Bankr. D. Minn. 1985); *Curtis v. Carey,* 393 S.W.2d 185 (Tex. Civ. App. 1965); *Marshall v. Bardin,* 169 Kan. 534, 220 P.2d 187 (Kan. 1950), all involving priority disputes with respect to aircraft between attachment or judgment creditors and antecedent purchasers and all according priority to the antecedent buyers.

The foregoing decisions in favor of antecedent purchasers are based on the premise that a judgment creditor's lien on personal property is merely derivative of the judgment debtor's rights and interest in such property. When property of the debtor has been sold prior to entry of a judgment against the debtor, and particularly if the property has been removed from the jurisdiction, a judgment lien cannot attach to it. *See In re Bellanca Aircraft Corp., supra,* 56 Bankr. at 378 ("Where personal property has been conveyed by the judgment debtor to a third party, therefore, no lienable interest in such property remains with the judgment debtor, and the judgment creditor must look elsewhere to satisfy his claim.") *Accord Marshall v. Bardin, supra,* 220 P.2d at 190 ("An attaching creditor acquires no greater right in the property seized than the defendant debtor in the attachment owned.") (Citations omitted.) *Cf. In re Gelking, supra,* 754 F.2d at 781 (concerning consensual security interests, "attachment, of course, is a necessary step in the perfection of a security interest * * *. (U.C.C. §§ 9-203(1), 303(1)). Further, in order for a security interest to attach, the debtor must have 'rights in the collateral.' * * * (U.C.C. § 9-203(1)(c)).")

As a judgment creditor can acquire no rights superior to those of its judgment debtor, Compass's judgment lien could not attach to the Cessna aircraft which had been sold by the judgment debtor in August 1979, resold twice, and then removed from Florida by Moore ten months before Compass sought its declaration of judgment lien in the Dade County, Florida Circuit Court. This result does not defeat the purpose of section 503(c) of the Federal Aviation Act as a judgment creditor is not the kind of innocent third party who engages in transactions in reliance on the FAA register intended to be protected by the Act. *See In re Bellanca Aircraft Corp., supra,* 56 Bankr. at 379 ("While it is clear that purchasers of or persons taking security interests in aircraft must and do significantly rely on the FAA register * * * the same cannot be said of general creditors * * *. Instead such creditors bargain on the basis that, in the absence of collection, a debt may be later reduced to judgment and thereby be enforced against the then existing property interests of the judgment debtor * * *. [They do] not deal on the faith of the FAA register to the extent that [they] should benefit from its protections.") *Accord Curtis v. Carey, supra,* 393 S.W.2d at 189; *Marshall v. Bardin, supra,* 220 P.2d at 190-91. Accordingly, we reverse the judgment of the district court and remand for further proceedings consistent with this opinion.

As indicated in the *Compass* case, the U.S. Court of Appeals ruled that a general creditor, such as *Compass,* can't attach a judgment lien to an aircraft that had already been sold by the judgment debtor. The court ruled that since general creditors don't "deal on the faith of the FAA register," they should not be allowed to benefit from its protections. This reinforces the view that the protection provided by the FAA Registry is for those innocent parties who rely on the register for protection.

The Basics of Recording Interests in Aircraft

By now, you should be a bit more sensitive to the importance of the federal registry in protecting ownership rights in your aircraft. Now let's take a look at a typical example of how the recording system works. After that you'll get some detailed guidance to ensure that you and your counsel properly follow the FAA's rules for recording interests in airplanes.

An illustration of the recording process

Perhaps the best way to review the mechanics of recording aircraft documents is through an example. Let's set the stage. Assume you (for purposes of this example, Joe Buyer) are investigating the purchase of a used Cessna 172. The seller's name is Seller Industries, Inc. (Seller). The aircraft sales price is $140,000. It has passed your mechanic's rigorous prepurchase inspection and you are ready to move forward on the transaction.

Now is a good time for you to contact a reputable title search company to perform a title search on the aircraft. Let's say that after contacting a title and escrow company, you are issued the following report as illustrated in Fig. 3-1.

This is the typical title report you might expect to receive. It indicates that the current registered owner of the aircraft is indeed Seller. That's a good start. It also lists a lien or encumbrance from the First National Bank of America (FNBA). That's probably not unusual. More than likely FNBA financed Seller's aircraft purchase and the amount payable from Seller to FNBA has not yet been paid off.

AIRCRAFT TITLE REPORT

Prepared for: **Search Date: 11/20/03**

Joe Buyer
2 North Avenue
Southtown, Maryland

Aircraft Description:

Registration No. N9999B
Make and Model: C-172
Serial No. 12345

Current Owner:

Seller Industries, Inc.
123 Broad Street
North Creek, Florida

Ownership Type: Individual

Current owner acquired through bill of sale dated 6/15/85, filed with FAA on 6/17/85;
recorded by FAA on 6/23/85 (FAA Document No. B2222).

Previous Owner:

Joe Conveyance
456 Registration Lane
Highway, Missouri

Liens or Encumbrances

Security agreement for $85,000.00 executed on 6/13/85, filed with FAA on 6/16/85,
recorded by FAA on 6/17/85 in favor of First National Bank of America.

Title Search by: _____

Figure 3-1 Specimen aircraft title report.

On the basis of the title report, you decide to go through with the purchase. You are also smart enough to know the value of an escrow agent, so you retain one to assist in this transaction.

In order to purchase the aircraft, you will have to finance $120,000 of the $140,000 purchase price. You have a loan commitment from the Second Bank of Maryland (SBM).

The escrow agent will be expecting receipt of a bill of sale from Seller. In appropriate form, the bill of sale should look something like the illustration in Fig. 3-2.

FORM APPROVED
OMB NO. 2120-0042

UNITED STATES OF AMERICA
U.S. DEPARTMENT OF TRANSPORTATION FEDERAL AVIATION ADMINISTRATION

AIRCRAFT BILL OF SALE

FOR AND IN CONSIDERATION OF $ THE UNDERSIGNED OWNER(S) OF THE FULL LEGAL AND BENEFICIAL TITLE OF THE AIRCRAFT DESCRIBED AS FOLLOWS:

UNITED STATES REGISTRATION NUMBER N 9999B

AIRCRAFT MANUFACTURER & MODEL C-172

AIRCRAFT SERIAL No. 12345

DOES THIS DAY OF 19
HEREBY SELL, GRANT, TRANSFER AND DELIVER ALL RIGHTS, TITLE, AND INTERESTS IN AND TO SUCH AIRCRAFT UNTO:

Do Not Write In This Block
FOR FAA USE ONLY

PURCHASER

NAME AND ADDRESS
(IF INDIVIDUAL(S), GIVE LAST NAME, FIRST NAME, AND MIDDLE INITIAL.)

Joe Buyer
2 North Avenue
Southtown, MD 99999

DEALER CERTIFICATE NUMBER

AND TO EXECUTORS, ADMINISTRATORS, AND ASSIGNS TO HAVE AND TO HOLD SINGULARLY THE SAID AIRCRAFT FOREVER, AND WARRANTS THE TITLE THEREOF.

IN TESTIMONY WHEREOF HAVE SET HAND AND SEAL THIS DAY OF 19

SELLER

NAME (S) OF SELLER (TYPED OR PRINTED)	SIGNATURE (S) (IN INK) (IF EXECUTED FOR CO-OWNERSHIP, ALL MUST SIGN.)	TITLE (TYPED OR PRINTED)
Seller Industries		
Inc.	Sam Seller	President

ACKNOWLEDGMENT (NOT REQUIRED FOR PURPOSES OF FAA RECORDING: HOWEVER, MAY BE REQUIRED BY LOCAL LAW FOR VALIDITY OF THE INSTRUMENT.)

ORIGINAL: TO FAA

AC Form 8050-2 (9/92) (NSN 0052-00-629-0003) Supersedes Previous Edition

Figure 3-2 Sample bill of sale for sale of aircraft.

The bill of sale is a very simple form. As with any other document recorded with the FAA it should include the following information to comply with FAR Section 49.43: make, model, and manufacturer's serial number. Notice that the president of Seller will sign the bill of sale [the signature must be in ink—see FAR Section 49.13(a)] and indicate clearly that he is signing the document in his capacity as the president of Seller Industries, Inc.

The escrow agent will also be looking for a release from Seller's bank, FNBA. FAR Section 49.17(d)(4) requires that this release come in the form of FAA AC Form 8040-41, Part II. This release is sent in anticipation of your payoff to FNBA so that FNBA's lien on the aircraft will be released. The release form issued by FNBA to the escrow agent is probably going to look something like the illustration provided in Fig. 3-3.

From you, the escrow agent will be expecting a completed application for aircraft registration. In this case, your registration application should be prepared as illustrated in Fig. 3-4.

Finally, the escrow agent will be expecting that your financing bank, SBM, will be recording a lien for the amount that it financed in this transaction. SBM may submit a security agreement on FAA AC Form 8040-98. SBM's security agreement is illustrated in Fig. 3-5.

Once all this paperwork has been filed, the escrow agent is ready to close this transaction once your $20,000 deposit and the $120,000 you financed has been wire transferred to the escrow account. Once your money and your bank's money hits the account, the bill of sale and registration application are immediately filed with the FAA to record your ownership interest in the aircraft. The escrow agent will then record the release of lien from the seller's bank and the security agreement from your bank. The end result is that you are now the proud owner of a Cessna 172 with a lien for $120,000 from your bank.

The mechanics of recording

With an example under your belt, we can now review some detailed guidance to make sure you get things right the first time when you submit documents for recording with the FAA. This discussion of mechanics may not be exciting, but it is important to make sure that your transaction goes smoothly through the recording process.

Getting the name right.   It's always a good idea to get the names right on any legal documents. However, it is absolutely critical in the case of recording documents with the FAA. FAA Advisory Circular Form 8040-93 ("AC 8040-93") advises the following when it comes to the use of names:

- Use the same name on all documents. If your name is listed as William A. Jones on one document, don't use Bill Jones on a subsequent document.

- Use full names for corporations (exactly as indicated on the corporation's articles of incorporation).

THIS FORM SERVES TWO PURPOSES
PART I acknowledges the recording of a security conveyance covering the collateral shown
PART II is a suggested form of release which may be used to release the collateral from the terms of the conveyance

PART I - CONVEYANCE RECORDATION NOTICE

NAME (Last name first) OF DEBTOR

Seller Industries, Inc.

NAME and ADDRESS OF SECURED PARTY/ASSIGNEE

First National Bank of America
12 Creditors Lane
Moneytown, MA 99999

NAME OF SECURED PARTY'S ASSIGNOR (if assigned)

Do Not Write In This Block
FOR FAA USE ONLY

FAA REGISTRATION NUMBER	AIRCRAFT SERIAL NUMBER	AIRCRAFT Mfd. (BUILDER) and MODEL
N9999B	12345	C-172

ENGINE MFR. AND MODEL	ENGINE SERIAL NUMBER(S)

PROPELLER MFR. and MODEL	PROPELLER SERIAL NUMBER(S)

THE SECURITY CONVEYANCE DATED __6/13/85__ COVERING THE ABOVE COLLATERAL WAS RECORDED BY THE FAA AIRCRAFT REGISTRY ON __6/17/85__ AS CONVEYANCE NUMBER __XXXX009__.

_____
FAA CONVEYANCE EXAMINER

PART II - RELEASE - (This suggested release form may be executed by the secured party and returned to the FAA Aircraft Registry when terms of the conveyance have been satisfied. See below for additional information.)

THE UNDERSIGNED HEREBY CERTIFIES AND ACKNOWLEDGES THAT HE IS THE TRUE AND LAWFUL HOLDER OF THE NOTE OR OTHER EVIDENCE OF INDEBTEDNESS SECURED BY THE CONVEYANCE REFERRED TO HEREIN ON THE ABOVE DESCRIBED COLLATERAL AND THAT THE SAME COLLATERAL IS HEREBY RELEASED FROM THE TERMS OF THE CONVEYANCE. ANY TITLE RETAINED IN THE COLLATERAL BY THE CONVEYANCE IS HEREBY SOLD, GRANTED, TRANSFERRED, AND ASSIGNED TO THE PARTY WHO EXECUTED THE CONVEYANCE, OR TO THE ASSIGNEE OF SAID PARTY IF THE CONVEYANCE SHALL HAVE BEEN ASSIGNED: PROVIDED, THAT NO EXPRESS WARRANTY IS GIVEN NOR IMPLIED BY REASON OF EXECUTION OR DELIVERY OF THIS RELEASE.

This form is only intended to be a suggested form of release, which meets the recording requirements of the Federal Aviation Act of 1958, and the regulations issued thereunder. In addition to these requirements, the form used by the security holder should be drafted in accordance with the pertinent provisions of local statutes and other applicable federal statutes. This form may be reproduced. There is no fee for recording a release. Send to FAA Aircraft Registry, P.O. Box 25504, Oklahoma City, Oklahoma 73125.

ACKNOWLEDGMENT (If Required By Applicable Local Law):

DATE OF RELEASE: _____

First National Bank of America
(Name of security holder)

SIGNATURE (in ink) _____

TITLE _____ Assistant Vice President _____

(A person signing for a corporation must be a corporate officer or hold a managerial position and must show his title. A person signing for another should see Parts 47 and 49 of the Federal Aviation Regulations 14 CFR)

Re-print 5/92

Figure 3-3   Sample release.

FORM APPROVED
OMB No. 2120-0042

UNITED STATES OF AMERICA DEPARTMENT OF TRANSPORTATION **FEDERAL AVIATION ADMINISTRATION-MIKE MONRONEY AERONAUTICAL CENTER** AIRCRAFT REGISTRATION APPLICATION	CERT. ISSUE DATE

UNITED STATES
REGISTRATION NUMBER **N** 9999B

AIRCRAFT MANUFACTURER & MODEL

C-172

AIRCRAFT SERIAL No.

12345

FOR FAA USE ONLY

TYPE OF REGISTRATION (Check one box)

☒ 1. Individual   ☐ 2. Partnership   ☐ 3. Corporation   ☐ 4. Co-owner   ☐ 5. Gov't.   ☐ 8. Non-Citizen Corporation

NAME OF APPLICANT (Person(s) shown on evidence of ownership. If individual, give last name, first name, and middle initial.)

● Joe Buyer

TELEPHONE NUMBER: (243) 555-1212

ADDRESS (Permanent mailing address for first applicant listed.)

Number and street: 2 North Avenue

Rural Route: P.O. Box:

CITY	STATE	ZIP CODE
Southtown	Maryland	99999

☐   **CHECK HERE IF YOU ARE ONLY REPORTING A CHANGE OF ADDRESS**
ATTENTION! Read the following statement before signing this application.
This portion MUST be completed.

A false or dishonest answer to any question in this application may be grounds for punishment by fine and / or imprisonment (U.S. Code, Title 18, Sec. 1001).

CERTIFICATION

● I/WE CERTIFY:

(1) That the above aircraft is owned by the undersigned applicant, who is a citizen (including corporations) of the United States.

(For voting trust, give name of trustee: _____), or:

CHECK ONE AS APPROPRIATE:

a. ☐ A resident alien, with alien registration (Form 1-151 or Form 1-551) No. _____

b. ☐ A non-citizen corporation organized and doing business under the laws of (state) _____ and said aircraft is based and primarily used in the United States. Records or flight hours are available for inspection at _____

(2) That the aircraft is not registered under the laws of any foreign country; and

(3) That legal evidence of ownership is attached or has been filed with the Federal Aviation Administration.

NOTE: If executed for co-ownership all applicants must sign. Use reverse side if necessary.

TYPE OR PRINT NAME BELOW SIGNATURE

	SIGNATURE	TITLE	DATE
EACH PART OF THIS APPLICATION MUST BE SIGNED IN INK.	Joe Buyer		
	SIGNATURE	TITLE	DATE
	SIGNATURE	TITLE	DATE

NOTE   Pending receipt of the Certificate of Aircraft Registration, the aircraft may be operated for a period not in excess of 90 days, during which time the PINK copy of this application must be carried in the aircraft.

AC Form 8050-1 (12/90) (0052-00-628-9007) Supersedes Previous Edition

Figure 3-4 Sample aircraft registration application.

Paperwork Reduction Act: This information is collected to provide evidence of security interest. The information is used by the aircraft registry in the recording of security interests. We estimate that it will take approximately 30 minutes to complete this form. The information is required to perfect a security interest in the described collateral. (This form or equivalent may be used.). This information is public information, and no confidentiality is provided. An agency may not conduct or sponsor, and a person is not required to respond to a collection of information unless it displays a currently valid OMB control number. The number associated with this collection is 2120-0042.

DEPARTMENT OF TRANSPORTATION
FEDERAL AVIATION ADMINISTRATION
CIVIL AVIATION REGISTRY
AIRCRAFT REGISTRATION BRANCH
P. O. Box 25504
Oklahoma City, Oklahoma 73125
AIRCRAFT SECURITY AGREEMENT

NAME & ADDRESS OF DEBTOR

Joe Buyer
2 North Avenue
Southtown, MD   99999

NAME & ADDRESS OF SECURED PARTY/ASSIGNOR

Second Bank of Maryland
25 Lender Lane
Promissory, MD   20000

ASSIGNED/NAME & ADDRESS OF ASSIGNEE

ABOVE SPACE
FOR FAA USE ONLY

Date: xx/xx/xx

A security interest is hereby granted to the secured party on the following described collateral:
AIRCRAFT (FAA registration number, manufacturer, model, and serial number):

NOTICE: ENGINES LESS THAN 750 HORSEPOWER AND PROPELLERS NOT CAPABLE OF ABSORBING 750 OR MORE RATED SHAFT HORSEPOWER ARE NOT ELIGIBLE FOR RECORDING.
ENGINES (manufacturer, model, and serial number):

PROPELLERS (manufacturer, model, and serial number):

SPARE PARTS LOCATIONS (air carrier's name, city, and state):

together with all equipment and accessories attached thereto or used in connection therewith, including engines of _____ horsepower, or the equivalent, and propellers capable of absorbing _____ rated takeoff shaft horsepower, described above, all of which are included in the term aircraft as used herein.
The above described aircraft is hereby mortgaged to the secured party for the purpose of securing in the order named:
FIRST: The payment of all indebtedness evidenced by and according to the terms of that certain promissory note, herein below described, and all renewals and extensions thereof.

Note bearing date of __xx/xx__ executed by the debtor and payable to the order of Second Bank of MD _____ in the aggregate sum of $ 120,000 with interest thereon at the rate of ___7.00___ per centum per annum, from date, payable in installments as follows:

The principal and interest of said note is payable in __120__ installments of $ __1,760__ each on the ___1st__ day of each successive month

beginning with the xx___ day of __xx_____. The last payment of $__1,760__ is due on the ___xx_____ day of xx____ .

SECOND: The prompt and faithful discharge and performance of each agreement of the debtor herein contained made with or for the benefit of the secured party in connection with the indebtedness to secure which this instrument is executed, and the repayment of any sums expended or advanced by the secured party for the maintenance or preservation of the property mortgaged hereby or in enforcing their rights hereunder. Said debtor hereby declares and hereby warrants to the said secured party that they are the absolute owner of the legal and beneficial title to the said aircraft and in possession thereof, and that the same is free and clear of all liens, encumbrances, and adverse claims whatsoever, except as follows: (If no liens other than this mortgage, indicate "none".)

AC Form 8050-98(7/00)

Figure 3-5 Sample aircraft security agreement.

It is the intention of the parties to deliver this instrument in the state of __Maryland__ .

Provided, however, that if the debtor, their heirs, administrators, successors, or assignees shall pay said note and the interest thereon in accordance with the terms thereof and shall keep and perform all and singular the terms, covenants, and agreements in this security agreement, then this security agreement shall be null and void.

Time is of the essence of this security agreement. It is hereby agreed that, if default be made in the payment of any part of the principal or interest of the promissory note secured hereby at the time and in the manner therein specified, or if any breach be made of any obligation or promise of the debtor herein contained or secured hereby, or if any or all of the property covered hereby be hereafter sold, leased, transferred, mortgaged, or otherwise encumbered without the written consent of the secured party may deem himself insecure, then the whole principal sum unpaid upon said promissory note, with the interest accrued thereon, or advanced under the terms of this security agreement, or secured thereby, and the interest thereon shall immediately become due and payable at the option of the secured party.

Upon default, secured party may at once proceed to foreclose this mortgage in any manner provided by law, or the secured party may at its option, and they are hereby empowered so to do, with or without foreclosure action, enter upon the premises where the said aircraft may be and take possession thereof; and remove and sell and dispose of the same at public or private sale, and from the proceeds of such sale retain all costs and charges incurred by secured party in the taking or sale of said aircraft, including any reasonable attorney's fees incurred; also all sums due him on said promissory note, under any provisions thereof, or advanced under the terms of this security, and interest thereon, or due or owing to the said secured party, under any provisions of this security agreement, or secured hereby, with the interest thereon, and any surplus of such proceeds remaining shall be paid to the debtor, or whoever may be lawfully entitled to receive the same. If a deficiency occurs, the debtor agrees to pay such deficiency forthwith.

Said secured party or his agent may bid and purchase at any sale made under this mortgage or herein authorized, or at any sale made upon foreclosure of this security agreement.

In witness whereof, the debtor has hereunto set _____ his _____ hand and seal on the day and year first above written.

ACKNOWLEDGMENT:
(If required by applicable local law)

NAME OF DEBTOR ____ Joe Buyer _____

SIGNATURE(S) (IN INK) _____
(If executed for co-ownership, all must sign)

TITLE _____
(If signed for a corporation, partnership, owner, or agent)

ASSIGNMENT BY SECURED PARTY

For value received, the undersigned secured party does hereby sell, assign, and transfer all right, title, and interest in and to the foregoing note and security agreement and the aircraft covered thereby, unto the assignee named on the face of this instrument at the address given, and hereby authorizes the said assignee to do every act and thing necessary to collect and discharge the same. The undersigned secured party warrants and agrees to defend the title of said aircraft hereby conveyed against all lawful claims and demands except the rights of the maker. The undersigned secured party warrants that the secured party is the owner of a valid security interest in the said aircraft. (A Guaranty Clause or any other provisions which the parties are desirous of making a part of this assignment should be included in the following space.)

Dated this _____ day of _____ .

ACKNOWLEDGMENT:
(If required by applicable local law)

NAME OF SECURED PARTY (ASSIGNOR) _____

SIGNATURE(S) (IN INK) _____
(If executed for co-ownership, all must sign)

TITLE _____
(If signed for a corporation, partnership, owner, or agent)

THIS FORM IS ONLY INTENDED TO BE A SUGGESTED FORM OF SECURITY AGREEMENT WHICH MEETS THE RECORDING REQUIREMENTS OF TITLE 49, UNITED STATES CODE, AND THE REGULATIONS ISSUED THEREUNDER. IN ADDITION TO THESE REQUIREMENTS, THE FORM OF SECURITY AGREEMENT SHOULD BE DRAFTED IN ACCORDANCE WITH THE PERTINENT PROVISIONS OF LOCAL STATUTES AND OTHER APPLICABLE FEDERAL STATUTES. THIS FORM MAY BE REPRODUCED.

SEND, WITH APPROPRIATE FEE, TO: AIRCRAFT REGISTRATION BRANCH
P.O. BOX 25504
OKLAHOMA CITY, OKLAHOMA 73125-0504

AC Form 8050-98(7/00)

Figure 3-5 *(Continued)*

- Don't use nicknames.

- Show "Jr." and/or "Sr." if appropriate.

- Women should use their own names (e.g., Janet A. Smith rather than Mrs. Raymond Smith).

- Trade names are not sufficient. Use the legal name of the business (e.g., individual owner, corporation, LLC, partnership, etc.).

While we're on the topic of names, you should be aware of what you need to do if your name changes after signing a recorded document. If your name changes or a corporate name changes, you should send the FAA the original or a true copy of the legal instrument approving the change (e.g., marriage license).

Signature requirements.   Another place where people make errors in recordable documents is the signature block. AC 8040-93 suggests the following rules for ensuring that your signature is correctly executed:

- If you are the sole owner of an aircraft, you should sign as an individual owner. Your title is "Owner." If you have a business with a "doing business as" name, that name should also be used. In such a case the signature block of a recordable document would read: "Joseph L. Kelly, d/b/a Kelly Enterprises, signed by Joseph L. Kelly."

- If you are part of a co-ownership, each of the co-owners must sign every document related to the aircraft. Each co-owner title is "Co-owner." The FAA also requires that if you have a corporate co-owner, any documents signed by the corporate co-owner must be signed by an officer identified as an officer of the corporation.

- For a partnership, one partner may sign documents to be recorded. She should indicate that she is signing as a partner (use the title "Partner") and show the full partnership name.

- If your airplane is owned through a corporation, the full name of the corporation must be shown on the document to be recorded. An officer should sign the document and indicate his or her title.

A related issue is the appropriate way to get someone else's signature to "count" for FAA purposes. Sometimes this is necessary because of death, disability, bankruptcy, or other reasons. Again, AC Form 8040-93 has some suggestions for you:

- Guardians must submit a certified true copy of the court order that appointed them as guardians. The FAA also requires that the names of the airplane owner and guardian must appear on the document, with the guardian's signature indicating that he/she is signing on behalf of the owner.

- If someone has a power of attorney, they should be prepared to present the FAA with a certified true copy of the power of attorney document, signed by the aircraft's owner.

- If an aircraft is owned by an estate, the executor or personal representative can execute aircraft ownership documents on behalf of the estate. The executor or personal representative should submit the document signed plus the written authority (Letters Testamentary or Letters of Administration) that person has to act on behalf of the estate. The executor should make it clear that he or she is signing as a representative of the estate (e.g., Joseph George, Personal Representative, Executor).

- A trustee must submit either a certified true copy of the court order appointing him or her as trustee or a certified true copy of the complete trust instrument. Trustees in bankruptcy must submit a certified true copy of their appointment by the court.

Finally, if you ever make an error in submitting a document for recording or a transaction is not completed after documentation is submitted, you may submit a statement signed by both parties explaining the erroneous filing. The statement should include the date the documents were submitted, the names of the parties, the FAA recording date and document number, and a $4.00 fee.

Special Recording Topics

In addition to the usual transfers of ownership in an aircraft, there are other recording issues that require the attention of aircraft owners. In certain circumstances, liens and encumbrances can be recorded against specified parts of your aircraft. There are also occasions where certain types of tax liens may validly exist on an aircraft you purchased and you will not be able to determine the existence of these tax liens by a search of FAA's records. It is not uncommon for mechanics to file a lien against an aircraft. Finally, there are circumstances where you may desire to gift all or a portion of your aircraft to an individual or charity. Each of these topics will be addressed below.

Recording interests in engines, propellers, appliances, and accessories

Beside the ability to record liens against aircraft, the FAA also provides for the recordation of liens and other encumbrances against engines and propellers. FAR Section 49.41(a) specifies the conditions that must be met before any encumbrances may be recorded against engines and/or propellers. First, the engine or propeller must be specifically identified by make, model, horsepower, and manufacturer's serial number. Second, if someone wants to record an interest against a specific engine, the engine must have 750 or more rated takeoff horsepower. A propeller must be capable of absorbing 750 or more rated takeoff shaft horsepower.

If these requirements are met, an encumbrance may be recorded against specifically identified engines or propellers. Releases on engines and propellers are recorded in the same manner as they are for aircraft. The creditor files FAA AC Form 8040-1 (or an equivalent form) describing the encumbrance and the engine(s) or propeller(s) released.

For air carriers, encumbrances may also be recorded against aircraft appliances and spare parts. FAR Section 49.41 provides for the recordation of encumbrances against an air carrier's engines and propellers (regardless of horsepower ratings), appliances, and spare parts. In order for the encumbrance to be recordable, the document submitted to the FAA must contain or be accompanied by a statement from the air carrier and specifically describe the location or locations of the spare parts covered by the encumbrance.

Does the *Philko* case apply to recording interests against propellers, engines, appliances, and spare parts? It appears from case law that the answer is clearly yes. In one case involving an encumbrance against an air carrier's spare parts, a federal bankruptcy court held that federal laws related to recording interests in aircraft apply to both aircraft and spare parts (see *In re Avair, Inc.* 96 B.R. 261).

Mechanic's liens

If a mechanic has performed work or provided materials in the process of repairing or maintaining your aircraft, most state laws will permit the mechanic to hold a lien against your aircraft for the value of those services and/or materials. In some states the lien can only be created and maintained by possession of your aircraft. In other states, a mechanic's lien can be created without the need for the mechanic to have possession of your aircraft.

The latest information published by the FAA indicates that the FAA Registry will accept the filing of a mechanic's lien from 32 states plus the Virgin Islands (see Table 3-1). In every other state, the FAA requires a court judgment before it will accept a lien.

Obviously, if you have had work done on your airplane and you have not paid for the work, you may be subject to the filing of a lien against your airplane. If the work on your airplane originated in any of the states or territories listed in Table 3-1, the mechanic does not have to seek or receive a court judgment against you to be able to successfully record the lien.

This means that the lien has the ability to place a "cloud" over your title. In other words, any potential buyer doing a title search on your aircraft will see the

TABLE 3-1 Mechanic's Lien States

Alaska	Indiana	Missouri	South Carolina
Arizona	Iowa	Nevada	South Dakota
Arkansas	Kansas	New Jersey	Tennessee
California	Kentucky	New Mexico	Texas
Connecticut	Maine	North Dakota	Virginia
Florida	Michigan	Ohio	Washington
Georgia	Minnesota	Oklahoma	Wyoming
Illinois	Mississippi	Oregon	Virgin Islands

recorded lien through a title search. If the potential buyer is prudent, he/she may be reluctant to purchase your aircraft until you get the lien cleared from your title report.

Getting the lien cleared may or may not be easy. If it is a simple matter of you owing money to a repair shop, you can pay the money owed and the lien should then be cleared. If you are disputing the amount owed or the validity of the lien, the dispute may require adjudication by the courts. This may cause you significant delays. No matter what, the FAA will generally not get involved in the dispute. The issue may only be resolved by turning to the appropriate state law for guidance.

Tax liens

One of the few holes in the national recording system is the IRS tax lien. FAR Section 49.17(a) recognizes the problem by expressly warning that notices of federal tax liens are not recordable at the FAA Registry because they are required to be filed elsewhere by federal law.

The federal law in question is 26 U.S.C. 6323, which states that liens against personal property (a category of property including aircraft) must be filed in a state or local government office in which the aircraft is located. The law goes on to specifically say that if the IRS wants to file an aircraft lien against you, it will not be subject to the legal requirements of filing under our current national filing system (26 U.S.C. 6323(f)(4)).

This obviously puts you at a disadvantage when you are trying to ensure that you will be getting clear title to an aircraft you may want to purchase. You may know the most recent location of the aircraft you are going to purchase. However, it will be very difficult for you to track down previous locations where the aircraft may have been based in the past.

The bottom line is that you may be subject to a lien that may be as difficult to locate as a needle in a haystack. This tax lien issue has been a nagging problem for aircraft owners for many years. It is difficult to understand why the IRS is given immunity from the FAA's rules for recording liens and encumbrances. In any case, the law is what it is. Thankfully, there have been few documented cases of an aircraft owner losing his/her aircraft to the IRS because of a tax lien created by a previous owner.

Is there any way to completely protect yourself from such an unpleasant surprise? At this point the best answer is no. However, there may be ways to reduce your exposure to harm. Some companies provide federal tax lien services. However, the service is often limited to searching the records of your seller. This may not be adequate if the lien was related to an earlier owner. Another possible source of protection is title insurance (discussed a bit later in this chapter). Recently, a number of title insurance companies have begun to offer title insurance that covers federal tax liens. The amount of coverage tends to be limited, but it can prove helpful if you are unlucky enough to find yourself with the IRS at your doorstep demanding the keys to your airplane.

Repossessed aircraft

If you are looking to purchase a repossessed aircraft, the chain of title will be a bit different than usual. With the typical repossession, the debtor who defaulted on his/her debt generally will not transfer the airplane by a bill of sale to the creditor repossessing the aircraft. The creditor repossesses the aircraft and then records a "Certificate of Repossession of Encumbered Aircraft" with the FAA. A sample of the FAA's suggested form for the certificate is found in Fig. 3-6.

In addition to a repossession certificate (or its equivalent), the FAA will also require the underlying security agreement or a true copy of the security agreement if it has not yet been recorded with the FAA. If the repossession was part of a foreclosure and you are attempting to register your aircraft, the FAA requires that the sheriff, auctioneer, or other authorized person in charge of the sale execute a bill of sale. That person must also state that the sale was made in compliance with local law.

Gifts

Sometimes aircraft owners decide to gift their aircraft to charities, friends, or family members. The best approach to handling this sort of transfer is to execute a deed of gift. An example of a deed of gift for an airplane is presented in Fig. 3-7. As indicated in the deed of gift, there may be occasions when you will have to determine whether there are U.S. Gift Tax considerations. Federal gift taxes will usually come into play when the dollar value of the amount gifted exceeds $11,000 (subject to periodic revision) in any given year. If this is the case, the taxes may be avoided by making a series of annual gifts of a partial interest in the aircraft.

Aircraft Title Insurance

You've probably got a title insurance policy out on your house. Why not take one out on that airplane you just purchased? Let's take a look at aircraft title insurance policies for general aviation aircraft owners so that you can better decide if a title insurance policy is for you.

First, you should be aware that there are aircraft title insurance policies for lenders and for aircraft owners. Of course, lenders are very interested in protecting title to the aircraft they have loaned money on. They often have more at stake than you do as the aircraft owner.

For purposes of this discussion, the focus will be on title insurance for aircraft owners. Although there is no one standard policy, there are a number of common threads running through many of the aircraft title insurance policies in force today.

In order to examine any type of insurance policy, it is important to identify what is covered and what is excluded from coverage. In that regard, aircraft title insurance policies are no different from other insurance policies. There

U.S. DEPARTMENT OF TRANSPORTATION FEDERAL AVIATION ADMINISTRATION

Aircraft Registration Branch
PO Box 25504
Oklahoma City, Oklahoma 73125-0504

CERTIFICATE OF REPOSSESSION OF ENCUMBERED AIRCRAFT

The undersigned hereby certifies that they are the true and lawful holder of a note or other evidence of indebtedness secured by a

_____ on the following described aircraft:
               (Type of Security Agreement)

Aircraft Manufacturer and Model   _____

Aircraft serial number   _____ FAA registration number   _____

Said Security agreement on the above aircraft bears the date of   _____ and was executed by

_____ to   _____

and assigned to   _____ . This Security

agreement was recorded under Title 49, United States Code, Section 44107, on the _____ day of _____, _____,

and was entered in the Civil Aviation Registry as document no.   _____ .

On the   _____ day of _____, _____, the aforesaid   _____ breached the

obligations and promises contained in the Security agreement. The undersigned certifies that the secured party has performed all

obligations imposed on it by the security agreement and applicable local laws; that in accordance with the terms of said security

agreement, and pursuant to the pertinent laws of the state of   _____ ,the undersigned repossessed the aircraft

described above and foreclosed on the   _____ day of   _____, _____ , and that pursuant to local law,

divested the said debtor, and any and all persons claiming by, through or under them, of any and all title they had or may have had,

and the secured party now owns the aforesaid aircraft, or the aircraft has been sold.

NOTE: If the agreement involved was not recorded with the Aircraft   NAME OF HOLDER OF SECURITY AGREEMENT
         Registration Branch, the original or certified true copy
         should accompany this certificate of repossession   _____
                                                                          SIGNATURE (IN INK)

                                                                  _____
                                                                                Title

ACKNOWLEDGMENT (Not required for purposes of FAA recording; however, may be required by local law for validity of the instrument.)
_____

AC Form 8050-4 (2-00) Supercedes previous editions

Figure 3-6   Certificate of repossession.

DEED OF GIFT

I, _____, hereby give, convey, and assign to _____ all of my rights, title,

and interest in the _____ model _____ aircraft bearing the serial number

_____ and the United States Registration Number _____. [If applicable--

The interest conveyed herein is a gift, the value of which does not exceed _____

dollars ($_____).]

Executed on _____, 200_

_____
_____

SUBSCRIBED AND SWORN to before me this _____ day of _____, 200_.

_____
Notary Public

My commission expires _____

Figure 3-7 Deed of gift.

is usually a list of covered items and sometimes an equally long list of items that will cause coverage to be excluded.

You'll typically find coverage for the following situations in an aircraft title policy:

- Ownership disputes where your ownership interest is being challenged because of errors or omissions in FAA records

- If it turns out that somewhere in your aircraft's chain of title there were forged documents and the true owner is now at your doorstep claiming his lawful right to own the aircraft you thought was yours

- If you unknowingly purchase a stolen aircraft and the rightful owner makes a claim to the airplane

- If you unknowingly purchase an aircraft when the aircraft's title was previously transferred under duress, fraud, or undue influence
- When your claim to your aircraft may be challenged because someone in the chain of title lacked contractual capacity (for instance, a minor or someone with a mental disability)
- If your title claim is being challenged because of defects in filing or recording of documents
- If after you purchase the title insurance policy someone forges a mortgage or security interest in your aircraft
- If there are state or federal tax liens on your aircraft (usually subject to dollar limitations)
- If there are mechanic's liens or other nonpossessory liens for labor or materials furnished before the policy date which you did not agree to or create
- If you run into problems because of the "blind spot" in the federal recording system (possibility of someone else recording an interest at the same time you are recording your interest and you are unable to discover their claim)

Perhaps just as important, an aircraft title insurance policy creates a duty to defend for the insurer. This means that if a claim is made against your title to the aircraft, and the claim is arguably covered under your policy, you are entitled to have the insurance company pay for your defense. The duty to defend is an important part of any policy and you will want to carefully ensure that the conditions triggering the insurer's duty meet your satisfaction.

As with any insurance policy, there are also exclusions from coverage that you should consider. Here's a typical list of circumstances where claims against your ownership interest might not be covered by your aircraft title insurance policy:

- If your aircraft title is placed in jeopardy because of your illegal use of the aircraft, including environmental protection laws
- If a government police authority confiscates or seizes your airplane
- If claims arise from outside the United States
- If you created, agreed to, or assumed a defect in title or lien or encumbrance
- If you knew of a potential title problem that could not be found by the title insurance company in its search of records, and you failed to disclose the problem before the policy became effective
- Where liens or encumbrances are created in the normal course of business after the policy becomes effective
- If someone else has possession of the aircraft at the time the policy becomes effective

Again, each policy is going to be a bit different. You and your legal counsel should review different policies to determine which, if any, will best fit your needs. Policy premiums also vary depending on the amount of coverage you are seeking. In most cases you will also find that the premium is payable in a lump sum when you first purchase the policy.

4

Aircraft Registration

The law mandates that you cannot operate a civil aircraft unless the owner
has registered the aircraft with the FAA. This chapter will cover:

- Eligibility for aircraft registration with the FAA
- Registration procedures
- The FAA registration certificate
- Aircraft registration markings
- Registration issues for import and export aircraft

Eligibility for Registration

FAR Section 47.3 requires that an aircraft is eligible for U.S. registration only if
it is owned by:

1. A U.S. citizen.
2. A permanent resident in the United States (green card holder).
3. A noncitizen U.S. corporation if the aircraft is based and used primarily in
   the United States (this means 60 percent or more of its flight time is in the
   United States).

Additionally, the aircraft cannot be registered under the laws of any other
country. In order to qualify as a U.S. citizen, you must meet any one of the following criteria established in FAR Section 47.2:

1. You are an individual who is a U.S. citizen or a citizen of a U.S. possession.
2. You are a member of a partnership in which each partnership member is an
   individual U.S. citizen.

3. You own all or part of a corporation or association created under the laws of the United States or any territory or possession of the United States and the corporation possesses all of the following characteristics:
   - The president is an individual who is a U.S. citizen.
   - Two-thirds or more of the directors (or trustees) of the corporation are U.S. citizens.
   - Two-thirds or more of the corporation's managing officers are U.S. citizens.
   - At least 75 percent of the voting rights in the corporation's stock are owned or controlled by persons who are citizens of the United States or one of its possessions.

If you are a U.S. permanent resident or you otherwise qualify as an individual, partnership, or corporate U.S. citizen, read on. If not, you should turn to Chap. 5 to determine the best approach to qualifying for U.S. registration. Once you've made that determination, you can return to this chapter for a discussion of registration basics.

FAR Section 47.5 (b) requires that an aircraft be registered only in the name of its legal owner. Does that mean that registration equals ownership? The law and regulations expressly answer that question with a clear no. FAR Section 47.5 (c) states that "Section 501(f) of the Act…provides that registration is not evidence of ownership in any proceeding in which ownership by a particular person is in issue." The regulation goes on to state, "The FAA does not issue any certificate of ownership or endorse any information with respect to ownership on a Certificate of Aircraft Registration." Essentially, the FAA is stating that it issues a registration certificate only to the person who appears to be the airplane's owner based on the evidence submitted to the FAA Registry. The FAA Registry will not make any determinations regarding the authenticity or legality of the documents submitted by registration applicants. Therefore, you cannot depend on your registration certificate as proof that you own your airplane—all the more reason to do everything you can to ensure that you have clear title through an aircraft title search and a professionally managed closing on your aircraft sale.

Registration Procedures

The FAA requires the use of FAA Form 8050-1 to apply for registration. A blank sample form is presented in Fig. 4-1 as a reference to the following discussion of the registration process.

The registration application has three copies attached. The first page is white. The second copy page is green. The third copy page is pink. The first and second pages are sent to the FAA. While the FAA is processing your application, you'll keep the pink copy and use it as a temporary registration. Now let's walk through the form to ensure that your application will be complete.

Figure 4-1 Aircraft registration application.

N-number designation

In most cases, this will be an easy chore. If you are purchasing a used or newly manufactured aircraft that has been previously registered in the United States, it already has an N-number assigned. As long as you are not looking for a special or different (customized) N-number, you simply insert the existing N-number in the box provided at the top of AC Form 8050-1.

If you are looking for a custom N-number, you will have to make a request for a special N-number. Your request will have to be submitted in writing. The FAA's website currently indicates that you should include your name, address, phone number, signature, and where appropriate (for instance, an aircraft owned by a corporation) the title of the person making the request. The current fee for this service is $10.00 (with a check made payable to the United States Treasury). A sample letter requesting a custom N-number is presented in Fig. 4-2.

The address used in the sample letter is the FAA's mailing address. If you want to get your request to the FAA by overnight mail you should send the request to: FAA, Aircraft Registration Branch, AFS-700, 6425 S. Denning, Registry Building 118, Oklahoma City, OK 73169.

In deciding the N-number you'd like to use for your airplane, you should keep the following guidelines in mind:

- You can't use N-numbers N1 through N99. These numbers are reserved for FAA use only.

- You can't use N-numbers beginning with NC, NX, NR, or NL (except as provided in FAR Section 45.22 for exhibition, antique, and other specially defined aircraft).

- The letters I and O cannot be used in an N-number.

- Your N-number cannot begin with a zero directly after the N prefix.

- Your N-number cannot exceed five characters in addition to the required prefix N.

- The characters can be one to five numbers (N54310), one to four numbers and one suffix letter (N5432A), or one to three numbers and two suffix letters (N321AB).

Under current procedures, the FAA will reserve your special N-number for one year. By the time one year is completed, you must either have the N-number assigned to a specific aircraft or renew the reservation in writing along with an additional $10.00 fee.

If the FAA approves your special N-number request, it will send you an Assignment of Special Registration Number (AC Form 8050-64). After you have placed the special N-number on the aircraft, you have 5 days to complete and mail the AC Form 8050-64 back to the FAA. You can keep a copy of the form (the form is sent in triplicate) and use that form until you get a permanent registration certificate from the FAA indicating your special N-number.

If your aircraft is an amateur-built aircraft and has never been registered anywhere, you will have to request a special N-number for your airplane. If

```
                         DEAN OWNER
                         1 WEST DRIVE
                      TOWNSHIP, OR 00009

Date: _____

FAA
Aircraft Registration Branch
P.O. Box 25504
Oklahoma City, OK  73125

Dear Sir or Madam:

I am writing to request an N-number assignment for my airplane.  Information regarding
the airplane is as follows:

Make/Builder: Cessna
Type: C-172
Model: M
Serial Number: 00001

This aircraft is currently registered as N_____.  I have enclosed my fee of
$10.00 for this request.

My choices for a registration number are as follows:

1st N123M
2nd N1AM
3rd N1234M
4th N321M
5th N213M


Sincerely,


Dean Owner
```

Figure 4-2   Sample letter to FAA requesting special N-number for aircraft.

your aircraft is being brought into the United States from a foreign country, you will have to submit evidence to the FAA that it was never registered in a foreign country. The FAA is not specific on what constitutes sufficient evidence. Presumably, your sworn affidavit indicating that the aircraft in question has never been registered in any other place should suffice.

Aircraft identification

As indicated in Fig. 4-1, the FAA's registration application requires that you provide the make, model, and serial number for your aircraft. In most cases, this will be a simple task.

Type of registration

As indicated on the FAA's registration application, you may register your aircraft in any one of the following ways:

- Individual
- Partnership
- Corporation
- Co-owner
- Government
- Noncitizen corporation

Actually, there is one additional type of registration permitted by the FAA that is not indicated on the current form. The FAA will permit you to register your aircraft under the name of a limited liability company (LLC). Details on how you should prepare an LLC's application for aircraft registration are provided below.

If you are registering your airplane as an individual, you simply use your name and sign with your name. Your title is "Owner."

If you register as a partnership, the names of each general partner must be shown on the application for registration. For FAA purposes, the name of the partnership is the name that your partnership uses to do business or the names of all the general partners. One general partner can sign for your partnership if the full partnership name is indicated on the registration application and the partner signs with the title "Partner." Keep in mind that if your partnership has a corporation as a partner, the aircraft is not eligible for U.S. registration. FAR Section 47.2 requires that the members of a partnership be individuals only.

Registration of a corporate-owned aircraft is relatively straightforward. You must show the full name of the corporation. A corporate officer or manager with authority must sign the registration application and indicate his or her title with the corporation.

If co-ownership owns your aircraft, each co-owner's name must be shown on the registration application along with each one's respective share of the aircraft (this should also be indicated on the bill of sale). Each co-owner should sign the application. The title for each co-owner is "Co-Owner." The fact that a trade name is used does not negate the need for each co-owner's name to appear on the registration application.

Government applications for registration must be signed by a person with capacity and authority to register an aircraft. This person must indicate his/her title on the registration application.

If you are registering your aircraft as a noncitizen corporation, please turn to Chap. 5 for detailed guidance on the requirements. This is a special form of registration that permits non–U.S. citizens to obtain U.S. aircraft registration if certain conditions are met.

As indicated above, you can also elect to register your aircraft in the name of an LLC. This has become a more common entity choice for aircraft ownership because of the relative ease of forming an LLC and the liability protections that may be obtained through an LLC. However, if you intend to place your aircraft in an LLC, you should be aware that the FAA has some special policies and rules that you will have to comply with in order to successfully register your airplane. The FAA's policies on LLCs are detailed in an FAA memorandum dated August 23, 1999. The FAA's memorandum indicates clearly that the primary concern of the FAA is ensuring that the citizenship of the LLC's members is established. As a result of the FAA's policy on LLCs, you will have to submit additional information along with your aircraft registration package if you wish to register your aircraft through an LLC. A sample letter is shown in Fig. 4-3. However, each situation will be different and you should seek the advice of counsel in preparing the letter. The letter should be sent to the FAA on your attorney's letterhead or the letterhead of your LLC.

In addition to the letter, it might be appropriate and helpful if you send the FAA certified and/or true copies of your LLC's articles of organization and/or operating agreement (if applicable). Although there does not appear to be any written guidance on the issue, you should check or "X" the box marked "Corporation" (box 3) under the "Type of Registration" if you intend to register your airplane under the name of an LLC. When you sign the registration application, you should sign with your name. Your title will be either "Member" or "Managing Member" depending on how your LLC has been structured.

Name, address, and certification

When you indicate the "Name of Applicant," be sure that you are using the identical name to the name shown on whatever evidence of ownership you are using (typically a bill of sale). You must also ensure that the type of registration you choose and the title of the signer agree. For instance, if you checked the "Corporation" box, the signature shown at the bottom of the form must be the signature of a corporate officer or manager identified as such.

When you fill in the address box, be certain that you give the FAA a physical address. A post office or mail drop address will be okay for mailing purposes, but the FAA also wants a physical location shown on the application.

Note that when you sign the application for registration you are making some important certifications. In most cases, you will be certifying that you are a U.S. citizen or an entity (partnership, corporation, LLC) that qualifies as a U.S. citizen.

In the less typical case, you may have to certify that you qualify for registration because you have established a voting trust (see Chap. 5), you are a resident alien, or you are a noncitizen corporation with an aircraft based and used primarily in the United States (see Chap. 5).

Finally, you will have to certify that (1) your aircraft is not registered in a any foreign country and (2) that legal evidence of your ownership is attached to the registration application or has already been filed with the FAA Registry. This legal evidence is typically an aircraft bill of sale.

_____, LLC
Address of LLC

March 4, 2003

Federal Aviation Administration
Aircraft Registration Branch
Mike Monroney Aeronautical Center
Oklahoma City, OK 73125

Re: Registration of N_____ to _____, LLC

Dear Sir or Madam:

In accordance with a memorandum dated August 23, 1999, from the Manager of the Aircraft Registration Branch, AFS-750, I've prepared the following as a statement in support of registration of the above-referenced aircraft in the name of a limited liability company.

1. The full name of the applicant is _____, LLC.

2. The state in which the LLC is lawfully organized is _____.

3. The effective date of the LLC is _____.

4. The name of each of the members of the LLC and the type of entity of each member is indicated below:

Name	Type of Entity	U.S. Citizenship Verification
_____	Individual	U.S. citizen

5. The LLC is managed by its members.

6. The members of the LLC may act independently.

Figure 4-3 Sample letter for registration of aircraft to LLC.

All of these certifications are important and they are made when you sign the registration application. The FAA takes these certifications seriously (as it should) and so does the law. Any false or misleading statements may subject you to criminal sanctions including fines and/or imprisonment.

The Registration Certificate

The first thing to remember is that when you forward your registration application to the FAA, you must keep the pink copy of the form. This copy will act

7. The name of each of the managers of the LLC and the type of entity of each manager is indicated below:

Name	Type of Entity	U.S. Citizenship Verification
_____	Individual	U.S. citizen

The undersigned certifies that the LLC is a U.S. citizen and eligible to register aircraft in its name in that at least 2/3 of the managers or managing members identified above are U.S. citizens within the meaning of 49 U.S.C. Section 40102 (a)(15) and at least seventy-five percent (75%) of the voting interest is owned or controlled by persons who are U.S. citizens or citizens of one of its possessions; that "Citizens of the United States" means (a) an individual who is a citizen of the United States or one of its possessions; or (b) a partnership of which each member is such an individual; or (c) a corporation or association created or organized under the law of the United States or of any state, territory, or possession of the United States of which the president and two-thirds or more of the board of directors or other managing officers thereof are such individuals and in which at least seventy-five percent (75%) of the voting interest is owned or controlled by persons who are citizens of the United States or one of its possessions.

I have attached a copy of the LLC's Certificate of Formation and the confirmation received by the State of _____ indicating that the filing was accepted.

Very truly yours,

_____
Member, _____, LLC

Figure 4-3 *(Continued)*

as your temporary registration. The pink slip is valid as a temporary registration for 90 days. There is no indication for a date on the registration application, other than the date that you sign the application. Presumably that will be the date that starts the 90-day clock in motion. It is very important to keep in mind that your temporary "pink slip" registration may not be used for operations outside the United States. If you must travel outside the United States with your aircraft shortly after its purchase, you will have to request expedited service from the FAA and file a Declaration of International Operations (see Chap. 5).

When the FAA is satisfied that you meet the requirements for registration, it will issue a Certificate of Registration on FAA AC Form 8050-3. If your aircraft was previously registered in the United States or was never registered anywhere, the effective date of your registration certificate will be the date the FAA received your registration application and supporting documents. If your aircraft was previously registered in a foreign country (see registration procedures in the "Import/Export Issues" section later in this chapter), your registration will be effective on the date the FAA issues your registration certificate.

How long is your registration certificate effective? FAR Section 47.41 states that a registration certificate is valid until the FAA suspends or revokes the certificate (which for most holders means indefinitely) unless one of the following happens:

- Your aircraft is registered under the laws of a foreign country.
- You make a written request to the FAA to cancel your aircraft's registration.
- Your aircraft is totally destroyed or scrapped.
- You transfer ownership of your aircraft.
- You, the trustee, or the entity holding ownership in the aircraft lose U.S. citizenship or permanent resident status.
- You have died and 30 days have elapsed since your death.
- Your aircraft is registered under ownership of a noncitizen corporation and the aircraft is no longer based and primarily used in the United States.

If any of these situations arises, either you or the administrator of your estate must send the FAA your aircraft's registration certificate with the reverse side completed. If the administrator of your estate is handling this matter, he or she should send in the certificate within 60 days after your death.

Another responsibility you will have as a registered aircraft owner is keeping the FAA up-to-date on your location. FAR Section 47.45 requires that you notify the FAA of any change in address within 30 days of your move. The change of address can be accomplished using the FAA's Aircraft Registration Application (see Fig. 4-1). Notice that in the middle of the form there is a box. Next to the box it states: "CHECK HERE IF YOU ARE ONLY REPORTING A CHANGE OF ADDRESS." If you have a change of address to report, you should check this box, write in your new address and the aircraft's N-number (at the top of the form) and send the form to the FAA's Aircraft Registry.

If your registration certificate is lost, destroyed, stolen, or just missing, you can request a duplicate certificate from the FAA. The fee for the duplicate registration certificate is currently $2.00. If you need to operate your aircraft immediately, you should request that the FAA issue you a collect telegram. You can use the telegram as a temporary registration for your aircraft as long as it is carried in the aircraft and until you receive your duplicate registration certificate from the FAA.

Your certificate will undergo periodic scrutiny by the FAA. As a general rule, the FAA will send you a Triennial Aircraft Registration Report every 36 months. The purpose of the report is to confirm that your aircraft is still eligible for registration with the U.S. registry. The report should be completed and returned to the FAA within 60 days after being issued by the FAA. If you fail to complete and return the triennial report, your aircraft's registration may be suspended or revoked.

Aircraft Registration Markings

In most cases, you will receive your aircraft with all of its registration markings already in place. If not, you will have to refer to FAR Part 45 for information regarding registration markings. As a general rule, the big issues in marking an aircraft for registration purposes are size, permanence, legibility, and placement. Each of these three features is discussed below.

Size

FAR Section 45.29 requires the following height requirements for registration markings:

1. Fixed-wing aircraft must display marks at least 12 inches in height (with the general exception of fixed-wing aircraft with at least 2-inch registration marks before November 1, 1981, and aircraft manufactured after November 2, 1981, but before January 1, 1983).

2. Gliders must display marks at least 3 inches in height.

3. Most amateur-built aircraft (180 knots CAS or less) must display registration marks at least 3 inches in height.

4. Airships and balloons must display registration marks at least 3 inches in height.

5. With certain exceptions, rotorcraft must display registration marks at least 12 inches in height.

Certain exceptions to these rules exist for antique, exhibition, and restored aircraft. Exhibit or airshow aircraft may be operated without any registration marks displayed. Older U.S. registered small aircraft built at least 30 years ago may generally be operated with 2-inch registration markings.

As far as width is concerned, the FAA requires that the letters/numbers in your registration markings be two-thirds as wide as they are high. Exceptions to this rule are the number 1 (which must be one-sixth as wide as it is high) and the letters M and W, which can be as wide as they are tall.

Your registration marks must also be spaced properly. The FAA requires that the space between your letters/numbers be at least one-fourth the width of the letters/numbers. Each character must also be formed by solid lines one-sixth as thick as the letter or number is high.

One final technical rule. Your registration marks for a fixed-wing aircraft must also be uniform. They must have the same height, width, thickness, and spacing on both sides of the aircraft fuselage or tail.

Permanence

FAR Section 45.21 (c)(1) requires that your registration mark "be painted on the aircraft or affixed by any other means insuring a similar degree of permanence." FAA Advisory Circular AC 45-2A can assist you in complying with this requirement. The AC states that it is acceptable to use a paint that would require thinners or strippers to remove. The AC also indicates that it would be okay to affix your registration marks in the form of decals.

The use of tape capable of being peeled off or water-base paint would not meet the FAA's requirements. FAR Section 45.21 (d) states that the only time registration marks can be applied with easily removable letters/numbers is when:

1. Your aircraft is intended for immediate delivery to a foreign buyer.
2. Your aircraft is operating with a temporary registration number.
3. You have marked the aircraft temporarily with 12-inch letters/numbers in order to legally operate in an ADIZ or DEWIZ.

Legibility

The regulations state that your registration marks must be legible, have no ornamentation, and contrast in color with the background. These somewhat subjective standards may make it somewhat difficult to implement the rules.

AC 45-2A provides some guidance on how to implement the regulations. The AC states that shading will not be considered ornamentation if it enhances the readability of your registration marks. Likewise, the AC states that a border around letters or numbers would also be satisfactory if it adds to readability.

On a bit more complicated note, the FAA states in AC 45-2A that if your registration marks meet the requirements of FAR Section 25.811(f)(2) "the FAA will consider the aircraft to also comply with the FAA's legibility requirements." This sounds great until you read FAR Section 25.811(f)(2). This section states:

> Each outside marking including the band, must have color contrast to be readily distinguishable from the surrounding fuselage surface. The contrast must be such that if the reflectance of the darker color is 15 percent or less, the reflectance of the lighter color must be at least 45 percent. "Reflectance" is the ratio of luminous flux reflected by a body to the luminous flux it receives. When the reflectance of the darker color is greater than 15 percent, at least a 30-percent difference between its reflectance and the reflectance of the lighter color must be provided.

The usefulness of this test is probably quite limited to most lay persons. Nonetheless, the FAA advises that the "reflectance" referenced in this test can be measured by "appropriate electro-optical instrumentation or by use of photometer card sets." It seems that the best bit of advice is to make sure that your registration markings are carefully selected and reviewed by a reputable paint shop that has experience with this kind of work.

One thing that is certain is the need to use Roman capital letters starting with the letter "N" in every case and followed by your aircraft registration number. Every suffix letter used in the registration number must also be a Roman capital letter.

Placement

The FAA is also very particular about where your aircraft registration markings are placed. The required location(s) of the markings will depend on whether your aircraft is a fixed-wing aircraft or a non-fixed-wing aircraft. The controlling regulations are found in FAR Sections 45.25 and 45.27.

If you own a fixed-wing aircraft, you must display your registration marks on the vertical tail surfaces or the sides of the fuselage. The only rare exception to this rule is where one side of the tail or fuselage is large enough to meet the size requirements for markings and the other side is not. If this is the case, you must place full-sized markings on the side that is large enough. If both sides are not large enough, marks as "large as practicable" (see FAR Section 45.29(f)) should be placed on the aircraft.

In most cases, your registration marks must be displayed horizontally. The only exception to this is the situation where 3-inch marks are permitted. If that is the case, the marks may be placed vertically on a tail surface. If you display the marks on your aircraft's fuselage, they must be placed between the trailing edge of the wing and the leading edge of the horizontal stabilizer. The FAA does indicate that if engine pods or other aircraft appurtenances are located within this area and are integral parts of the fuselage, you can place your marks on the pods or appurtenances.

For non-fixed-wing aircraft, the FARs require the following:

- If you own a rotorcraft, you must display registration marks horizontally on both surfaces of the cabin, boom, fuselage, or tail.

- For airships you must place your registration markings horizontally on (1) the upper surface of the right horizontal stabilizer and on the surface of the left horizontal stabilizer with the beginning of the mark placed on the leading edge of each stabilizer and (2) each side of the bottom half of the vertical stabilizer.

- For spherical balloons, your registration marks must be placed in opposite positions near the point where the balloon is widest.

- For nonspherical balloons, the FAA states that you must position your marks on each side of the balloon near its maximum cross section and immediately above either the rigging band or the points of attachment of the basket or cabin suspension cables.

Import/Export Issues

The FAA has special registration rules you should be aware of if you intend to export an aircraft. Some practical guidance for complying with these rules is provided below.

Importing an aircraft

If you want to register an aircraft that was previously registered in a foreign country, here are the items the FAA will need to process your registration:

1. A bill of sale or other persuasive evidence showing a transfer of title from a foreign seller to you.
2. A statement by an official of the foreign country's aircraft registry that the aircraft's registration has been terminated or is invalid in that country. The statement should show the official's name and title and describe the aircraft by make, model, and serial number.
3. An FAA application for aircraft registration (AC Form 8050-1).
4. A $5.00 registration fee (payable by check or money order).

The FAA has indicated that it would be helpful if you placed all of the above in an envelope marked with the word "IMPORT" in red when you send your documents for processing.

Exporting an aircraft

FAR Section 47.47 governs the necessary paperwork for when you wish to export an aircraft that is currently registered in the United States. The first step is to fill out the reverse side of your aircraft's registration certificate. The information you fill out on the reverse side of the registration certificate, plus the information already on the certificate, should provide the FAA with the following information regarding your aircraft:

1. Make, model, and serial number
2. U.S. registration number
3. Country where your aircraft will be exported
4. Name of the person or entity to whom you are transferring ownership of your aircraft

The FAA will also need satisfaction that there are no outstanding liens on record or other encumbrances on the aircraft. If such liens or encumbrances exist, the holders of the liens or encumbrances must consent to the transfer.

Once you have submitted the required information to the FAA, the FAA will send a notice to the country to which your aircraft will be exported. The notice will go by ordinary mail or airmail. If you want the notice to travel any faster, you will have to pay for the expedited service. You should indicate "EXPORT" in red ink on the outside of your envelope when you send your request to the FAA.

5

Aircraft Registration for Non–U.S. Citizens

This chapter deals with the special issues that a non–U.S. citizen must address if she or he wishes to register an aircraft under the registry of the United States. As indicated in Chap. 4, in order to be eligible for U.S. registration, an aircraft must be owned by an individual, co-ownership, partnership, limited liability company, or corporation that qualifies as "U.S. citizens." The law also permits individual U.S. permanent residents to own U.S.–registered aircraft. Therefore, if you do not initially qualify under the FAA definition for a U.S. citizen, you must either do something to qualify your aircraft for U.S. registration or you won't be able to get your airplane registered with an N-number.

In this chapter you will receive detailed guidance and solutions if you are a non–U.S. citizen seeking to register your aircraft in the United States. The solutions covered by this chapter include the formation of noncitizen U.S. corporations, the use of trusts, and the use of voting trusts. Some specimen legal documents and FAA forms you and your counsel may wish to review are included throughout this chapter and in the CD that accompanies this book.

Noncitizen U.S. Corporations

If you are a non–U.S. citizen and you plan to base and use your airplane primarily within the United States, a relatively simple solution is at hand. FAR Section 47.9 specifically permits a non–U.S. citizen corporation to obtain an N-number on the condition that 60 percent or more of your airplane's flight hours are accumulated within the United States.

So how does this work in your case? If you are the typical non–U.S. citizen addressing this question, you are most likely visiting the United States on a visa (usually work-related). You are a pilot and you'd like to continue your flying activities during your temporary stay in the United States.

In order for this regulatory solution to be available to you, you must first determine if you will accumulate at least 60 percent of your flight hours in the United

States (the "60 percent rule"). If the answer is "no," this is not the solution for you. You may have to review the discussion regarding trusts and voting trusts below. If you can answer this question with a "yes," then this is the easiest and least expensive way to get your airplane registered as a U.S. civil aircraft.

The first step is to establish a corporation. Remember that under this FAA rule, you can be the sole shareholder, the sole director, and the sole officer of the corporation and still comply with the law—even though you do not qualify as a U.S. citizen. The only condition is that your use of the aircraft must comply with the 60 percent rule. You should take note that in order to establish registration under FAR Part 47.9 you must establish a corporation. No other form of aircraft ownership will be acceptable to the FAA for this purpose.

If you have not already purchased an aircraft, it may be best to suspend the purchase until you have had an opportunity to get your corporation established. If you are tight on time, most states have expedited service for processing your articles of incorporation if you pay additional fees. If you have already purchased the aircraft in your own name, you can simply transfer the aircraft from yourself to your newly formed non–U.S. citizen corporation by transferring title through a bill of sale. Remember that the bill of sale is your aircraft's evidence of ownership, required by the FAA for registration.

Once you have a bill of sale that has transferred aircraft title to your corporation, you are ready to register your aircraft. FAR Section 47.9 requires that along with your bill of sale, you must submit a certified copy of the articles of incorporation for your corporation. Certified copies are typically easy to obtain. Often, the best thing to do is to request a certified copy at the same time you submit articles of incorporation for filing.

FAR Section 47.9 also requires that you submit:

1. A certification that your corporation is qualified to do business in one or more states in the United States

2. A certification that your corporation's airplane will be based and primarily used in the United States

3. The location where flight hour records for your corporation's aircraft will be maintained

These three items are addressed on the bottom portion of the FAA aircraft registration application. A sample filled-in registration application is illustrated in Fig. 5-1.

While you are waiting for your permanent registration certificate, you may fly your aircraft using the pink copy of the registration application as long as you keep the airplane within the United States. Once you receive your permanent registration certificate, you can operate the aircraft outside the United States, but you must remain vigilant to ensure that you are meeting the 60 percent rule.

The FAA will track your airplane use every 6 months starting with the month of your aircraft's registration. This tracking is accomplished every 6 months by use of FAA Form 8050-117 (Flight Hours for Corporations Not U.S. Citizens). A sample filled-in form is illustrated for your reference in Fig. 5-2.

FORM APPROVED
OMB No. 2120-0042

UNITED STATES OF AMERICA DEPARTMENT OF TRANSPORTATION
FEDERAL AVIATION ADMINISTRATION-MIKE MONRONEY AERONAUTICAL CENTER
AIRCRAFT REGISTRATION APPLICATION

CERT. ISSUE DATE

UNITED STATES
REGISTRATION NUMBER **N** 9876A

AIRCRAFT MANUFACTURER & MODEL
Piper PA-21

AIRCRAFT SERIAL No.
54321

FOR FAA USE ONLY

TYPE OF REGISTRATION (Check one box)

☐ 1. Individual   ☐ 2. Partnership   ☒ 3. Corporation   ☐ 4. Co-owner   ☐ 5. Gov't.   ☐ 8. Non-Citizen Corporation

NAME OF APPLICANT (Person(s) shown on evidence of ownership. If individual, give last name, first name, and middle initial.)

● Aviation Unlimited, Inc.

TELEPHONE NUMBER: (270) 555-1212

ADDRESS (Permanent mailing address for first applicant listed.)

Number and street: 17 East Highway

Rural Route: P.O. Box:

CITY	STATE	ZIP CODE
Kingsbrooke	South Carolina	99999

☐ **CHECK HERE IF YOU ARE ONLY REPORTING A CHANGE OF ADDRESS**
ATTENTION! Read the following statement before signing this application.
This portion MUST be completed.

A false or dishonest answer to any question in this application may be grounds for punishment by fine and / or imprisonment (U.S. Code, Title 18, Sec. 1001).

● **CERTIFICATION**

I/WE CERTIFY:

(1) That the above aircraft is owned by the undersigned applicant, who is a citizen (including corporations) of the United States.

(For voting trust, give name of trustee: _____), or:

CHECK ONE AS APPROPRIATE:

a. ☐ A resident alien, with alien registration (Form 1-151 or Form 1-551) No. _____

b. ☒ A non-citizen corporation organized and doing business under the laws of (state) South Carolina and said aircraft is based and primarily used in the United States. Records or flight hours are available for inspection at see address above

(2) That the aircraft is not registered under the laws of any foreign country; and
(3) That legal evidence of ownership is attached or has been filed with the Federal Aviation Administration.

NOTE: If executed for co-ownership all applicants must sign. Use reverse side if necessary.

TYPE OR PRINT NAME BELOW SIGNATURE

	SIGNATURE	TITLE	DATE
EACH PART OF THIS APPLICATION MUST BE SIGNED IN INK.		President	
	SIGNATURE	TITLE	DATE
	SIGNATURE	TITLE	DATE

NOTE   Pending receipt of the Certificate of Aircraft Registration, the aircraft may be operated for a period not in excess of 90 days, during which time the PINK copy of this application must be carried in the aircraft.

AC Form 8050-1 (12/90) (0052-00-628-9007) Supersedes Previous Edition

Figure 5-1 Sample registration application for noncitizen corporation.

AGENCY DISPLAY OF ESTIMATED BURDEN
The Federal Aviation Administration estimates that the average burden for this report form is .5 hour per response. You may submit any comments concerning the accuracy of this burden estimate or any suggestions for reducing the burden to the Office of Management and Budget (OMB). You may also send comments to the Federal Aviation Administration, Civil Aviation Registry, P. O. Box 25504, Oklahoma City, OK 73125-0504. Attention: OMB number 2120-0042.

U.S. Department of Transportation
Federal Aviation Administration

Civil Aviation Registry
P.O. Box 25504
Oklahoma City, Oklahoma 73125-0504

October 17, 2002

Aviation Unlimited, Inc.
17 East Highway
Kingsbrooke, SC 99999

Flight Hours For Corporations Not U.S. Citizens N 9876A

Records on file with the Federal Aviation Administration (FAA) Civil Aviation Registry indicate that your aircraft was registered pursuant to Section 501(b)(1)(A)(ii) of the Federal Aviation Act of 1953, as amended, and Section 47.9 of the Federal Aviation Regulations which relate to registration of aircraft by "Corporations not U. S. citizen". Section 47.9(f) requires that you submit a statement regarding flight hours accumulated every six months. If the aircraft is operated outside the United States during the reporting period, the statement must set forth the total time in service of the airframe during that period and the total flight hours accumulated by the aircraft within the United States during that period. To continue to be eligible for registration, at least 60 percent of the total flight hours must be accumulated within the United States. If the aircraft was operated exclusively within the United States during the reporting period, a statement to that effect will meet the requirements of the regulation.

Your failure to submit the required report may result in action being taken to revoke the aircraft registration certificate.

INSTRUCTIONS

This form is provided to assist you in furnishing the information required by regulation for the period of APRIL 01, 2002 to SEPTEMBER 30, 2002 for aircraft:

N-number 9876A , Serial number 54321

Make/model Piper PA-21

Please read carefully. Complete and sign Section A if the aircraft accumulated flight hours outside the United States; or, sign the certification in Section B if the aircraft was operated exclusively within the United States.

Section A	Section B
_____ Total flight hours accumulated during reporting period.	
_____ Total flight hours accumulated within United States during reporting period.	I/we the undersigned do hereby certify N 9876A was operated exclusively within the United States.
I/we the undersigned do hereby certify the flight hours listed are accurate and comply with the Federal Aviation Regulations.	
Signature	Signature
Corporate Title	Corporate Title
Date	Date

ADDRESS CHANGE

If the location of your records for flight hours within the United States has changed, please enter the new address.

Street	City	State	Zip

AC Form 8050-117 (5-95)

Figure 5-2 Sample flight hours form.

Usually, the FAA will fill in all the necessary information regarding the dates and aircraft identification. Your job is to fill in the appropriate certifications. You will use Section A near the bottom of the form only if your aircraft has accumulated flight hours both outside and inside the United States. If you have operated your aircraft exclusively within the United States, you will use Section B.

Note that the form requires you to keep the FAA up-to-date on where flight hour records for your aircraft may be located. It is imperative that you keep the FAA posted on any change of address for your corporation and the location of flight hour records (see the very bottom portion of Form 8050-117).

Trusts

If you do not qualify as a U.S. citizen and you intend to operate your aircraft more than 40 percent or more of the time outside the United States, FAR Section 47.9 won't be much help to you. One alternative you and your counsel may consider is the use of a trust. If structured properly, a trust may assist you in legally qualifying as a U.S. citizen and, in turn, getting your aircraft registered under the registry of the United States.

This is a more complex solution than the solution offered to non–U.S. citizen corporations that will base and primarily use their aircraft in the United States. The trust solution will most certainly require the assistance of counsel experienced in this area in order to ensure that you are acting in accordance with the law.

Trust basics

Here's how this solution might work for you. First, you must understand the basic nature of a trust. Under the law, a trust is a recognized legal tool that allows a "creator," "settlor," or "trustor" (usually the aircraft owner) to place property (including an airplane) "in trust" for the benefit of a designated "beneficiary." It is okay for the trustor and beneficiary to be the same person—they often are. The property in trust is entrusted to a party legally known as a "trustee."

Here's the tricky part for nonlawyers. The magic of a trust is that it allows title to an aircraft to be broken down into two parts.

The first part is called "legal" title. The trustee gets legal title. This essentially means that the government and other third parties will view the trustee as the legal owner of the aircraft.

The second part is called "equitable" title. The beneficiary gets equitable title. Equitable title means the use, enjoyment, and benefit of the property held in trust.

The relationship between the trustee and the beneficiary is known as a fiduciary relationship. That means that the trustee has a legal obligation to preserve the trust property and to ensure that it is only owned and managed in accordance with the trust's instructions.

Figure 5-3 illustrates the nature of a trust.

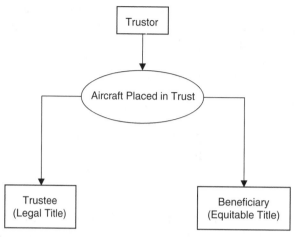

Figure 5-3 Trusts.

Finding a trustee

With that background in mind, we can get back to your immediate problem. The object is to ensure that legal title to your aircraft is held by a trustee who is a U.S. citizen as defined by the FAA.

This may be accomplished by finding an individual, partnership, limited liability company, or corporation who is a U.S. citizen and willing to act as a trustee. If you have a friend or relative who is willing to act in this capacity, you should consider yourself quite fortunate. As you might guess, the friend or relative who agrees to act as trustee takes on potential legal exposure for property or personal injury caused by the aircraft owned by the trust. There is scant case law on this matter when it comes to aircraft trusts, but there is certainly exposure to lawsuits. That potential exposure is a burden the trustee must be willing to accept. Therefore, the trustee should be very careful to ensure that he or she is named in any insurance policy as an insured. The trustee should also be comfortable with the level of insurance coverage obtained.

If you do not have the luxury of a friend or relative who is willing to act as a trustee, there are people and entities that are in the business of acting as trustees. You can consult with your lawyer to locate these entities. Of course, they will charge you for the setup of your trust. In most cases, they will also charge you an annual or monthly fee for their services as trustee of the trust.

The trust agreement

Once you have found someone willing to act as a trustee, you must prepare a written trust document that will be acceptable to the FAA. Under FAR Section 47.11, the regulations require that if you want to name a trustee as the owner of your aircraft, the FAA must receive a "complete and true" copy of the trust agreement or instrument creating the trust. A very simple specimen agreement can be found in the CD that comes with this book. A review of the basic provisions of the agreement follows.

Recitals ("witnesseth" clause).   An aircraft trust agreement should have some background information included in the form of recitals. The recitals should identify the airplane involved using make, model, serial number, and N-number (if available). To be fully descriptive, the recitals should indicate that the purpose of the trust is to ensure that the aircraft involved is made eligible for U.S. registration. It is also important to expressly indicate that the trustee (often called the "Owner Trustee") is willing to serve in that capacity.

Creation of trust.   There should be language in the agreement clearly stating that you are transferring your aircraft to the owner trustee. In accordance with basic trust law, the trustee should expressly state that he or she is holding the aircraft for your use or benefit in accordance with the trust agreement.

Owner trustee provisions.   Your trust agreement should include provisions that, at a minimum, govern the following issues:

1. When a trustee may be removed
2. How the trust will deal with the resignation of a trustee
3. Filling vacancies left by trustees

You and your counsel should carefully consider how you want to deal with each of these issues. There is no one "canned" approach that will permit a smooth transition from one trustee to another in the case of removal, resignation, or vacancy for any reason.

Nonetheless, there are a few things that you must be aware of in order to have the FAA's blessing for your trust agreement. Perhaps one of the most important issues is the removal of a trustee. The FAA wants assurance that you will not be able to remove a trustee without cause. Therefore, it is important to carefully word the agreement to indicate that the trustee may only be removed "for cause." Some lawyers will desire to go further and list specific reasons why a trustee might be removed. It is also important that your agreement specifically state that an owner trustee automatically ceases to continue as owner trustee if he/she/it is no longer a U.S. citizen. Provisions regarding resignation and the filling of vacancies should allow, to the greatest extent possible, flexibility and speed in getting a new owner trustee in place.

Limitations on trustee and liability of trustee.   From the FAA's standpoint, this is another critical provision in your trust agreement. The FAA is most concerned that the owner trustee has real control over your aircraft. Essentially, this means that the owner trustee has the same power and discretion that you would have.

One limitation that the FAA will permit relates to the gross negligence or willful misconduct of the owner trustee. The FAA will permit removal of a trustee for cause if the cause involves gross negligence or willful misconduct.

This is also a good time for you and your counsel to refine the scope of the owner trustee's responsibilities. Most owner trustees will only want to be legally responsible for matters relating to ownership and operation of your aircraft. Further, an owner trustee may insist that he or she not be held liable for any advice or lack thereof from accountants or attorneys.

Insurance coverage for trustee.   Your trustee may justifiably be concerned with potential exposure to liability. If this is the case, you may have to include a provision indicating that the owner trustee will be a named insured on your liability policy. This is not a difficult request for most insurers to deal with. Insurer's can also provide the owner trustee with a certificate of insurance confirming coverage and the right to be notified if the insurance policy is canceled for any reason.

Miscellaneous provisions.   Just as in any agreement, there are certain provisions that you will want to include to ensure that if there are future problems or modifications to the trust agreement, they can be dealt with in an orderly fashion. Typically, you will want to ensure that no changes can be made without a written modification signed by both parties. You may also want to designate a court or arbitrator to oversee any disputes.

Affidavit of citizenship.   FAR Section 47.7(c)(2)(iii) requires that if any beneficiary under the trust is not a U.S. citizen or resident alien, the trustee must prepare an affidavit of citizenship. The regulation specifically requires that the trustee state that he or she is not aware of "any reason, situation, or relationship (involving beneficiaries or other persons who are not U.S. citizens or resident aliens) as a result of which those persons together would have more than 25 percent of the aggregate power to influence or limit the exercise of the trustee's authority." Essentially, the affidavit has to confirm to the FAA that the owner trustee has true decision-making capacity that is effectively unfettered by the noncitizen owner/beneficiary of the trust.

Getting your aircraft registered

Now that you've got your agreement in place, the next step is to get your airplane registered as a U.S.–registered aircraft. In the typical transaction, you might be interested in purchasing an aircraft from a broker or private party. If this is the case, you may decide to purchase the aircraft and file a bill of sale with the FAA indicating that you are the aircraft's owner. Of course, if you are not a U.S. citizen, you may still record your interest in the aircraft. However, you will not be able to get the aircraft registered with an N-number. The following set of steps is designed to help you and your counsel through the registration process:

1. Purchase the aircraft in your name.
2. Select an owner trustee.

3. Transfer legal title to the aircraft to the owner trustee.

4. Submit an application for registration to the FAA with the owner trustee listed as owner.

Remember that while the FAA is examining your application for registration, you may not operate your aircraft outside the United States. The "pink slip" portion of the registration application representing your temporary registration is not acceptable for international flights.

If you are in a hurry to operate the aircraft outside the United States, you may consider requesting expedited review of your application. Many of the title and escrow companies make this service available for an extra fee. The FAA will also require a statement indicating the specifics of your planned international operation. See Fig. 5-4 for a sample Declaration of International Operations.

When submitting documents to the FAA or a title and escrow service, it is always a good idea to include a brief, explanatory cover letter outlining your document submission for registration. A typical cover letter would look something like the letter illustrated in Fig. 5-5. This sample letter also serves as a good checklist to ensure that you've covered all bases in your submission.

Once the FAA accepts the registration application, the plane is officially registered under the U.S. registry and the owner trustee is listed as the registered owner. If you ever desire to sell the aircraft, you will have to instruct the owner trustee to transfer title to the aircraft back to you. You may then sign a bill of sale transferring title to the aircraft to your buyer.

An overview of the entire process (from aircraft purchase to aircraft sale) of registration by trust for non–U.S. citizens is illustrated in Fig. 5-6.

Voting Trusts

The FAA regulations specifically provide for voting trusts as a vehicle for qualifying domestic corporations as U.S. citizens for purposes of aircraft registration. You can find the details of the FAA's rules for voting trusts in FAR Section 47.8.

However, before delving into the FARs, it is advisable that you have a fundamental understanding of how a voting trust works. Once the basics are reviewed, you'll have a much better picture of how a voting trust operates to qualify your aircraft for U.S. registration.

To get a feel for how a voting trust works, you first have to understand some fundamentals about owning shares of corporate stock. When you own stock, you basically own two things—a proportional share of the corporation's net assets and the right to vote on corporate matters. A voting trust permits you to legally separate these two aspects of stock ownership—the ownership rights and the voting rights. See Fig. 5-7 for an illustration of how a voting trust works.

So how does this help you qualify your aircraft for U.S. registration? To answer this question, you first have to go back to the definition of U.S. citizen in FAR Section 47.2. Specifically, FAR Section 47.2 defines a U.S. citizen as: "A

DECLARATION
OF
INTERNATIONAL OPERATIONS

_____, as the owner trustee of a _____ aircraft with manufacturer's serial number _____, hereby declares that this aircraft is scheduled to depart the United States from _____ with a destination of _____ on or around _____.

Expedited registration is requested in support of the scheduled international operations of this aircraft.

_____
_____
As Owner Trustee

Dated: _____

Telephone: _____

Figure 5-4 Declaration of International Operations.

corporation or association organized under the laws of the United States of any State, Territory, or possession of the United States, of which the president and two-thirds or more of the board of directors and other managing officers thereof are such individuals and in which at least *75 percent of the voting interest* is owned or controlled by persons who are citizens of the United States or of one of its possessions" (emphasis added).

To qualify for registration under this definition, your aircraft must:

1. Be owned by a domestic corporation.
2. Two-thirds of the corporation's managing officers must be U.S. citizens.

<u>VIA OVERNIGHT MAIL</u>

March 4, 2003

XYZ Aircraft Title and Escrow Services
FAA Road
Oklahoma City, OK 99999

Attn.: Title and Escrow Agent
Re: Registration Application for _____

Dear_____:

I've enclosed the following executed documents along with a check for $5 to cover aircraft registration fees:

1. FAA Aircraft Registration Application;

2. Bill of Sale transferring legal title to the aircraft from [Seller] to [Trustor];

3. Bill of Sale transferring legal title to the aircraft from [Trustor] to [Owner Trustee], as Owner Trustee;

4. Trust Agreement dated _____between [Trustor] and [Trustee];

5. Affidavit of Citizenship signed by [Trustee] and witnessed by a notary public; and

6. Declaration of International Operations

This application requires expedited service. Please bill any processing charges to my office.

Thank you for your assistance. Please don't hesitate to give me a call if you have any questions. My office number is _____.

Very truly yours,

Aviation Lawyer

cc: [Trustor and Trustee]

Figure 5-5 Sample letter to FAA with registration application for trusts.

3. Two-thirds of the corporation's board of directors must be U.S. citizens.

4. U.S. citizens must own at least 75 percent of the voting interest in the corporation's stock.

The first three parts of the FAA's requirements are clear enough. First, you must form a domestic U.S. corporation. Next, you must fill the managing officer

Figure 5-6 Overview of trusts in aircraft transactions.

and director positions at least two-thirds with a person or persons who qualify as U.S. citizens.

The final requirement gets to the heart of the voting trust. It requires that you turn over at least 75 percent of the voting interest in the corporation to a person who is a U.S. citizen. FAR Section 47.8(a)(2) indicates that the voting trustee must be able to establish by affidavit that:

1. She or he is a citizen of the United States.

2. She or he is not a past, present, or prospective director, officer, employee, attorney, or agent of any other party to the trust agreement.

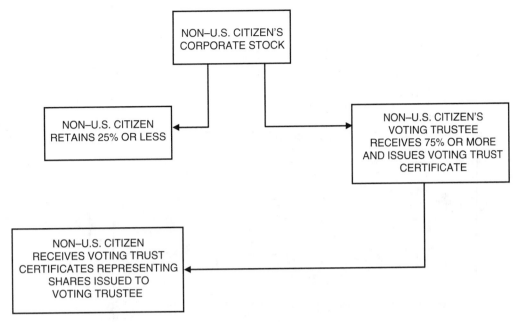

Figure 5-7 Voting trusts.

3. She or he is not a present or prospective beneficiary, creditor, debtor, supplier, or contractor of any other party to the trust agreement.

4. She or he is not aware of any reason, situation, or relationship under which any other party to the agreement might influence the exercise of her or his totally independent judgment under the voting trust agreement.

Essentially, the FAA requires the trustee to be able to make independent decisions on how to exercise her or his voting rights for the corporation. If all of these requirements can be met, your aircraft may be eligible for registration under the voting trust rules.

The regulations also require that you submit a voting trust agreement to the FAA. Specimen voting trust documents (from a very simple voting trust agreement) are provided as an illustration for you and your counsel in the accompanying CD. Notice that the voting trust documents are broken down into three components:

1. Agreement

2. Exhibit A with Restriction Language

3. Voting Trust Certificate

4. Exhibit B with Affidavit by Voting Trustee

The agreement lays out the mechanics for separating stock voting rights from ownership rights. FAR Section 47.8 (b) also requires that the agreement

FORM APPROVED
OMB No. 2120-0042

UNITED STATES OF AMERICA DEPARTMENT OF TRANSPORTATION
FEDERAL AVIATION ADMINISTRATION-MIKE MONRONEY AERONAUTICAL CENTER
AIRCRAFT REGISTRATION APPLICATION

CERT. ISSUE DATE

UNITED STATES
REGISTRATION NUMBER **N** 234GW

AIRCRAFT MANUFACTURER & MODEL
Beech Baron

AIRCRAFT SERIAL No.
98765

FOR FAA USE ONLY

TYPE OF REGISTRATION (Check one box)

☐ 1. Individual   ☐ 2. Partnership   ☒ 3. Corporation   ☐ 4. Co-owner   ☐ 5. Gov't.   ☐ 8. Non-Citizen Corporation

NAME OF APPLICANT (Person(s) shown on evidence of ownership. If individual, give last name, first name, and middle initial.)

VoteTrust, Inc.

TELEPHONE NUMBER: (301) 877-5555

ADDRESS (Permanent mailing address for first applicant listed.)

Number and street: 3034 Hone Ave

Rural Route: P.O. Box:

CITY	STATE	ZIP CODE
NY	NY	12345

☐   **CHECK HERE IF YOU ARE ONLY REPORTING A CHANGE OF ADDRESS**
     ATTENTION! Read the following statement before signing this application.
     This portion MUST be completed.

A false or dishonest answer to any question in this application may be grounds for punishment by fine and / or imprisonment (U.S. Code, Title 18, Sec. 1001).

CERTIFICATION

I/WE CERTIFY:

(1) That the above aircraft is owned by the undersigned applicant, who is a citizen (including corporations) of the United States.

(For voting trust, give name of trustee: Tom Citizen _____), or:

CHECK ONE AS APPROPRIATE:

a. ☐ A resident alien, with alien registration (Form 1-151 or Form 1-551) No. _____

b. ☐ A non-citizen corporation organized and doing business under the laws of (state) _____ and said aircraft is based and primarily used in the United States. Records or flight hours are available for inspection at _____

(2) That the aircraft is not registered under the laws of any foreign country; and

(3) That legal evidence of ownership is attached or has been filed with the Federal Aviation Administration.

NOTE: If executed for co-ownership all applicants must sign. Use reverse side if necessary.

TYPE OR PRINT NAME BELOW SIGNATURE

	SIGNATURE	TITLE	DATE
EACH PART OF THIS APPLICATION MUST BE SIGNED IN INK.		President	
	SIGNATURE	TITLE	DATE
	SIGNATURE	TITLE	DATE

NOTE Pending receipt of the Certificate of Aircraft Registration, the aircraft may be operated for a period not in excess of 90 days, during which time the PINK copy of this application must be carried in the aircraft.

AC Form 8050-1 (12/90) (0052-00-628-9007) Supersedes Previous Edition

Figure 5-8 Sample registration application for voting trust.

provide for the succession of a voting trustee in the event of death, disability, resignation, termination of citizenship, or any other event leading to the replacement of the voting trustee. The agreement should also acknowledge the independent nature of the voting trustee.

Exhibit A provides language that must be placed on each shareholder's stock certificate. This language will notify the holder or potential holder of the stock that the transfer of stock is restricted by the voting trust agreement. It should also provide a location for obtaining a copy of the voting trust agreement.

The Voting Trust Certificate is evidence of the voting trustee's voting rights in the corporation. Here's the way it works. The voting trustee gets the actual shares of stock issued by the corporation. The non–U.S. citizen shareholder gets the voting trust certificate as evidence of his/her ownership rights in the corporation. The actual mechanics of all this should be spelled out in the voting trust agreement and should be administered by qualified legal counsel.

The final piece of documentation necessary is the voting trustee's affidavit. As indicated above, the primary purpose of this affidavit is to assure the FAA that the voting trustee is able to act independently in the exercise of his or her duties.

Once you are ready to submit your aircraft registration application, you should provide the name of the voting trustee as indicated on line 1 of the certification portion of FAA Form AC 8050-1. A sample filled-in registration application is provided in Fig. 5-8.

Aircraft Storage and Maintenance

6

Hangar and Tie-Down Agreements

One inescapable cost of aircraft ownership is the cost of storage. As a rule, there are generally two ways to store an aircraft that you intend to use on a regular basis. The first approach is to find a covered space called a hangar. The second approach is to tie down your aircraft on the ramp at the field where you wish to base your aircraft.

In order to either hangar or tie down your aircraft, you will usually be required to agree to certain conditions by a landlord who will be renting you the hangar or tie-down space. Typically, the conditions are spelled out in a written agreement drafted by the landlord. Before you sign any agreements, you should be aware of the legal ramifications and pitfalls of the agreement.

This chapter is designed to acquaint you with common legal issues that might come up in your hangar or tie-down agreement. We'll start with a review of hangar agreements and then move to a discussion of leasing a tie-down space.

Hangar Leases

At most airports that have hangars, the landlord will have a hangar (or T-hangar) lease agreement that has already been drafted by the landlord or the landlord's attorneys. Before you sign on the dotted line, you should be aware of what you are signing. Perhaps the best way to get a feel for the issues you might encounter is to walk through an actual hangar lease agreement that has provisions similar to the lease you are presented by your hangar landlord. The lease agreement we'll use as an illustration is found in Fig. 6-1. Comments and analysis follow and refer back to the lease illustration.

As indicated earlier, the landlord or the landlord's attorney will usually have drafted the hangar lease you will be presented. That means that it is most likely that the lease will be drafted to the landlord's advantage. You have to be vigilant to ensure that you are not agreeing to anything that may cause you legal problems down the road. An analysis of the hangar lease is presented below with labeled comments on the various sections of the lease.

Figure 6-1   Specimen T-hangar lease agreement.

T-HANGAR LEASE AGREEMENT

This T-HANGAR LEASE AGREEMENT (the "Agreement") is entered into as of this 3rd day of May 2002, by and between _____ ("Landlord"), whose address is _____, and Joe Tenant ("Tenant"), whose address is _____.

In consideration of the mutual covenants and agreements contained herein to be performed by the respective parties, and for other good and valuable consideration, the parties hereby agree as follows:

1. *Lease of the Hangar:* Landlord hereby leases to Tenant Hangar No. _____ (the "Hangar") in Building No. _____ located at the _____ (the "Airport") in the State of _____.

The Hangar shall be used and occupied by Tenant solely for the storage/parking of the following described aircraft:

Make/Model/Color: _____

Registration No. _____ (the "Aircraft"), or any other similar Aircraft owned or leased by Tenant (the "Substitute Aircraft"), provided Tenant has obtained the prior written consent of Landlord to store the Substitute Aircraft in the Hangar. In the event Tenant is permitted to store Substitute Aircraft in the Hangar, all provisions of the Agreement applicable to the Aircraft shall also be applicable to the Substitute Aircraft.

2. *Term:* The term of this Agreement shall commence _____ and shall continue in effect until _____.

3. *Rent:* For the use of the Hangar, Tenant shall pay Landlord the amount of Three Hundred Dollars ($300.00) per month, plus applicable sales tax, for a monthly total of $321.00, payable in advance on or before the first day of each month. Monthly rent and all additional charges shall be paid promptly when due, without notice or demand and without deduction, diminution, abatement, counterclaim or setoff of any amount or for any reason whatsoever, to Landlord at its address set forth above.

A five percent (5%) late charge not less than $50.00 will be assessed on all rent not received by the 5th of each month. If payment remitted by check or note or other instrument is presented for payment to the appropriate institution or individual, and if said funds are insufficient, payment will be subject to a $35.00 reprocessing fee.

All sums of money required to be paid by Tenant to Landlord under this Lease shall bear interest at the highest rate permitted by law from the date same was due until the date same is paid in full.

4. *Security Deposit:* Tenant has deposited with Landlord a security deposit in the amount of $300.00. The security deposit will be held by Landlord without interest as security for the full and faithful performance by Tenant of its obligation hereunder, and may be co-mingled with other monies of Landlord. In the event of default by Tenant, Landlord may use all or any part of the security deposit for the payment of any unpaid rent or for any other monies owed by Tenant to Landlord. Upon the termination of this Lease, any portion of the security deposit not so used or applied shall be returned to Tenant, provided Tenant faithfully performs its obligations hereunder, by mail within a reasonable time after the termination of this Lease. The security deposit shall not be applied by Tenant toward the last month's rent.

5. *Obligations of Landlord:* Landlord will maintain the structural components of the Hangar, including doors and door mechanisms, and Landlord will provide restroom facilities, water and normal building maintenance without additional cost to Tenant, provided, however, that Landlord reserves the right to assess an additional fee for consumption of utilities by Tenant beyond nominal requirements by Landlord.

Figure 6-1 *(Continued)*

6. *Obligations of Tenant*

6a. Storage: The Hangar shall be used only for storage/parking of the above-identified Aircraft unless otherwise approved in writing by Landlord. No commercial activity of any kind whatsoever shall be conducted by Tenant in, from or around the Hangar.

6b. Electrical Service: Tenant, at its expense, will be responsible for obtaining electrical service to the Hangar.

6c. Building Maintenance and Repair: Tenant shall maintain the Hangar in a neat and orderly condition, and shall keep the Hangar floor clean and clear of debris, oil, grease and/or toxic chemicals. No hazardous or flammable materials will be stored within or about the Hangar unless stored within an EPA or local Fire Marshall approved container/cabinet. No boxes, crates, rubbish, paper or other litter that could cause or support combustion shall be permitted to accumulate within or about the Hangar.

Tenant shall be responsible for all damage to the leased premises caused by Tenant's acts or negligence, not to exclude repair of apron in front of Tenant's hangar due to fuel spillage. In the event Tenant does not promptly repair any damage to the premises or property for which Tenant is responsible, Landlord reserves the right, in addition to any other rights or remedies available to Landlord, to make such repairs, at Tenant's expense, the cost of which shall become due and payable as part of Tenant's next monthly rental payment.

6d. Use of Hangar: The Hangar is not to be used as a commercial repair/maintenance shop. No maintenance on the Aircraft shall be performed in the Hangar without the prior written approval of Landlord except such minor maintenance as would normally be performed by an aircraft owner without the benefit of an aircraft mechanic. Painting is strictly prohibited. Tenant shall take steps to ensure that the performance of such maintenance work shall not damage the Hangar. Tenant shall control the conduct and demeanor of its employees and invites, and of those doing business with it, in and around the Hangar and shall take all steps necessary to remove persons whom Landlord may for good and sufficient cause deem objectionable.

6e. Compliance with Laws: In utilizing the Hangar during the term of this Agreement, Tenant agrees to and shall comply with all applicable ordinances, rules and regulations established by any federal, state or local government agency exercising jurisdiction over the premises, including FAA regulations and rules and regulations of the Airport. Non-aviation use of the Hangar is not permitted.

It is expressly understood that no storage of airplanes, boats, cars, trucks, trailers or mobile homes is permitted outside of the Hangar or anywhere on Landlord's property except designated parking areas for such vehicles or aircraft. No pets or other animals are allowed on the property.

Tenant is not permitted to physically occupy the Hangar with both sliding doors in the fully closed position.

Tenant is to use common courtesy when opening the sliding hangar doors as to ensure there is no blockage of the adjacent (neighboring) hangars from the sliding doors.

6f. Expiration or Termination of This Lease: Upon the expiration or termination of this Lease, Tenant shall surrender the Hangar to Landlord in substantially the same condition as the Hangar was in at the beginning of this Lease and in good and clean condition, reasonable wear and tear excepted. Tenant shall be liable for any and all damage to the Hangar caused by Tenant's use, including, but not limited to, bent or broken interior walls, damage to unsealed floors and/or apron immediately adjacent to Hangar due to fuel or oil spillage, contamination from hazardous materials or damage to doors due to Tenant's improper or negligent operation. Should Tenant remain in possession of the Hangar after the

Figure 6-1 (*Continued*)

expiration of the term or earlier termination of this Lease, with or without the consent of Landlord, express or implied, such holding over shall, in the absence of a written agreement to the contrary, be deemed to have created and be construed to be a tenancy at sufferance terminable on written notice by Landlord to Tenant, at double the rent installments (prorated on a monthly basis) in effect during the lease year immediately preceding the expiration of the term of this Lease, and otherwise subject to all of the other terms, covenants and conditions of this Lease insofar as the same may be applicable to a tenancy at sufferance, without prejudice to any remedy which Landlord may have against Tenant for holding over unlawfully, provided, however, that if Tenant holds over with the prior written consent of Landlord, the rent installments will not be doubled as hereinabove provided.

6g. Tenant's Responsibility for Conduct: All Tenant employees will obtain and display proper identification in accordance with prevailing Airport regulations for all areas of the Airport where required. All costs incurred in obtaining such required identification badge authorizations or endorsements shall be borne solely by Tenant.

Tenant shall indemnify and hold harmless Landlord (inclusive of its subsidiaries, affiliates and parent company, as now or hereafter constituted) and its officers, directors, agents, Tenants, customers, contractors, subcontractors, invites and employees from and against any and all fines, penalties, damages or legal actions which may be imposed by the Airport, FAA or any other agency having jurisdiction at or on the Airport as a result of Tenant's or its officers, directors, agents, contractors, subcontractors, invites or employees failure to comply and adhere to any and all federal state, local or Airport regulations in effect as of the effective date of this Lease or promulgated from time to time thereafter.

7. *Safe Use of Premises:* Tenant at their expense shall provide and keep in good working order a fire extinguisher on the premises which must be an "ABC-type" or other type acceptable to Landlord. Tenant agrees to make no unlawful, offensive or noxious use of the Hangar. In addition, no explosives, firearms, volatile or flammable chemicals or any other property which would materially increase the hazard of fire shall be stored at the Hangar. Inspection fees assessed by the City and/or Airport Authority shall be paid by Tenant.

8. *Primary Lease:* It is expressly understood and agreed that if the primary lease between the Owner of the Airport and Landlord, which covers the Hangar and adjacent areas, is terminated, canceled or for any reason abated as to any portion of the hangar or adjacent areas, such termination, cancellation or abatement will operate as a cancellation of this Agreement, and Landlord will be relieved of liability for any and all damages Tenant may sustain as a result therefrom.

9. *Sublease/Assignment:* Tenant agrees not to sublease the Hangar or to assign this Agreement without the prior written approval of Landlord, which approval may be withheld in Landlord's sole discretion. The parking of aircraft other than the Aircraft in the Hangar shall constitute a sublease.

Neither an assignment of Tenant's interest in this Lease nor a subletting, occupancy or use of the Hangar Space or any part thereof by any person or entity other than Tenant, nor the collection of rent by Landlord from any person or entity other than Tenant as provided in this provision, nor the application of any such rent shall, in any circumstances, relieve Tenant from its obligation fully to observe and perform the terms, covenants and conditions of this Lease on Tenant's part to be observed and performed.

10. *Condition of Premises:* Tenant has inspected the Hangar and Tenant hereby accepts the Hangar in its present condition without any liability or obligation on the part of Landlord to make any alterations, improvement or repairs of any kind on or about said Hangar.

11. *Alterations:* Tenant covenants and agrees not to install any fixtures or make any alterations, additions or improvements to the Hangar without the prior written approval of Landlord. All fixtures installed or additions and improvements made to the Hangar shall, upon completion of such additions and improvements, become Landlord's property and shall remain in the Hangar at the termination of this Agreement, however terminated, without compensation or payment to Tenant.

Figure 6-1   *(Continued)*

12. *Insurance*

    12a. Insurance Coverages: Tenant shall obtain and maintain at all times during the term of this Lease, from a financially solvent insurance carrier(s) authorized to conduct business in the State of _____, the following types and minimum amounts of insurance:

       General Liability Insurance, inclusive of aircraft liability and premises liability; with a combined single limit of $2,000,000 per occurrence insuring Tenant's liability against bodily injury to persons, invites, including passengers and damage to property;

       All-Risk Property Insurance, with coverage to be at Full Replacement Value for Aircraft, furnishings, equipment, spare parts and all other content for the Hangar.

       Automobile Liability Insurance, with the coverage of $2,000,000 per occurrence, and an express representation specifying the applicability of such insurance on airport premises.

    12b. Certificates of Insurance: All Tenant insurance is to be carried by one or more insurance companies licensed to do business in the State where this Lease is effective and approved by Landlord. Landlord shall be named as an additional insured under the applicable liability policies and furnished duly executed certificate(s) of all required insurance, together with satisfactory evidence of the payment of the premiums therefore, on the date Tenant first occupies the Hangar and, upon renewals of such policies, no less than thirty (30) days prior to the expiration of the term of such coverage. The insurance policies of Tenant shall further provide at least thirty (30) days advance written notice to Landlord and Tenant of any material changes, cancellation, non-renewal or changes adverse to the interests of Landlord or Tenant.

       It is expressly understood by Tenant that the receipt of any required insurance certificate(s) by Landlord hereunder does not constitute agreement that the insurance requirements of this Lease have been fully met or that the insurance policies indicated on the certificate are in compliance with all requirements of this Lease. Further, the failure of Landlord to obtain certificates or other evidence of insurance from Tenant shall not be deemed a waiver by Landlord. Nonconforming insurance shall not relieve Tenant of its obligation to provide the insurance specified herein. Nonfulfillment of the insurance conditions by Tenant hereunder may constitute a material breach of this Lease and Landlord retains the right to suspend the Lease until proper evidence of insurance is provided or, in the continued absence of such insurance evidence, terminate this Lease, in Landlord's sole discretion.

    12c. Waiver of Subrogation: All policies shall expressly waive the underwriters' and insurance carriers' right of subrogation against Landlord and/or its insurance carriers.

    12d. Primary Insurance: Consistent with the indemnification provisions of this Lease, Tenant's insurance policies will respond on a primary basis, with any insurance carried by Landlord to be construed as secondary or excess insurance.

    12e. Tenant's Liability Not Limited: NOTWITHSTANDING THE PROVISIONS OF THIS SECTION, FOR PURPOSES OF THIS LEASE, TENANT ACKNOWLEDGES THAT ITS POTENTIAL LIABILITY IS NOT LIMITED TO THE AMOUNT OF LIABILITY INSURANCE COVERAGE IT MAINTAINS NOR TO THE LIMITS REQUIRED HEREIN.

    12f. Invalidation or Conflict with Existing Insurance Policies: Tenant shall not do, permit or suffer to be done any act, matter, thing or failure to act in respect to the Hangar that will (a) invalidate or be in conflict with any insurance policies covering the Hangar or any part thereof; or (b) increase the rate of insurance on the Hangar or any property located therein. If, by reason of the failure of Tenant to comply with

(Continued)

Figure 6-1 *(Continued)*

the provisions of this Lease, the insurance rate shall at anytime be higher than it otherwise would be, then Tenant shall reimburse Landlord and any other Hangar Tenants, on demand, for that part of all premiums for any insurance coverage that shall have been charged because of such actions by Tenant.

13. *Casualty:* In the event the Hangar, or the means of access thereto, shall be damaged by fire or any other cause, the rent payable hereunder shall not abate provided that the Hangar is not rendered unTenantable by such damage. If the Hangar is rendered unTenantable and Landlord elects to repair the Hangar, the rent shall abate for the period during which such repairs are being made, provided the damage was not caused by the acts or omissions of Tenant, its employees, agents or invites, in which case the rent shall not abate. If the Hangar is rendered unTenantable and Landlord elects not to repair the Hangar, this Agreement shall terminate.

14. *Indemnity; Force Majeure:* Tenant agrees to release, indemnify and hold Landlord, its officers and employees harmless from and against any and all liabilities, damages, business interruptions, delays, losses, claims, judgments, of any kind whatsoever, including all costs, attorneys' fees, and expenses incidental thereto, which may be suffered by, or charged to, Landlord by reason of any loss of or damage to any property or injury to or death of any person arising out of or by reason of any breach, violation or non-performance by Tenant or its servants, employees or agents of any covenant or condition of the Agreement or by any act or failure to act by those persons. Landlord shall not be liable for its failure to perform this Agreement or for any loss, injury, damage or delay of any nature whatsoever resulting therefrom caused by any Act of God, fire, flood, accident, strike, labor dispute, riot, insurrection, war or any other cause beyond Landlord's control.

15. *Environmental Responsibilities of Tenant*

15a. Environmental Removal and Disposal: Tenant shall be responsible for the proper removal and disposal of all Hazardous and Regulated Substances, as defined herein, generated by Tenant as a result of Tenant's activities in, on and from the Hangar. Such removal and disposal shall include, but not be limited to Tenant's manifesting such regulated substances under Tenant's assigned Environmental Protection Agency (EPA) identification number and ensuring that removal of such regulated materials from the Hangar and Landlord's leasehold is accomplished in accordance with Airport, local, state and federal guidelines. Additionally, environmental contamination, which impacts Landlord's Airport leasehold as a result of Tenant's improper handling, disposal, release or leakage of any regulated substances while utilizing the Hangar, shall be the sole responsibility of Tenant. Tenant shall also be responsible for the safe and proper removal of all regulated substances it generates in conjunction with its use and occupancy of the Hangar upon termination of this Lease. For purposes of this provision, "Hazardous and Regulated Substances" shall mean any hazardous or toxic substances, materials or wastes, pollutants or contaminants, as defined, listed or regulated now or in the future by any federal, state or local law, rule, regulation, ordinance, statute or order or by common law decision, including without limitation, petroleum products or by-products.

15b. Environmental Indemnification: Tenant shall indemnify, defend and hold harmless Landlord (including Landlord's subsidiaries, affiliates and parent company, as now or hereafter constituted), the Airport and their respective officers, directors, agents, customers, Tenants, contractors, subcontractors, invites, guests and employees from and against any and all claims (including, without limitation, third party claims from bodily injury or real or personal property damage), actions, administrative proceedings (including informal proceedings), judgments, damages, punitive damages, penalties, fines, taxes and assessments, liabilities (including sums paid in settlement of claims), interest, impairments, losses, fees and expenses (including attorneys' fees and expenses incurred in enforcing this provision or collecting any sums due hereunder), consultant and expert fees, together with all other costs and expenses of any kind or nature, including any and all expenses of cleaning up or disposing of any such Hazardous and Regulated Substances (collectively, "Environmental Damages"), that arise directly or indirectly in connection with the presence, suspected presence, release or suspected or threatened release of any Hazardous and Regulated Substances arising from or caused by Tenant's use of the Hangar or Tenant's failure to perform the covenants of this

Figure 6-1 *(Continued)*

Paragraph 15. Tenant shall have no responsibility for any Environmental Damages, which preceded Tenant's initial date of use of the Hangar. For the purpose of this provision, the Parties mutually agree Tenant's initial date of Space use was _____.

The obligations, covenants and agreements of Tenant contained in this Paragraph 15 shall survive termination of this Lease for any reason.

16. *Subordination and Attornment*

16a. Subordination and Nondisturbance: This Lease is subject and subordinate to any and all mortgages which may now or hereafter encumber the airport's fee simple interest or the Landlord's leasehold interest in the real property of which the demised premises is a part thereof, and to all renewals, modifications and extensions thereof; provided that with respect to any future mortgage, the subordination described above shall be subject to the following: in the event that any proceedings are brought by the mortgagee (a) to foreclose the mortgage or any renewal, modification, consolidation, replacement or extension thereof, for any reason whatsoever, or (b) to succeed to the interest of Landlord by foreclosure, deed in lieu thereof or otherwise, and provided no uncured event of default under this lease shall have occurred and be continuing, Tenant's possession of the demised premises and Tenant's rights and privileges under this lease and any extension or extensions thereof shall not be diminished, interfered with or disturbed by the mortgagee as a result of such foreclosure under the mortgage or by any such attempt to foreclose or to succeed to the interest of Landlord by foreclosure, deed in lieu thereof, or otherwise. Tenant shall, within seven (7) days after request of Landlord, execute any subordination documents which Landlord or any mortgagee of the demised premises may reasonably request, but no such document shall be required to effectuate said subordination. Any such subordination documents shall, as to any future mortgage, comply with the terms of this Paragraph.

16b. Attornment: Notwithstanding any other provision of this lease, all rights of Landlord, including property rights and the rights in this lease, are freely saleable, transferable and conveyable. Tenant agrees that in the event of a sale, transfer or assignment of Landlord's interest in the demised premises, or in the event any proceedings are brought for the foreclosure of or for the exercise of any power of sale under any mortgage made by Landlord encumbering the demised premises, to attorn to and to recognize such transferee, purchaser or mortgagee as Landlord under this lease.

17. *Disclaimer of Liability:* LANDLORD HEREBY DISCLAIMS, AND TENANT HEREBY RELEASES LANDLORD FROM, ANY AND ALL LIABILITY, WHETHER IN CONTRACT OR TORT (INCLUDING STRICT LIABILITY AND NEGLIGENCE) FOR ANY LOSS, DAMAGE OR INJURY OF ANY NATURE WHATSOEVER SUSTAINED BY TENANT, ITS EMPLOYEES, AGENTS OR INVITES DURING THE TERM OF THIS AGREEMENT, INCLUDING BUT NOT LIMITED TO LOSS, DAMAGE. OR INJURY TO THE AIRCRAFT OR OTHER PROPERTY OF TENANT THAT MAY BE LOCATED OR STORED IN THE HANGAR, UNLESS SUCH LOSS, DAMAGE OR INJURY IS CAUSED BY LANDLORD'S GROSS NEGLIGENCE OR INTENTIONAL WILLFUL MISCONDUCT. THE PARTIES HEREBY AGREE THAT UNDER NO CIRCUMSTANCES SHALL LANDLORD BE LIABLE FOR INDIRECT, CONSEQUENTIAL, SPECIAL OR EXEMPLARY DAMAGES, WHETHER IN CONTRACT OR TORT (INCLUDING STRICT LIABILITY AND NEGLIGENCE).

18. *Rights Reserved to Landlord:* Landlord reserves the following rights with respect to the Hangar:

18a. Upon prior written notice to Tenant and during reasonable hours (except in situations of emergency where no notice shall be required), to enter the Hangar for inspections or repairs to the Hangar or adjacent premises.

18b. Upon prior written notice to Tenant and during reasonable hours, to exhibit the Hangar to lenders or prospective lenders, or to prospective purchasers of the leasehold; and, during the last six (6)

(Continued)

Figure 6-1 (*Continued*)

months of the term, to prospective new Tenants; provided, however, that Landlord shall exercise its rights under this section in such a manner so as to minimize the interference with Tenant's use and occupancy of the Hangar.

18c. Upon reasonable notice to Tenant (except in situations of emergency where no notice shall be required), to enter the Hangar for any reasonable purpose whatsoever related to the safety, protection, preservation or improvement of the Hangar without being deemed guilty of an eviction or disturbance of Tenant's use and possession of the Hangar.

18d. Removal of Aircraft during Repair: Consistent with Landlord's obligations as set forth in Paragraph 5 to keep the Hangar in proper repair, should Hangar maintenance or repairs, whether preventive or non-routine, be required to be undertaken by Landlord and such maintenance or repair, in Landlord's sole judgment, shall compromise or potentially compromise the safety or integrity of the Aircraft, Landlord's personnel or other Hangar Tenant's aircraft and/or personnel, then Landlord, concurrent with notification of Tenant, shall remove the Aircraft from the Hangar until such time as the maintenance/repair work is completed or, in Landlord's sole determination, the maintenance/repair work is completed to the extent that no further danger exists to the Aircraft or personnel. Landlord agrees to pro-rate the following month's Hangar rent of Tenant for all days the Aircraft cannot be accommodated in the Hangar. The foregoing proration of rent shall be the sole compensation due Tenant as a result of Landlord's actions herein.

19. *Default:* This Agreement shall be breached if: (a) Tenant shall default in the payment of any rental payment within five days of its due day; (b) Tenant shall default in the performance of any other covenant herein, and such default shall continue for ten (10) days after written notice thereof from Landlord; (c) Tenant shall cease to do business as a going concern; (d) a petition is filed by or against Tenant under the Bankruptcy Act or any amendment thereto (including a petition for reorganization or an arrangement); or (e) Tenant assigns its property for the benefit of creditors.

In the event of this Agreement by Tenant, Landlord shall, at its option, and with notice to Tenant, have the right to terminate this Agreement and to remove the Aircraft and any other property of Tenant from the Hangar using such force as maybe reasonably necessary, without being deemed guilty of trespass, breach of peace or forcible entry and detainer, and Tenant expressly waives the service of any notice. Exercise by Landlord of either or both of the rights specified above shall not prejudice Landlord's right to pursue any other remedy available to Landlord in law or equity.

20. *Governing Law:* This Agreement shall be construed in accordance with the laws of the State of _____.

21. *Relationship of Parties:* The relationship between Landlord and Tenant shall always and only be that of Landlord and Tenant. Tenant shall never at anytime during the term of this Agreement become the agent of Landlord, and Landlord shall not be responsible for the acts or omissions of Tenant, its employees or agents.

22. *Remedies Cumulative:* The rights and remedies with respect to any of the terms and conditions of this Agreement shall be cumulative and not exclusive, and shall be in addition to all other rights and remedies available to either party in law or equity.

23. *Notices:* All notices to be given hereunder shall be in writing and shall be sent by certified mail to the addresses shown on the front page of this Lease, or to such other address as either party may have furnished by prior written notice sent pursuant hereby. Any notices permitted or required to be given by the terms of this Lease shall be effective upon mailing and shall be deemed sufficient if mailed by United States mail, with proper postage and address affixed thereto.

24. *Integration:* This Agreement constitutes the entire agreement between the parties, and supersedes all prior independent agreements between the parties related to the leasing of the Hangar. Any change or modification hereof must be in writing signed by both parties.

Figure 6-1 *(Continued)*

25. *Waiver:* The waiver by either party of any covenant or condition of this Agreement shall not thereafter preclude such party from demanding performance in accordance with the terms hereof. No failure by Landlord to insist upon the strict performance of any term, covenant, agreement, provision, condition or limitation of this Lease or to exercise any right or remedy consequent upon a breach thereof, and no acceptance by Landlord of full or partial rent during the continuance of any such breach or application of the security deposit in light of any breach, shall constitute a waiver of any such breach or of any such term, covenant, agreement provision, condition, limitation, right or remedy. No term, covenant, agreement, provision, condition or limitation of this Lease to be kept, observed or performed by Landlord or by Tenant, and no breach thereof, shall be waived, altered or modified except by a written instrument executed by Landlord or by Tenant, as the case may be. No waiver of any breach shall affect or alter this Lease, but each and every term, covenant, agreement, provision, condition and limitation of this Lease shall continue in full force and effect with respect to any other then existing or subsequent breach thereof.

26. *Successors Bound:* This Agreement shall be binding on and shall inure to the benefit of the heirs, legal representatives, successors and assigns of the parties hereto.

27. *Severability:* If a provision hereof shall be finally declared void or illegal by any court or administrative agency having jurisdiction over the parties of this Agreement, the entire Agreement shall not be void, but the remaining provisions shall continue in effect as nearly as possible in accordance with the original intent of the parties.

28. *Time Is of the Essence:* Time is of the essence in the performance of all of Tenant's obligations under this Lease.

29. *Taxes, Assessments and Fees:* Tenant shall be solely responsible for the payment of all taxes, assessments, license fees or other charges that may be levied or assessed during the Term of this Lease upon or against any personal property or equipment located within or upon the Hangar and/or the Space which is owned by, leased to or in the care, custody and control of Tenant.

IN WITNESS THEREOF, the parties have executed this Agreement as of the day and year first above written.

WITNESSES:　　　　　　　　LANDLORD:

_____

_____　　　　　　　BY: _____

　　　　　　　　　　　　　　_____

　　　　　　　　　　　　　　TENANT:

WITNESSES:　　　　　　　　_____

_____　　　　　　　BY: _____

_____

Identification of the parties

You can't overlook the simple things. It is always a good idea to know the person or entity you are dealing with in a transaction as substantial as an aircraft hangar lease. On many occasions, the landlord may be a state, county, or local government. In other settings you may be doing business with a privately owned fixed-base operator (FBO). Although it is not true in every case, you are likely to encounter great difficulty getting a governmental body to modify any terms of the lease that you find troublesome. In any case, you should ensure that you know who the other party is—it will make the reasons for some of the lease provisions a bit clearer. In the case of our specimen hangar lease, we are dealing with a fictional FBO. It could be that the FBO holds title to the hangars or simply leases them all from a government owner and sublets the space to aircraft owners.

Hangar/aircraft identification

Again, there is a fairly obvious need for this provision. You need to know the hangar that has been assigned to you and the landlord needs to identify your aircraft by type and model to ensure it is an airplane that will be suitable for the hangar space provided. Notice that this hangar agreement permits you to store the identified aircraft and any other "similar" aircraft (substitute aircraft). However, if you want to store a substitute aircraft, you must obtain the written consent of the landlord. This is not very problematic. However, to protect yourself from an arbitrary landlord, you may want to request that language be added that indicates that the landlord will not be able to unreasonably withhold consent for a substitute aircraft. Most landlords will be honest and deal with you in good faith. However, you don't want to be burned just because you replaced an older aircraft (especially when the hangar space can accommodate both types of aircraft with room to spare).

Term

The term of the lease is obviously a very critical part of your deal. Although there may be advantages to a long-term lease, it could prove problematic if you sell the aircraft before the term is up or you change airport bases. Any such move could subject you to paying rent to the landlord until the end of the lease term for a hangar that you are not using. The lease depicted in Fig. 6-1 is for a little over 1 year. It is not unusual to see 1-year leases. However, you and your landlord can agree to any term for your lease.

Rent payments

The section outlining payments in the lease we are reviewing is pretty straightforward. If a lease is for an extended period, it may include rent escalation provisions that increase the rent over the latter years of the lease. Just make sure you know what you will be paying at any given time for the lease of your hangar space. The late charge assessment is also predictable and in this case, fairly reasonable.

Security deposits

Paragraph 4 of our specimen lease requires a security deposit of $300.00. The requirement for a security deposit is something you might expect. Typically, the deposit will equal 1 or 2 months' rent. In some states the law may cap the amount of the security deposit. Although these caps often apply to residential leases only, that does not meant that you can't negotiate a deposit amount you can live with. This lease predictably states that the security deposit can be used to compensate the landlord for unpaid rent or damages to the hangar that you may have caused. One thing that may catch your eye in this provision is the fact that it allows the landlord to comingle the security deposit with the landlord's funds. Depending on who the landlord is, this may be a problem. Suppose the landlord goes bankrupt? Under the terms of this lease, you are just another unsecured creditor, who will probably see only pennies on the dollars owed you, if you see anything at all. It would be comforting to see a provision that states that the landlord must set aside your security deposit in a trust account that bears interest for you. In some states a separate trust account and/or interest-bearing security deposits may be a requirement. However, as usual, this protection is often built in for residential tenants only. You should also note the common provision in this lease indicating that you cannot use your security deposit to pay for your last month of rent.

Landlord's obligations

As expected, our specimen lease does not require much of the landlord. However, the key items in paragraph 5 include the landlord's responsibility to maintain the hangar structure and provide restroom facilities. This is a very standard provision outlining the landlord's limited obligations.

Tenant's obligations

Paragraph 6 outlines the tenant's obligations under the lease. In this lease, the provisions are fairly predictable. A few points should be made so that the lease we're reviewing may be compared to other leases:

1. Paragraphs 6(c) and 6(e) are largely concerned with compliance with applicable laws and ordinances, including fire and hazardous material regulations. You are probably well aware of FAA requirements for the safe operation of your aircraft. However, if your landlord is holding you to knowledge of any local codes, you should be provided with a copy of the applicable codes for your reference. It is inadvisable to agree to comply with any laws or ordinances unless you know what you are getting yourself into. If costs such as inspection fees may be assessed to you, you should investigate to determine how much they may cost you.

2. Paragraph 6(d) restricts the tenant's ability to have maintenance performed (other than FAA-approved maintenance by a pilot) without express written approval. With a provision like this, you may want to request a modification indicating that the landlord's approval will not be unreasonably withheld.

This may be especially important when the landlord is an FBO that may want to lock you into doing business with its maintenance shop.

3. Paragraph 6(f) requires that you surrender the leased hangar in much the same condition that you found it when you started your occupancy. One thing you might consider doing to protect yourself is to take strategically placed photographs of your hangar before you start your tenancy. By doing this, you may prevent any misunderstandings about whether damage was preexisting upon your arrival at the hangar. This also ties in with paragraph 10 of the specimen lease. Make sure you are satisfied with the condition of the hangar before you sign the lease. If you find something you don't like after the lease is executed, your landlord may not have a legal obligation to make repairs or alterations.

Primary lease

As indicated earlier, it is good to know your landlord and whether the landlord is leasing your hangar from someone else. In this case it appears that the landlord is leasing the hangar areas from the airport owners. When that lease is terminated, your rights to the hangar may be terminated. According to this provision, you will not have any right to pursue the landlord if this happens. There may not be a lot you can do about this provision. However, you should at least inquire about the termination date of the primary lease and of the possibility of it being renewed if the termination date falls within the time frame of your lease.

Sublease/assignment

Many aircraft owners tend to gloss over this type of provision. Later on they may regret that they did not examine it more carefully. What happens if you get a job transfer during the course of your lease and you need to relocate? What happens if your financial circumstances change and you sell your aircraft? Regardless of whether you have an airplane in the hangar or not, you are still obligated to pay rent under this lease. One thing you might be able to do is to sublease or assign your hangar to another party in order to reduce your financial burden of carrying the hangar. As written, this provision does not prohibit your sublet or assignment. However, it does require the prior written approval of the landlord, which may be withheld "in Landlord's sole discretion." This could be a problem. This is another case where you may want the language to be changed to indicate that the landlord's approval may not be unreasonably withheld. This provision also states that neither a sublet nor an assignment of your lease will relieve you of the obligations under the lease. Therefore, you may be liable if your subtenant or assignee fails to pay rent or damages the hangar in any way.

One additional note on this provision—there is a difference between a sublease and an assignment. An assignment of your hangar lease will occur if you transferred the entire remaining portion of your lease to a third party. A sublease would occur if you transferred only a portion of your remaining hangar lease to a third party. For instance, if 12 months remain on your hangar lease and you transfer the lease to another aircraft owner for the remaining 12 months, you would be assigning the lease. On the other hand, if 12 months remain and you are transferring only the next 3 months, you would be subleasing the hangar.

If an assignment occurs, the third party you assigned the lease to assumes your role as the original hangar tenant. However, if your assignee fails to pay rent, the landlord can pursue both the assignee and you. On the other hand, if you sublease the hangar and the subtenant does not pay rent to you, the landlord can pursue only you for the rent. Your only recourse is to go after the subtenant for the rent.

Alterations

It is very important that you carefully read and understand what this provision is saying. This type of provision can be the source of bad endings to a hangar lease. The first thing to note is that you cannot make any improvements or additions to your hangar without obtaining the approval of your landlord. The second thing that should jump out at you is that any improvements or additions you make will become the property of the landlord when your lease is terminated.

This provision bites a number of aircraft owners who have made their hangars a sort of home away from home. Often, they have built workbenches, cabinets, and shelves on their own time and expense. If you have the typical sort of hangar lease, you've got to get permission before you make any improvements or add anything that is attached to the hangar in any way. Next, you've got to come to grips with the fact that all your work and money to make the improvements may be enjoyed at no additional cost by the next hangar tenant.

Insurance

Paragraph 12 of our specimen hangar agreement discusses required insurance coverage. You may ask yourself why your hangar landlord is so interested in your insurance coverage. Essentially, the reason boils down to risk shifting. Your landlord and just about every aircraft hangar landlord want to be able to shift as much liability risk to you as possible. However, you've got to be careful. Sometimes attorneys and local contracting officers who know very little about general aviation draft these agreements. Therefore, they may attempt to shift more of the liability risk than you can realistically or practically handle.

Our agreement may present an example of overreaching by the landlord. Here are some of the bigger issues presented in paragraph 12 of the specimen agreement:

1. The amounts of insurance called for in Section 12a may be a problem for aircraft owners. The aircraft liability section calls for a single limit of $2,000,000 per occurrence. Unfortunately, this level of coverage may be unattainable for many aircraft owners. The automobile coverage of $2,000,000 with specifying

applicability on airport premises may also be unattainable at an affordable cost. You could not sign this agreement until you had an understanding of whether or not you could comply with these provisions and meet the amounts called for in the lease agreement.

2. The requirement that the landlord be named as an additional insured is commonplace and your insurance carrier would probably not balk at this requirement. The 30-day notice requirements for cancellation, nonrenewal, or changes may be unrealistic. Most insurance companies will give only 10 days' notice.

3. In paragraphs 12c and 12d the landlord's risk-shifting efforts become obvious. Paragraph 12c calls for you to convince your insurance carrier to expressly waive its customary right to subrogate against the landlord. Part 4 of this book contains a full discussion of subrogation and insurance. However, in the context of this lease it means that if the landlord negligently harms your property, your insurance company will have to pay you for the damages and your insurance carrier will not be able to go after your landlord for the damages. This type of scenario is not difficult to conjure up. For instance, let's say that a tractor lawn mower operated by the landlord's employees negligently rams the side of your aircraft causing serious damage. If you signed this agreement as is, your insurance company would perhaps be unable to recover the amount it paid you by going after the landlord for its negligence. You had better check with your insurance carrier before you agree to this provision. The same holds true for paragraph 12d. Your insurance carrier may not be willing to agree that it will be first in line to pay for any damage claims. Make sure your insurance company expressly agrees to any such provision before moving forward on the agreement.

4. Paragraph 12f also warrants some careful consideration. This paragraph seems to require that you gain familiarity with your landlord's insurance coverage on the hangars. How else would you know if any action you take will cause the invalidation of your landlord's insurance policies? This provision also begs the question of how you would be able to trace any increase in your landlord's insurance policy to some action or inaction on your part. Overall, this is a difficult provision to wade through. If you are faced with something like this, you should request a full explanation before agreeing to comply.

Casualty

This provision basically states that if your hangar or the access to your hangar is damaged, you will still have to pay rent as long as the hangar is not rendered "untenantable" by the damage. The question left open is what is meant by "untenantable." A search through standard and law dictionaries fails to produce a definition. Therefore, before you agree to the terms of this agreement, it would be advisable to request that a definition of this term be included in the agreement. Clearly, you don't want to be paying rent on a hangar that is severely damaged or inaccessible.

Indemnity

This is an interesting provision and it tends to show up in varying forms in just about every aircraft hangar agreement. The rather long first sentence holds you responsible for just about anything that can possibly go wrong if you should breach the terms of the hangar lease agreement. It is a very broad provision and the liability created may exceed the scope of any insurance coverage that you may have in place. The very next sentence states that the landlord will not be liable for its failure to perform the terms of the lease agreement. This can leave you in a pretty tough situation. After all, the landlord has at least agreed to maintain the hangar structure and doors. If the landlord now fails to follow through on its promises, this provision precludes you from any recourse against the landlord. Can you live with that? Landlords can tend to get pretty heavy-handed with this sort of provision. You'll need to decide whether you can live with the risk created.

Environmental responsibilities of tenant

Paragraph 15 of the specimen hangar lease agreement imposes some sweeping responsibilities and liabilities on you. As background to this provision, you should keep in mind that most landlords at airports are subject to scrutiny by local, state, and/or federal environmental agencies. There are strict penalties for violations of environmental protection rules. In fairness, landlords don't want to be subject to strict penalties for problems that you may have created. However, the broad responsibilities placed on you by these provisions will require that you at least have a fundamental understanding of the term "Hazardous and Regulated Substances" as it is used in this provision. For all practical purposes, the real issue will typically be your proper handling of fuel, oil, and hydraulic fluids. If you can ensure that there are no leaks or contamination caused by these fluids, you should be able to protect your interests. Compliance with this type of provision becomes even more important when you consider the fact that you will probably not be able to procure insurance to cover you for the type of risks addressed in this type of provision.

Subordination and attornment

Paragraph 16a of the specimen lease addresses the issue of subordination. Subordination in the context of this hangar lease means that your leasehold interest ranks lower in priority than any mortgage currently on the airport property or the landlord's present leasehold interest in the hangars. If any future mortgages are taken out on the underlying property, your leasehold interest in the hangar will be preserved. This provision also requires you to acknowledge your subordinate interest in the hangar at the reasonable request of the landlord or someone holding a mortgage on the property where your hangar is located.

Paragraph 16b addresses the issues of attornment. Attornment is a practical matter in any landlord/tenant relationship. It essentially requires that if your current landlord transfers or assigns the leased property to another person, you promise to recognize that other person as the landlord.

Disclaimer of liability

The paragraph labeled "Disclaimer of Liability" (paragraph 17) is perhaps one of the more problematic provisions in this hangar lease. Again, this is a fairly typical provision for hangar leases. Distilled to its essence, it states that you are releasing the landlord from any liability for harm that may come to you, your guests, or your employees, except if the harm is caused by the gross negligence or intentional willful misconduct of the landlord. This paragraph also states that even if the landlord acted with gross negligence or willful misconduct, you still could not pursue the landlord for indirect, consequential, or special damages.

So what does all this mean in plain English? Perhaps the best way to explain is with an example. Suppose your airplane was parked in its hangar after you rolled in from a flight. The hangar doors are open. You are nearby in the landlord's FBO purchasing some new charts. While you are in the pilot shop, you hear a crash and the unpleasant sound of metal bending. When you step outside, your worst fears are confirmed. A line person driving a fueling truck accidentally backed up into your hangar. Your airplane is substantially damaged.

Well, at least you've got insurance coverage. Or do you? After you make your claim, your insurance company may want to pursue the landlord for damages (that's the subrogation right we were discussing earlier). However, when your insurance company pursues the matter, it will discover that you have released the landlord from all liability unless the landlord was grossly negligent or engaged in willful misconduct. The truth of the matter is that this was most likely a case of simple negligence, not gross negligence, and your insurance company may be effectively blocked from stepping into your shoes and suing your landlord. By impeding your insurance carrier's right to subrogate, you may have also provided your insurance carrier with a way of denying your claim. Many insurance policies will prohibit you from waiving their right to subrogate. You could now be in the worst of all possible worlds, with no ability to collect an insurance recovery and no ability to pursue the landlord because you signed this release.

There is no simple solution to this very common clause in hangar leases. However, your best bet may be to formally present the entire lease agreement to your insurance carrier for review. You'll want your insurance company's express and written verification that your agreement to this provision and any of the other risk-shifting provisions in your hangar agreement will not compromise your insurance coverage.

Rights reserved to landlord

Paragraph 18 discusses certain rights reserved to the landlord. Upon closer examination, it is apparent that this provision addresses the landlord's right to access your hangar and your aircraft under certain circumstances. None of these provisions is particularly unusual. However, you might be wise to insist on a specific number of hours' or days' notice of any landlord action that will allow access to your aircraft or hangar (obviously, the notice provisions will have to be waived in case of a true emergency).

Standard contract provisions

Paragraphs 19 to 29 are typical boilerplate provisions that you would expect to see in any agreement and certainly in an aircraft hangar lease. You may find the paragraph on governing law to be of particular interest in case you and the landlord are from different states. Litigating disputes is an expensive proposition. If they must be litigated, it may be best done from the convenience of your home state's court system.

Tie-Down Agreements

Tie-down agreements are usually a lot simpler than hangar agreements. Because you are no longer renting a building or enclosed structure, many of the issues that existed with the hangar lease are no longer relevant.

Again, it may be best to review a typical, real-life tie-down agreement to review some of the major points likely to be encountered with your tie-down agreement. The following illustration in Fig. 6-2 comes from an actual FBO tie-down agreement that has been deidentified, but not edited in any manner. Let's review the agreement and determine if there are any issues that warrant special attention. After that, we'll take a look at the question of who is liable in case your aircraft is damaged while being stored at a tie-down facility.

Term of agreement

This agreement starts off by describing the term of the agreement. In this case, the lease term is month to month. That's a term of art in the legal world of landlord/tenant relations. It typically means that you are able to keep your airplane in the landlord's tie-down spots on a monthly basis. Either you or the landlord may cancel the lease. Customarily, a 1-month notice of cancellation is required on a month-to-month lease. This lease appears to follow that custom.

Payment and services

The payment provisions of this tie-down agreement are very straightforward. It's not at all unusual to see a different rate for grass or hard surface tie-downs. It is also not unusual for the landlord to expressly remind you that you are responsible for the security and protection of your own aircraft.

Insurance

Similar to the hangar agreement, this landlord is requiring "all-risk" insurance on your aircraft. What is different here is that no minimum amount of insurance is required. This provision is probably no big deal. Insurance is a must no matter what the requirements of this agreement.

Figure 6-2 Specimen tie-down agreement.

TIE-DOWN AGREEMENT

This TIE-DOWN AGREEMENT (the "Agreement") is entered into as of this _____ day of _____ 2003, by and between _____, a _____ Limited Liability Company (hereinafter referred to as "LANDLORD"), and _____ (hereinafter referred to as "Client").

1. This will confirm that Client shall be granted the non-exclusive permission to park an aircraft (hereinafter collectively referred to as the "Aircraft") on or about the demarcated tie-down area located on or about the LANDLORD facility. Such permission shall commence on _____ and continue month to month. This agreement shall be terminable by either party by giving thirty (30) days' written notice.

The tie-down spots are allocated and utilized on a first come and first served basis and there are no assigned tie-down spots. LANDLORD is permitted to relocate the Aircraft about the tie-down area in its sole and unfettered discretion. Likewise, LANDLORD is permitted to relocate the entire tie-down area on or about its facility incident to its overall development plan.

2. In consideration for such parking of Aircraft No. _____, Type _____, Client shall pay to LANDLORD the sum of $30.00 per month for grass tie-down or $60.00 per month for hard surface tie-down. Such fee shall be paid via credit card type (circle one) Exxon/Visa/MasterCard/American Express, No. _____, Expiration Date _____, and LANDLORD is authorized to debit such credit card on the first day of each month for the parking fee.

3. LANDLORD shall provide no services or materials to Client incident to the parking or relocation of the Aircraft and Client is solely responsible for securing, covering, maintaining, protecting and locking of the Aircraft.

4. As a condition of this Agreement, Client agrees to maintain, at its sole expense, "all-risk" insurance on the Aircraft. This insurance policy shall name LANDLORD as a named insured. Client shall deliver to LANDLORD copies of certificates evidencing the existence of the insurance.

5. Client hereby releases LANDLORD from any and all liability, whether in contract, tort or otherwise, for any loss, damage or injury of any nature sustained by Client or the Aircraft. Client agrees to indemnify, defend and hold LANDLORD harmless from any and all claims, demands, suits and judgments arising out of the parking, utilization and storage of the Aircraft on or about the LANDLORD facility.

6. As additional consideration to LANDLORD, Client agrees to purchase any and all aviation fuel and supplies from LANDLORD when such purchases are made at Witham Field.

7. This Agreement is immediately terminable by LANDLORD as a result of Client's creation and/or maintenance of a nuisance on or about the _____ or a condition and/or situation which otherwise disrupts the orderly operation of the _____. Client shall properly dispose of any and all aviation fuel and oil and same should not be improperly discarded or dumped about LANDLORD's facility. It is further agreed that any and all activities shall be in compliance and conformity with any and all rules, regulations, orders and laws promulgated and adopted by _____ County, the state of _____, the Federal Aviation Administration (FAA) or any Federal, State and local administrative, law enforcement or regulatory agency having jurisdiction of Client's use.

IN WITNESS THEREOF, the parties have executed this Agreement as of the day and year first above written.

LANDLORD   CLIENT

_____.   _____

_____   _____
FOR LANDLORD   FOR CLIENT

Indemnification

Paragraph 5 of this agreement may be a problem. Just as we discussed in the hangar agreement section, this type of hold harmless and indemnification provision may impede your ability to recover a claim from your insurance carrier. Again, what happens if the landlord is negligent and dings your aircraft with its equipment? Do you really want to let the landlord off the hook? Will your insurance company pay up if you do? You should have this provision reviewed by your insurance carrier before you commit to this type of provision in a tie-down agreement.

"Purchase" provision

Although it is not highly unusual, you don't see provisions like paragraph 6 in every tie-down agreement. Signing this agreement requires you to purchase all aviation fuel and supplies from the landlord. It can probably be assumed that this provision is only enforceable as long as the lease is in place.

Compliance with applicable laws

This type of provision is expected (just as with a hangar lease). It would be reasonable to request that the landlord attach any local or state provisions that you need to comply with so that you will be put on notice. It may be difficult for you to do the research and unearth all the codes and regulations you may have to comply with under this agreement.

Lease or bailment?

If your aircraft is damaged while parked at a tie-down facility, the question of legal liability will quickly become an issue. Are you liable or is the tie-down facility liable for the damages incurred to your aircraft?

This question often turns on whether the arrangement with the tie-down facility (usually an FBO or aircraft operator) is a bailment. A bailment is a legal arrangement where you deliver personal property (e.g., your aircraft) to another in trust for a specific purpose (e.g., storage). With a bailment, there is an implied or express agreement that your property will be returned to you intact.

If your arrangement is legally classified as a bailment, the tie-down facility owes you the duty to exercise ordinary and reasonable care for the preservation of your aircraft. On the other hand, if your arrangement is classified as a lease or license, no such duty would exist and you would have to prove that the tie-down facility was negligent in order for the tie-down facility to be held liable for damages to your aircraft. Case 6-1 provides a good example of the legal reasoning that will have to be employed to determine whether a bailment exists.

Case 6-1

Aerowake Aviation, Inc. v. Clifford M. Winter, Jr. and
Avemco Insurance Company
Supreme Court of Alabama
423 So. 2d 165 (1982)

OPINION BY: Almon, J.

OPINION: Aerowake Aviation, Inc. (Aerowake) appeals from a judgment against it
in favor of Clifford M. Winter (Winter) and his subrogated insurance carrier, Avemco
Insurance Company (Avemco). Winter sued Aerowake for damage to his airplane
while it was parked at Aerowake's airport facility. Winter sued on a theory of bailment, alleging that Aerowake breached its duty to exercise ordinary and reasonable
care for the protection, preservation, and return of his airplane. The trial court
entered judgment on a jury verdict of $7,840 for Avemco and $8,660 for Winter, and
denied Aerowake's motion for judgment notwithstanding the verdict or, in the alternative, for a new trial.

Aerowake argues that the court erred in denying its motion for new trial because
the verdict is contrary to the great weight of the evidence in that no bailment relationship existed between the parties. We disagree, and affirm.

Winter, a retired lieutenant colonel in the United States Air Force, constructed an
experimental two-man aircraft over the course of several years in the late 1970's. In the
summer of 1979 he went to the North Huntsville Airport, operated by Aerowake, and
talked with Jim Wakefield, the president and a chief stockholder of Aerowake, about
arrangements for keeping his plane at the airport. Aerowake provided services including fuel, employee attendance from 8:00 a.m. to 8:00 p.m., flight school, food and
beverages, car rental, airplane maintenance, tie-downs, and hangars. In August 1979
Winter contracted with Aerowake to store his airplane at the airport.

He initially used a tie-down spot in an open grassy area, but he later took a space
in a covered plane port. The space had three tie-down rings and Aerowake provided
ropes for Winter to tie his plane down. The plane port had walls on both ends but
was open to the front and rear of the airplanes. Winter testified that he was particularly glad to have this space because it was right next to the trailer in which Leroy
Trulson (Trulson), Aerowake's mechanic, lived with his family.

Aerowake provided the trailer for Trulson to live in free of charge. Although
Wakefield testified that living on the premises was not a condition of Trulson's
employment, he did admit that Trulson's presence at the airport at night was a benefit
to the airport and the owners of airplanes parked there.

On September 5, 1980, Ron Sunholm (Sunholm) drove his dune buggy into Winter's
airplane and caused substantial damage to it. Sunholm was related by marriage to
Trulson's son, and the two of them rode the dune buggy in the fields around the airport. Sunholm stuck his dune buggy in a pond several days before the accident. He
borrowed a rope from Trulson in order to pull it out. Apparently after freeing the
dune buggy, he cranked it up and ran into Winter's airplane. Sunholm testified that
the throttle stuck and the brakes would not work because they had mud in them.

Trulson had seen Sunholm driving his dune buggy on the airport premises numerous times prior to the day it got stuck in the mud. Sunholm asked Trulson for help
in freeing the dune buggy, but he refused and only consented to let Sunholm borrow
a rope; nor did he supervise Sunholm's efforts, being busy with work on an airplane.

The dispute at trial and on appeal concerns the nature of the parties' arrangements for storage of Winter's airplane—whether it was a bailment or merely a lease

or license. If it was a bailment, Aerowake owed a duty as bailee to exercise ordinary and reasonable care for the protection of the airplane. If it was a lease or license, there would be no such duty on Aerowake.

The Court of Civil Appeals recently adopted the definition of bailment found at 8 Am. Jur. 2d Bailments § 2 (1980):

> The delivery of personal property by one person to another in trust for a specific purpose, with a contract, express or implied, that the trust shall be faithfully executed, and the property returned or duly accounted for when the special purpose is accomplished, or kept until the bailor reclaims it. [Footnotes omitted.]

Farmer v. Machine Craft, Inc., 406 So. 2d 981, 982 (Ala. Civ. App. 1981).

The parties argue principally from the Alabama cases regarding automobile parking lots: *Lewis v. Ebersole,* 244 Ala. 200, 12 So. 2d 543 (1943); *Ex parte Mobile Light & R. Co.,* 211 Ala. 525, 101 So. 177 (1924); *Mobile Parking Stations, Inc. v. Lawson,* 53 Ala. App. 181, 298 So. 2d 266 (Ala. Civ. App. 1974); *Kravitz v. Parking Service Co.,* 29 Ala. App. 523, 199 So. 727 (1940). Aerowake would have us hold that the facts here are more like those in *Ex parte Mobile Light & R. Co., supra,* and *Mobile Parking Stations, supra,* in that the alleged bailee did not have sufficient control over the property for the relationship to constitute a bailment.

We do not find this argument compelling. The circumstances of an airplane parking facility are not exactly analogous to those of an automobile parking lot. Some protection of the airplane from the weather is always provided—if not shelter, at least tie-down equipment to keep the airplane secure in the wind. Airplanes can usually be moved without the ignition keys. In fact, Winter's airplane did not require keys to start, so the primary indication of possession in automobile cases is lacking.

Aerowake states that it did not have Winter's permission to move or start his airplane. This is not determinative here for several reasons. Should an emergency arise such as a fire in the plane port, Aerowake would surely have a duty to extinguish the fire or move airplanes out of danger. Trulson testified that when a storm approached he inspected the tie-downs and made sure no loose objects such as garbage cans were around to be blown into the planes. Thus, the requisite amounts of possession and control and assumption of a duty to use reasonable care to protect the property were present.

In most cases of this type the defendant fixed-base operator, such as Aerowake, admits the existence of a bailment and argues that it met its duty of ordinary care. *Aircraft Sales & Service, Inc. v. Bramlett,* 254 Ala. 588, 49 So. 2d 144 (1950); *Naxera v. Wathan,* 159 N.W. 2d 513 (Iowa 1968); *Alamo Airways, Inc. v. Benum,* 78 Nev. 384, 374 P.2d 684 (1962). Where a fixed-base operator or an airport has challenged the existence of a bailment, most courts have found a bailment to exist. *City of Jackson v. Brummett,* 224 Miss. 501, 80 So. 2d 827 (1955); *Hendren v. Ken-Mar Airpark, Inc.,* 191 Kan. 550, 382 P.2d 288 (1963); *Meyer v. Moore,* 329 P.2d 676 (Okla. 1958); *Clack-Nomah Flying Club v. Sterling Aircraft, Inc.,* 17 Utah 2d 245, 408 P.2d 904 (1965). In *Meyer v. Moore, supra,* the facts were held to establish "a bailment for hire under which it was the duty of the defendants to use ordinary care in safeguarding the plaintiffs' airplane during the time it was stored at the airport," even though "the owner or pilot of the plane was responsible for tying down his own airplane." *Id.,* 329 P.2d at 679.

In *Central Aviation Company v. Perkinson,* 269 Ala. 197, 112 So. 2d 326 (1959), the Court found an agreement on the part of the company to tie down the airplane, and that the company breached this agreement by not providing ropes of sufficient strength. The case was not tried on an issue of bailment, but does go to show that airplane parking facilities provide more than a mere parking space.

Aerowake does not raise on appeal the question of whether it breached any duty it might have had nor whether the damage caused by Sunholm was foreseeable. The general supervision of the premises by Trulson, however, coupled with his admission that he had seen Sunholm driving his dune buggy on the property, indicate to this Court that the verdict and judgment are not manifestly unjust.

For the reasons stated above, the judgment of the trial court and its denial of Aerowake's motion for new trial are affirmed.

AFFIRMED.

Notice that the court held that aircraft storage arrangements are different from automobile parking lots. The court noted that aircraft are capable of being moved without keys. The court also cited numerous cases where tie-down facilities have unsuccessfully challenged the finding of a bailment.

One thing you may consider is the inclusion of language in your tie-down agreement in which the tie-down facility clearly acknowledges the bailment arrangement. This may be difficult to negotiate, but it may help clarify each party's duties in the case of a severe windstorm or unexpected mishap.

Legal Basics of Aircraft Maintenance

If you're an aircraft owner, you know (or you will soon learn) that one of your biggest responsibilities is aircraft maintenance. In this chapter, we'll attempt to tackle some of the legal basics of aircraft maintenance that will affect you and your airplane. To make the task a bit more manageable, this chapter will be broken down into several segments. After a brief overview of FAA's maintenance requirements, there will be separate discussions of:

- Aircraft inspections
- Maintenance record keeping
- Inoperative equipment
- Airworthiness directives (more commonly referred to as ADs)
- Returning aircraft to service
- Maintenance contract issues

Overview

FAR Section 91.403 plainly states that if you are an aircraft owner, you are responsible for maintaining your aircraft in an airworthy condition. It also reminds you that your responsibilities also include compliance with rules related to ADs in FAR Part 39.

The vast majority of aircraft owners are not trained aviation mechanics. Because of this, you will probably have to rely in large part on the services of trained maintenance personnel to ensure that your aircraft is properly maintained. However, the regulations do not permit you to abdicate all of your legal responsibility to FAA–certified maintenance providers. In essence, the regulatory scheme sets up a sort of partnership between you and certified maintenance providers to work together in an effort to keep your aircraft safe for flight. You are responsible for the oversight of your aircraft's maintenance and

FAA–certified maintenance providers are responsible for properly implementing the maintenance needed for your aircraft.

FAR Sections 91.403, 91.405, and 91.407 set out your basic maintenance responsibilities. These responsibilities require that you:

1. Ensure your aircraft is inspected as required by the regulations (see "Inspection Requirements," below) and receives any necessary maintenance between required inspection intervals.

2. Comply with all maintenance record-keeping requirements (see "Record Keeping" later in this chapter).

3. Comply with FAA rules for inoperative equipment (see "Inoperative Equipment" later in this chapter).

4. Comply with applicable ADs (see "Airworthiness Directives" later in this chapter).

5. Ensure the right person has performed your maintenance and has properly approved it for return to service (see "Getting Your Aircraft Back in Service" later in this chapter).

Inspection Requirements

FAR Sections 91.409, 91.411, and 91.413 lay out mandatory inspection programs for most general aviation aircraft. In this segment, we'll take a look at the basic inspection requirements you will have to comply with under the current FAA rules.

General rules

For the typical general aviation aircraft, the rule is that you cannot operate an aircraft unless you have had an annual inspection within the preceding 12 months. If your aircraft is used to carry persons (other than crew members) for compensation or hire or for paid flight instruction, a 100-hour inspection is also required by FAR Section 91.409. In order to get your aircraft to the place where a 100-hour inspection may be performed, you may exceed the 100-hour requirement by no more than 10 hours while you are on your way to the place where the inspection will be performed. However, any time in excess of the 100 hours is counted toward the next 100-hour inspection. Therefore, if you went over your last 100-hour interval by 2 hours in order to get to the place where your inspection would be performed, your next 100-hour inspection will be required in 98 hours.

The requirements for performing an annual inspection and/or a 100-hour inspection are found in FAR Part 43, appendix D. A copy of that appendix is reproduced in Fig. 7-1.

Beside the annual/100-hour inspection route, you also have the option of choosing a progressive inspection program for your aircraft. If the FAA

APPENDIX D TO PART 43 -- SCOPE AND DETAIL OF ITEMS (AS APPLICABLE TO THE PARTICULAR AIRCRAFT) TO BE INCLUDED IN ANNUAL AND 100-HOUR INSPECTIONS

(a) Each person performing an annual or 100-hour inspection shall, before that inspection, remove or open all necessary inspection plates, access doors, fairing, and cowling. He shall thoroughly clean the aircraft and aircraft engine.

(b) Each person performing an annual or 100-hour inspection shall inspect (where applicable) the following components of the fuselage and hull group:

(1) Fabric and skin -- for deterioration, distortion, other evidence of failure, and defective or insecure attachment of fittings.

(2) Systems and components -- for improper installation, apparent defects, and unsatisfactory operation.

(3) Envelope, gas bags, ballast tanks, and related parts -- for poor condition.

(c) Each person performing an annual or 100-hour inspection shall inspect (where applicable) the following components of the cabin and cockpit group:

(1) Generally -- for uncleanliness and loose equipment that might foul the controls.

(2) Seats and safety belts -- for poor condition and apparent defects.

(3) Windows and windshields -- for deterioration and breakage.

(4) Instruments -- for poor condition, mounting, marking, and (where practicable) improper operation.

(5) Flight and engine controls -- for improper installation and improper operation.

(6) Batteries -- for improper installation and improper charge.

(7) All systems -- for improper installation, poor general condition, apparent and obvious defects, and insecurity of attachment.

(d) Each person performing an annual or 100-hour inspection shall inspect (where applicable) components of the engine and nacelle group as follows:

(1) Engine section -- for visual evidence of excessive oil, fuel, or hydraulic leaks, and sources of such leaks.

(2) Studs and nuts -- for improper torquing and obvious defects.

Figure 7-1 FAA requirements for annual and 100-hour inspections.

(3) Internal engine -- for cylinder compression and for metal particles or foreign matter on screens and sump drain plugs. If there is weak cylinder compression, for improper internal condition and improper internal tolerances.

(4) Engine mount -- for cracks, looseness of mounting, and looseness of engine to mount.

(5) Flexible vibration dampeners -- for poor condition and deterioration.

(6) Engine controls -- for defects, improper travel, and improper safetying.

(7) Lines, hoses, and clamps -- for leaks, improper condition, and looseness.

(8) Exhaust stacks -- for cracks, defects, and improper attachment.

(9) Accessories -- for apparent defects in security of mounting.

(10) All systems -- for improper installation, poor general condition, defects, and insecure attachment.

(11) Cowling -- for cracks and defects.

(e) Each person performing an annual or 100-hour inspection shall inspect (where applicable) the following components of the landing gear group:

(1) All units -- for poor condition and insecurity of attachment.

(2) Shock absorbing devices -- for improper oleo fluid level.

(3) Linkages, trusses, and members -- for undue or excessive wear fatigue, and distortion.

(4) Retracting and locking mechanism -- for improper operation.

(5) Hydraulic lines -- for leakage.

(6) Electrical system -- for chafing and improper operation of switches.

(7) Wheels -- for cracks, defects, and condition of bearings.

(8) Tires -- for wear and cuts.

(9) Brakes -- for improper adjustment.

(10) Floats and skis -- for insecure attachment and obvious or apparent defects.

(f) Each person performing an annual or 100-hour inspection shall inspect (where applicable) all

Figure 7-1 *(Continued)*

components of the wing and center section assembly for poor general condition, fabric or skin deterioration, distortion, evidence of failure, and insecurity of attachment.

(g) Each person performing an annual or 100-hour inspection shall inspect (where applicable) all components and systems that make up the complete empennage assembly for poor general condition, fabric or skin deterioration, distortion, evidence of failure, insecure attachment, improper component installation, and improper component operation.

(h) Each person performing an annual or 100-hour inspection shall inspect (where applicable) the following components of the propeller group:

(1) Propeller assembly -- for cracks, nicks, binds, and oil leakage.

(2) Bolts -- for improper torquing and lack of safetying.

(3) Anti-icing devices -- for improper operation and obvious defects.

(4) Control mechanisms -- for improper operation, insecure mounting, and restricted travel.

(i) Each person performing an annual or 100-hour inspection shall inspect (where applicable) the following components of the radio group:

(1) Radio and electronic equipment -- for improper installation and insecure mounting.

(2) Wiring and conduits -- for improper routing, insecure mounting, and obvious defects.

(3) Bonding and shielding -- for improper installation and poor condition.

(4) Antenna including trailing antenna -- for poor condition, insecure mounting, and improper operation.

(j) Each person performing an annual or 100-hour inspection shall inspect (where applicable) each installed miscellaneous item that is not otherwise covered by this listing for improper installation and improper operation.

Figure 7-1 *(Continued)*

approves your progressive inspection program, you can avoid the annual inspection and 100-hour (if applicable) inspections.

In order to get your progressive inspection program approved, you will have to submit a written request to your local FAA Flight Standards District Office. FAR Section 91.409 (d) indicates that your plan will have to include:

- A certified A&P mechanic with an inspection authorization, a certificated airframe repair station, or the manufacturer of the aircraft to supervise the program
- A current set of detailed inspection procedures in the form of a manual
- Appropriate physical facilities
- Appropriate technical information on the aircraft

The FAA will review your plan to ensure that it provides for the complete inspection of the aircraft every 12 months. Because of the obvious cost and effort required for a progressive inspection program, most general aviation aircraft owners tend to pass on this option.

In addition to the inspection requirements outlined above, the FAA also requires inspections for altimeter systems, altitude reporting equipment, and ATC transponders.

For all of these items, the inspections must be performed within the preceding 24 months in order to be able to legally operate your aircraft. For the altimeter system and altitude reporting equipment, the manufacturer of the aircraft or a certificated repair station meeting the qualifications of FAR Section 91.411(b)(2) must perform the required tests. A certificated mechanic with an airframe rating may perform the altimeter static pressure system tests and inspections only. For most general aviation aircraft, the ATC transponder test may only be conducted by a certificated repair station meeting the requirements of FAR Section 91.413(c)(1) or the manufacturer of the aircraft on which the transponder to be tested is installed, if the transponder was installed by that manufacturer.

Exceptions

As indicated above, you won't need an annual inspection or a 100-hour inspection if you have an approved progressive inspection program in place for your aircraft. FAR Section 91.409(c)(1) also exempts experimental aircraft, aircraft with a provisional airworthiness certificate, and aircraft carrying a special flight permit. Of these three, the most commonly experienced exception is the special flight permit exception.

The special flight permit is often referred to in the aviation community as the "ferry permit." Essentially, it allows you, often with certain conditions attached, to operate your aircraft when it might otherwise be unairworthy. In order to get a ferry permit, you have to make a special request to the FAA, usually at the FAA Flight Standards District Office nearest to where your aircraft is based. To get a ferry permit, you must complete an application on FAA Form 8130-6, Application for Airworthiness Certificate. For a ferry permit, you will only need to fill out sections II and VII of the form and sign at section VII F. A sample application for a ferry permit is illustrated in Fig. 7-2.

As indicated in section II of the form, a ferry permit is typically issued for the following reasons:

- To ferry an aircraft for repairs, alterations, maintenance, or storage
- To evacuate an aircraft where there may be impending danger
- To operate an aircraft in excess of maximum certificated take-off weight
- To deliver an aircraft or export an aircraft
- To perform a production test-flight of your aircraft
- To demonstrate your aircraft to customers

The FAA will also want to know your itinerary for your flight, the crew required to operate your aircraft, a list of the ways (if any) that your aircraft is not airworthy, and any other information necessary for the issuance of a special flight permit.

The FAA may require that you have appropriate inspections made before the special flight permit may be released to you. If this is the case, an A&P mechanic or certificated repair shop will probably have to inspect the aircraft before it is flown.

Record Keeping

If you take a look at the Standard Airworthiness Certificate (FAA Form 8100-2) for your aircraft, it states the following: "Unless sooner surrendered, suspended, revoked, or a termination date is otherwise established by the Administrator this airworthiness certificate is effective as long as the maintenance, preventive maintenance and alterations are performed in accordance with Parts 21, 43 and 91 of the Federal Aviation Regulations as appropriate and the aircraft is registered in the United States." This language summarizes the regulatory requirements for maintaining your aircraft in an airworthy condition. For better or for worse, the only way you will be able to establish compliance with the relevant regulations, and the only way you are going to be able to confirm that you are operating an airworthy aircraft, is through complete and accurate records that comply with the FARs.

What records are required?

FAR Section 91.417(a) is pretty specific about the records that must be kept. The regulation requires you to keep records of the following for each aircraft (including the airframe), engine, propeller, rotor, and appliance:

Form Approved
O.M.B. No. 2120-0018

APPLICATION FOR AIRWORTHINESS CERTIFICATE

U.S. Department of Transportation
Federal Aviation Administration

INSTRUCTIONS - Print or type. Do not write in shaded areas; these are for FAA use only. Submit original only to an authorized FAA Representative. If additional space is required use an attachment. For special flight permits complete Sections II, VI and VII as applicable.

I. AIRCRAFT DESCRIPTION

1. REGISTRATION MARK	2. AIRCRAFT BUILDER'S NAME *(Make)*	3. AIRCRAFT MODEL DESIGNATION	4. YR MFR	FAA CODING
N7654G	Cessna	172-C	1965	

5. AIRCRAFT SERIAL NO.	6. ENGINE BUILDERS NAME *(Make)*	7. ENGINE MODEL DESIGNATION	
00002	Lycoming	120-I-542	

8. NUMBER OF ENGINES	9. PROPELLER BUILDER'S NAME *(Make)*	10. PROPELLER MODEL DESIGNATION	11. AIRCRAFT IS *(Check if applicable)*
One	Hartzell	9875	☐ IMPORT

II. CERTIFICATION REQUESTED

APPLICATION IS HEREBY MADE FOR: *(Check applicable items)*

A 1 ☐ STANDARD AIRWORTHINESS CERTIFICATE (Indicate category)   ☐ NORMAL   ☐ UTILITY   ☐ ACROBATIC   ☐ TRANSPORT   ☐ COMMUTER   ☐ BALLOON   ☐ OTHER

B ☐ SPECIAL AIRWORTHINESS CERTIFICATE (Check appropriate items)

7 ☐ PRIMARY

2 ☐ LIMITED

5 ☐ PROVISIONAL (Indicate class)   1 Class I   2 Class II

3 ☐ RESTRICTED *(Indicate operation(s) to be conducted)*
1	AGRICULTURE AND PEST CONTROL	2 ☐	AERIAL	3 ☐	AERIAL ADVERTISING
4	FOREST *(Wildlife conservation)*	5 ☐	PATROLLING	6 ☐	WEATHER CONTROL
0	OTHER *(Specify)*				

4 ☐ EXPERIMENTAL *(Indicate operation(s) to be conducted)*
1	RESEARCH AND DEVELOPMENT	2 ☐	AMATEUR BUILT	3 ☐	EXHIBITION
4	AIR RACING	5 ☐	CREW TRAINING	6 ☐	MARKET SURVEY
0	TO SHOW COMPLIANCE WITH THE CFR	7 ☐	OPERATING *(Primary Category)* KIT BUILT AIRCRAFT		

8 ☒ SPECIAL FLIGHT PERMIT *(indicate operation to be conducted then complete Section VI or VII as applicable on reverse side)*
1 ☒	FERRY FLIGHT FOR REPAIRS, ALTERATIONS, MAINTENANCE, OR STORAGE
2	EVACUATE FROM AREA OF IMPENDING DANGER
3	OPERATION IN EXCESS OF MAXIMUM CERTIFICATED TAKE-OFF WEIGHT
4	DELIVERING OR EXPORTING 5 ☐ PRODUCTION FLIGHT TESTING
6	CUSTOMER DEMONSTRATION FLIGHTS

C 6 ☐ MULTIPLE AIRWORTHINESS CERTIFICATE (Check ABOVE: "Restricted Operation" and "Standard" or "Limited" as applicable)

III. OWNER'S CERTIFICATION

A. REGISTERED OWNER *(As shown on certificate of aircraft registration)*   IF DEALER, CHECK HERE ➤ ☐

NAME	ADDRESS
Odie Owner	12 Edenwald Avenue, NY, NY 10002

B. AIRCRAFT CERTIFICATION BASIS *(Check applicable blocks and complete items as indicated)*

☐ AIRCRAFT SPECIFICATION OR TYPE CERTIFICATE DATA SHEET *(Give No. and Revision No.)*

☐ AIRWORTHINESS DIRECTIVES *(Check if all applicable AD's complied with and give the number of the last AD SUPPLEMENT available in the biweekly series as of the date of application)*

☐ AIRCRAFT LISTING *(Give page number(s))*

☐ SUPPLEMENTAL TYPE CERTIFICATE *(List number of each STC incorporated)*

C. AIRCRAFT OPERATION AND MAINTENANCE RECORDS

☐ CHECK IF RECORDS IN COMPLIANCE WITH 14 CFR section 91.417	TOTAL AIRFRAME HOURS	3 EXPERIMENTAL ONLY *(Enter hours flown since last certificate issued or renewed)*

D. CERTIFICATION - I hereby certify that I am the registered owner (or his agent) of the aircraft described above, that the aircraft is registered with the Federal Aviation Administration in accordance with Title 49 of the United States Code 44101 *et seq.* and applicable Federal Aviation regulations, and that the aircraft has been inspected and is airworthy and eligible for the airworthiness certificate requested.

DATE OF APPLICATION	NAME AND TITLE *(Print or type)*	SIGNATURE

IV. INSPECTION AGENCY VERIFICATION

A. THE AIRCRAFT DESCRIBED ABOVE HAS BEEN INSPECTED AND FOUND AIRWORTHY BY *(Complete these sections only if 14 CFR part 21.183(d) applies)*

2 ☐ 14 CFR PART 121 CERTIFICATE HOLDER *(Give Certificate No.)*	3 ☐ CERTIFICATED MECHANIC *(Give Certificate No.)*	6 ☐ CERTIFICATED REPAIR STATION *(Give Certificate No.)*

5 ☐ AIRCRAFT MANUFACTURER *(Give name or firm)*

DATE	TITLE	SIGNATURE

V. FAA REPRESENTATIVE CERTIFICATION

(Check ALL applicable blocks in items A and B)

A. I find that the aircraft described in Section I or VII meets requirements for
☐ THE CERTIFICATE REQUESTED
4 ☐ AMENDMENT OR MODIFICATION OF CURRENT AIRWORTHINESS CERTIFICATE

B. Inspection for a special flight permit under Section VII was conducted by:
☐ FAA INSPECTOR   ☐ FAA DESIGNEE
☐ CERTIFICATE HOLDER UNDER   ☐ 14 CFR part 65   ☐ 14 CFR part 121 or 135   ☐ 14 CFR part 145

DATE	DISTRICT OFFICE	4 DESIGNEE'S SIGNATURE AND NO.	1 FAA INSPECTOR'S SIGNATURE

FAA Form 8130-6 (5-01) Supercedes Previous Edition   NSN: 0052-00-024-7006

Figure 7-2 Sample application for ferry permit.

VI. PRODUCTION FLIGHT TESTING

A. MANUFACTURER

NAME

ADDRESS

B. PRODUCTION BASIS *(Check applicable items)*

☐ PRODUCTION CERTIFICATE *(Give production certificate number)*

☐ TYPE CERTIFICATE ONLY

☐ APPROVED PRODUCTION INSPECTION SYSTEM

C. GIVE QUANTITY OF CERTIFICATES REQUIRED FOR OPERATING NEEDS ———————➤

DATE OF APPLICATION	NAME AND TITLE *(Print or type)*	SIGNATURE

VII. SPECIAL FLIGHT PERMIT PURPOSES OTHER THAN PRODUCTION FLIGHT TEST

A. DESCRIPTION OF AIRCRAFT

REGISTERED OWNER	ADDRESS
Odie Owner	12 Edenwald Avenue, NY, NY 10002
BUILDER *(Make)*	MODEL
Cessna	172-C
SERIAL NUMBER	REGISTRATION MARK
00002	N7654G

B. DESCRIPTION OF FLIGHT   CUSTOMER DEMONSTRATION FLIGHTS ☐ *(Check if applicable)*

FROM	TO
FOK	DMW

VIA	DEPARTURE DATE	DURATION
	xx/xx	4 hours

C. CREW REQUIRED TO OPERATE THE AIRCRAFT AND ITS EQUIPMENT

☒ PILOT   ☐ CO-PILOT   ☐ FLIGHT ENGINEER   ☐ OTHER *(Specify)*

D. THE AIRCRAFT DOES NOT MEET THE APPLICABLE AIRWORTHINES REQUIREMENTS AS FOLLOWS

Past Due Annual Inspection

E. THE FOLLOWING RESTRICTIONS ARE CONSIDERED NECESSARY FOR SAFE OPERATION *(Use attachment if necessary)*

VFR Only

F. CERTIFICATION - I hereby certify that I am the registered owner (or his agent) of the aircraft described above: that the aircraft is registered with the Federal Aviation Administration in accordance with Title 49 of the United States Code 44101 et seq. and applicable Federal Aviation Regulations, and that the aircraft has been inspected and is airworthy for the flight described.

DATE	NAME AND TITLE *(Print or type)*	SIGNATURE

VIII. AIRWORTHINESS DOCUMENTATION (FAA/DESIGNEE use only)

☐ A. Operating Limitations and Markings in compliance with 14 CFR section 91.9 as Applicable

☐ B. Current Operating Limitations Attached

☐ C. Data, Drawings, Photographs, etc. (Attach when required)

☐ D. Current Weight and Balance Information Available in Aircraft

☐ E. Major Repair and Alteration, FAA Form 337 *(Attached when required)*

☐ F. This inspection Recorded in Aircraft Records

☐ G. Statement of conformity, FAA Form 8130-9 *(Attach when required)*

☐ H. Foreign Airworthiness Certification for Import Aircraft *(Attached when required)*

☐ I. Previous Airworthiness Certificate issued in Accordance with

14 CFR Section_____ CAR_____ *(Original attached)*

☐ J. Current Airworthiness Certificate Issued in Accordance with

14 CFR Section_____ *(Copy attached)*

FAA Form 8130-6 (5-01) Supercedes Previous Edition

NSN: 0052-00-024-7006

Figure 7-2 *(Continued)*

- *Maintenance.* FAR Part 1 defines maintenance as "inspection, overhaul, repair, preservation, and the replacement of parts, but excludes preventive maintenance."

- *Preventive maintenance.* Preventive maintenance is defined in FAR Part 1 as "simple or minor preservation operations and the replacement of small standard parts not involving complex assembly operations." As discussed above, in certain instances, an aircraft owner may perform this type of maintenance as long as the owner is at least a private pilot.

- *Alteration(s).* The term "alteration" is not defined in FAR Part 1. However, the term "major alteration" is defined as "an alteration, not listed in the aircraft, aircraft engine, or propeller specifications (1) [t]hat might appreciably affect the weight, balance, structural characteristics, or other qualities affecting airworthiness; or (2)[t]hat is not done according to accepted practices or cannot be done by elementary operations."

- *All required inspections (including 100-hour, annual, and progressive inspections).* You are responsible for ensuring that an appropriately rated mechanic conducts your aircraft inspections and makes the required entries. Note that your records should include an identification of the program used, identify the portion or segment of the inspection program accomplished, and contain a statement that the inspection was performed in accordance with the instructions and procedures for that program. If you keep separate records for your airframe, engine(s), propeller(s), and appliances, you must have a separate entry for each placed in the appropriate maintenance logs. However, the annual inspection is required to be entered only into the airframe record. Any inspector performing a required inspection may also present you with a discrepancy list or list of defects. This list will indicate any unairworthy items discovered during the inspection performed on your aircraft. These discrepancy lists become part of your aircraft's maintenance record and must be retained in accordance with FAR 91.417(b)(3).

- *The total time in service of each airframe, each engine, each propeller, and each rotor.* FAR Part 1 defines "time in service." It states, "Time in service, with respect to maintenance time records, means the time from the moment an aircraft leaves the surface of the earth until it touches down at the next point of landing." FAR Section 43.9 does not require that you maintain this record as part of entries made for maintenance, preventive maintenance, rebuilding, or alterations. However, FAR Section 43.11 requires that persons performing maintenance make it a part of their entries for Part 91 inspections. You may not use recording tachometers to substitute for time-in-service record keeping.

- *The current status of life-limited parts of each airframe, engine, propeller, rotor, and appliance.* This requirement may be met by recording the total time of the airframe, engine, propeller, rotor, or appliance affected at the same time the life-limited part (and its current time in service) is recorded.

- *The time since the last overhaul of all items installed on the aircraft that must be overhauled on a specified time basis.* You can meet this requirement by following the same procedures described above for life-limited parts.

- *The current inspection status of the aircraft, including the time since the last inspection required by the inspection program under which the aircraft and its appliances are maintained.* You can meet this requirement by recording the applicable total time in service at each inspection interval.

- *The current status of applicable airworthiness directives (including for each the method of compliance, the AD number and revision dates).* If the AD is recurring, record the time and date when the next action is required. Although it is not required, a separate AD record may be kept for your aircraft's airframe, engine(s), propeller(s), rotor, and appliances. Having a separate AD record for each makes transfers to other parties a bit easier and for that reason alone, separate logs for these items may be advisable.

- *Records of tests and inspections for altimeter systems, altitude reporting equipment, and air traffic control transponders.* These inspections are recorded in the same manner as other inspections and maintenance.

- *Copies of FAA Form 337s issued for major repairs and alterations must become a part of your aircraft's maintenance records.* In some instances, the manufacturers will indicate directly in their service letters, bulletins, and other documents whether a change constitutes a major repair or alteration. In other cases, this determination is left to the person accomplishing the maintenance on your aircraft. A copy of a blank FAA Form 337 is shown in Fig. 7-3 for your reference.

The regulation does not require that a separate record be kept for airframe, engines, propellers, rotors, and appliances. However, it is common practice for aircraft owners to keep separate records or logs for these items. In many cases, keeping separate logs may be more practical. This may be especially true if there is a possibility that you will transfer an engine or propeller at a later date.

What should my maintenance records contain?

FAR Section 91.417 (a)(1) requires that your aircraft maintenance records include the following:

- *A description of the work performed.* Any work description should allow a person unfamiliar with the work to understand what was done and how it was done. If you had a lot of work done on your airplane, this description could get unwieldy. Therefore, the regulations allow reference to technical data acceptable to the administrator in lieu of a long-winded description of work. The technical data may consist of manufacturers' manuals, service letters, bulletins, work orders, or FAA advisory circulars. Any other documents must be copied and made a part of the maintenance record and retained in accordance with the regulations.

US Department of Transportation Federal Aviation Administration	MAJOR REPAIR AND ALTERATION (Airframe, Powerplant, Propeller, or Appliance)	Form Approved OMB No. 2120-0020
		For FAA Use Only
		Office Identification

INSTRUCTIONS: Print or type all entries. See FAR 43.9, FAR 43 Appendix B, and AC 43.9-1 (or subsequent revision thereof) for instructions and disposition of this form. This report is required by law (49 U.S.C. 1421). Failure to report can result in civil penalty not to exceed $1,000 for each such violation (Section 901 Federal Aviation Act of 1958).

1. Aircraft	Make		Model	
	Serial No.		Nationality and Registration Mark	
2. Owner	Name (As shown on registration certificate)		Address (As shown on registration certificate)	

3. For FAA Use Only

4. Unit Identification / 5. Type

Unit	Make	Model	Serial No.	Repair	Alteration
AIRFRAME	———————— (As described in Item 1 above) ————————				
POWERPLANT					
PROPELLER					
APPLIANCE	Type				
	Manufacturer				

6. Conformity Statement

A. Agency's Name and Address	B. Kind of Agency	C. Certificate No.
	U.S. Certificated Mechanic	
	Foreign Certificated Mechanic	
	Certificated Repair Station	
	Manufacturer	

D. I certify that the repair and/or alteration made to the unit(s) identified in item 4 above and described on the reverse or attachments hereto have been made in accordance with the requirements of Part 43 of the U.S. Federal Aviation Regulations and that the information furnished herein is true and correct to the best of my knowledge.

Date	Signature of Authorized Individual

7. Approval for Return To Service

Pursuant to the authority given persons specified below, the unit identified in item 4 was inspected in the manner prescribed by the Administrator of the Federal Aviation Administration and is ☐ APPROVED   ☐ REJECTED

BY	FAA Flt. Standards Inspector	Manufacturer	Inspection Authorization	Other (Specify)
	FAA Designee	Repair Station	Person Approved by Transport Canada Airworthiness Group	
Date of Approval or Rejection	Certificate or Designation No.	Signature of Authorized Individual		

FAA Form 337 (12-88)

Figure 7-3   FAA Form 337.

- *The date the work was completed.*   Normally, this is the date that your aircraft is approved for return to service. It may be possible that the work on your aircraft was performed by one person and approved by another. If this is the case, the dates may be different and two signatures may appear.

- *The signature and certificate number of the person approving your aircraft for return to service.*   However, in practice you will also see the type of certificate held by the person authorizing your aircraft for return to service. Typically, the designations you will see will be A for airframe, P for power plant, A&P for airframe and power plant, IA for inspection authorization, or CRS for certificated repair station. While FAR Section 91.417 does not require this information on the certificate held, its rough counterpart in FAR Section 43.9(a)(4) requires this information. Knowing the type of certificate held by the person approving your aircraft for return to service will allow you to better meet your responsibilities for compliance with maintenance record requirements.

How long should I keep maintenance records?

The FAA requires the following record retention periods for maintenance records:

- For maintenance, preventive maintenance, alterations, and inspections you should keep records until the work is repeated or superseded by other work or for 1 year after the work is performed.

- Virtually all other records should be kept perpetually and transferred with your aircraft when it is sold or title changes hands some other way (i.e., gift).

- If you are presented with a list of discrepancies as a result of a required inspection, you should keep the list of discrepancies until the defects are repaired and the aircraft is returned to service.

Where should I keep my aircraft maintenance records?

The regulations require that you keep all maintenance records and make them available for inspection (upon reasonable request) to a representative of the FAA or NTSB (National Transportation Safety Board). This does not mean that you must keep your maintenance records on board your aircraft (in fact, keeping such records with your aircraft is probably a bad idea because of the possibility of loss or damage). It simply suggests that they should be kept in a safe place. If the FAA or NTSB makes a reasonable request to see your aircraft's maintenance logs, you should be in a position to get them to that person in a reasonable amount of time.

However, if you have a fuel tank installed on your aircraft within the passenger or baggage compartment, you must keep the required FAA Form 337 on board the aircraft. You should be prepared to present the form upon request to a local, state, or federal law enforcement agent.

What about records on rebuilt engines?

FAR Section 91.421 permits you to use a new maintenance record, without a previous operating history, for any aircraft engine that has been rebuilt by the manufacturer or by an agency approved by the manufacturer. The regulation defines a "rebuilt engine" as:

> a used engine that has been completely disassembled, inspected, repaired as necessary, reassembled, tested, and approved in the same manner and to the same tolerances and limits as a new engine with either new or used parts. However, all parts used in it must conform to the production drawing tolerances and limits for new parts or be of approved oversized or undersized dimensions for a new engine.

In order to grant your engine the status of "zero time," the following items must be entered in a signed statement in your maintenance records:

- The date the engine was rebuilt

- Each change made as required by applicable airworthiness directives

- Each change made in compliance with service bulletins (if the service bulletin requires an entry to be made)

You should note that FAR Section 43.2(b) prohibits the use of the term "rebuilt" unless the requirements above have been met.

What happens if my maintenance records are lost or destroyed?

This is not the most pleasant prospect, but it does happen from time to time. The FAA has provided some practical guidance on how to deal with this situation in paragraph 12 of FAA Advisory Circular 43-9C.

The FAA suggests that your first duty is to attempt to establish the total time in service of the airframe. It suggests that other records can lend clues to finding this number. Specific suggestions include searching records maintained by repair facilities and individual mechanics. If you've exhausted all reasonable efforts to find your answer and you are still coming up short, you can insert a notarized statement in your aircraft's new maintenance records noting the loss of the original records and establishing a best-estimate time-in-service number on the basis of your research to date.

Other items such as AD compliance, current status of life-limited parts, time since last overhaul, current inspection status, and a list of major alterations may create an insurmountable problem. You may even be forced to comply a second time with certain ADs.

All in all, this is the type of experience you would rather avoid. It can be expensive and take up huge chunks of your time. Keep your records safe and current. A fireproof safe is often the best place for storing originals and copies.

Transferring your maintenance records

When you sell your aircraft, you will be required to transfer your aircraft's maintenance records to the buyer (see FAR Section 91.419). If you are an aircraft seller, it may be advisable for you to keep a copy of the more relevant records. This could prove helpful in case of any future disputes over the condition of the airplane you are selling.

Inoperative Equipment

FAR Sections 91.405 (c) and (d) refer to the issue of inoperative instruments or equipment. These sections indicate that if your aircraft has any inoperative instruments or equipment that is permitted to be inoperative under FAR Section 91.213, you must repair, replace, remove, or inspect it at your aircraft's next required inspection. It also adds that if listed discrepancies include inoperative instruments or equipment, you must install a placard clearly marking the instrument or equipment as "inoperative."

These sections in the maintenance regulations lead us to take a closer look at FAR Section 91.213, which details the FAA's rules for inoperative equipment. FAR Section 91.213 is a rather unusual regulation. As a general rule, it requires that you cannot take off in an aircraft with inoperative instruments or equipment. That means that if anything at all is not functioning properly on your aircraft, whether it is necessary for flight or not, your aircraft cannot be taken off the ground. Presumably, this could mean that if you are planning a 15-minute flight in the pattern on an extreme VFR day, you could not take off if the aircraft's ADF receiver was not operable.

There are two exceptions to this rather harsh rule. The first one states that you may operate your aircraft with inoperable instruments or equipment if you have an approved "minimum equipment list" (MEL) and the inoperable item is not on the list. This sounds like a pretty easy fix for an otherwise difficult rule. However, most light general aviation aircraft (especially single-engine aircraft) do not have MELs in place. To make matters worse, getting your MEL may not be so simple a matter. It requires application to your local FAA office and undoubtedly some time and negotiation to get a MEL approved especially for your aircraft.

Most aircraft owners prefer to rely on another exception that allows you to deactivate and placard the inoperable item as "inoperative." The FAA leaves the determination as to whether your aircraft can fly safely without the inoperable item to the pilots operating the aircraft. Thankfully, there have not been many cases involving the issue of inoperative equipment. However, as a rule, it still remains a part of your responsibility for aircraft maintenance.

Airworthiness Directives

If you are new to the world of aircraft ownership, you may not have heard of the term airworthiness directive. In the aviation community, airworthiness directives are more commonly referred to as ADs. However, no matter what name you give them, they can often mean expensive trouble for you and your airplane. ADs will, for better or for worse, be a part of your life as an aircraft owner. This segment of our review of the legal basics of aviation maintenance will focus on ADs and how they may affect you and your aircraft.

Overview of AD regulations and processes

The FAA issues ADs when it observes an unsafe condition in an aircraft, aircraft engine, propeller, rotor, or appliance. Under FAR Part 39, the FAA has the authority to notify aircraft owners of unsafe conditions and, perhaps more important, to require the correction of the unsafe condition(s) by the aircraft owner. ADs will prescribe the conditions and limitations, including inspection, repair, or alteration, under which you will be able to continue the operation of your aircraft or aircraft part(s). Notice in the discussion on inspections that the FAA requires that your maintenance records show compliance with all applicable ADs.

Typically, ADs are published in the *Federal Register*. The *Federal Register* is a daily compilation of proposed rules by all U.S. regulatory agencies, including the FAA. Usually, the AD is published in the *Federal Register* as a proposed AD (in the legal community we call this a Notice of Proposed Rulemaking, or NPRM) and people in the aviation community (or for that matter anyone) are invited to comment on the appropriateness and/or necessity of the proposed AD.

If an AD is adopted, it is later published in its final version in the *Federal Register* and distributed by first class mail to registered owners and aircraft owners affected by the AD. A final AD may look something like the example in Fig. 7-4.

In circumstances where the FAA believes that an AD requires urgent compliance, it will issue an emergency AD. An emergency AD usually becomes effective within 30 days of its publication in the *Federal Register*. The FAA notifies aircraft owners about emergency ADs via first class mail, telegram, or other electronic methods. It also notifies aviation groups and associations, other government agencies, and the aviation authorities in other countries.

As indicated in Fig. 7-4, most ADs are applicable to the make and model stated, regardless of the classification or category of the airworthiness certificate issued for the aircraft. If a limitation on applicability exists, it may be indicated when the AD specifies aircraft serial numbers that are subject to the AD.

As discussed earlier, ADs may be applied to aircraft products such as engines, propellers, rotors, or appliances. This is where you have to be particularly vigilant. The FAA may not have a problem tracking you down as an aircraft owner of a particular make and model aircraft. However, it may have a problem identifying you as the owner of a particular propeller or engine. In order to ensure that you are kept up-to-date on these matters, you may want to subscribe to FAA's Advisory Circular 36-9 which provides a biweekly review of airworthiness directives. Much of this information is now available on FAA's website and you may wish to review it directly from time to time to make sure that no new ADs have become applicable to your aircraft or aircraft products.

2002-21-02 Cirrus Design Corporation: Amendment 39-12908; Docket No. 2002-CE-41-AD.

(a) *What airplanes are affected by this AD?* This AD applies to the following airplane models and serial numbers that are certificated in any category:

{Private}Model	Serial numbers
SR20 SR22	1005 through 1241, except 1235, 1237, and 1238. 0002 through 0333, except 0309, 0322, 0323, and 0328.

(b) *Who must comply with this AD?* Anyone who wishes to operate any of the airplanes identified in paragraph (a) of this AD must comply with this AD.

(c) *What problem does this AD address?* The actions specified by this AD are intended to prevent loss of the self-locking retaining nut on the roll and yaw trim cartridges during flight, which could result in jamming of the corresponding flight control system. Such jamming could lead to loss of control of the airplane.

(d) *What must I do to address this problem?* To address this problem, you must accomplish the following actions:

{Private}Actions	Compliance	Procedures
(1) Replace the self-locking retaining nut on the yaw trim cartridge and the roll trim cartridge with a new self-locking retaining nut, part number MS21044N3.	Within the next 10 hours time-in-service after November 8, 2002 (the effective date of this AD), unless already accomplished.	In accordance with Cirrus Alert Service Bulletin SB A20–27–06, Issued: September 20, 2002, and Cirrus Alert Service Bulletin SB A22–27–03, Issued: September 20, 2002, as applicable.
(2) Do not install any self-locking retaining nut on the yaw trim cartridge or the roll trim cartridge that is not part number MS21044N3.	As of November 8, 2002 (the effective date of this AD).	Not applicable.

(e) *Can I comply with this AD in any other way?* You may use an alternative method of compliance or adjust the compliance time if:

(1) Your alternative method of compliance provides an equivalent level of safety; and
(2) the Manager, Chicago Aircraft Certification Office (ACO), approves your alternative. Submit your request through an FAA Principal Maintenance Inspector, who may add comments and then send it to the Manager, Chicago ACO.

Note: This AD applies to each airplane identified in paragraph (a) of this AD, regardless of whether it has been modified, altered, or repaired in the area subject to the requirements of this AD. For airplanes that have been modified, altered, or repaired so that the performance of the

Figure 7-4 Sample FAA airworthiness directive.

requirements of this AD is affected, the owner/operator must request approval for an alternative method of compliance in accordance with paragraph (e) of this AD. The request should include an assessment of the effect of the modification, alteration, or repair on the unsafe condition addressed by this AD; and, if you have not eliminated the unsafe condition, specific actions you propose to address it.

(f) *Where can I get information about any already-approved alternative methods of compliance?* Contact Gregory J. Michalik, Aerospace Engineer, FAA, Chicago ACO, 2300 East Devon Avenue, Des Plaines, IL 60018; telephone: (847) 294-7135; facsimile: (847) 294-7834.

(g) *What if I need to fly the airplane to another location to comply with this AD?* The FAA can issue a special flight permit under sections 21.197 and 21.199 of the Federal Aviation Regulations (14 CFR 21.197 and 21.199) to operate your airplane to a location where you can accomplish the requirements of this AD.

(h) *Are any service bulletins incorporated into this AD by reference?* Actions required by this AD must be done in accordance with Cirrus Alert Service Bulletin SB A20-27-06, Issued: September 20, 2002, and Cirrus Alert Service Bulletin SB A22-27-03, Issued: September 20, 2002. The Director of the Federal Register approved this incorporation by reference under 5 U.S.C. 552(a) and 1 CFR part 51. You can get copies from Cirrus Design Corporation, 4515 Taylor Circle, Duluth, MN 55811; telephone: (218) 727-2737; or electronically at the following address: *http://www.cirrusdesign.com/sb.* You may view this information at FAA, Central Region, Office of the Regional Counsel, 901 Locust, Room 506, Kansas City, Missouri, or at the Office of the Federal Register, 800 North Capitol Street NW, Suite 700, Washington, D.C.

(i) *When does this amendment become effective?* This amendment becomes effective on November 8, 2002.

Figure 7-4 *(Continued)*

Complying with ADs

Once ADs are adopted, they become part of the FARs. As regulations, they are enforceable under the law. Therefore, no person may operate an aircraft unless it complies with all applicable ADs. FAR Part 1 states that "[o]perate, with respect to aircraft, means use, cause to use, or authorize to use aircraft, for the purpose...of air navigation including the piloting of aircraft with or without the right of legal control (as owner, lessee, or otherwise)." This means that anyone using or authorizing the use of your aircraft (obviously including you as the owner) is responsible for ensuring that ADs applicable to your aircraft have been satisfied. If you have lessees, you should ensure that ADs are communicated to each lessee. Remember that they will not get the same notice of ADs that you get because they are usually unknown to the FAA.

There is no single rule for the timing of compliance with an AD. Sometimes, ADs must be complied with immediately and before any future flights. Other times, ADs may be accomplished at the next 100-hour or annual inspection. You will have to review each AD you face one at a time to determine when it requires compliance.

Documenting compliance

As an aircraft owner, you are primarily responsible for ensuring that your aircraft is in compliance with applicable ADs. The way you will evidence compliance with ADs is through good record keeping. You are responsible for ensuring that appropriate maintenance entries have been entered in your logs. The entries should indicate the description of work performed, the date the work was completed, and the name of the person approving the work on your aircraft (and the name of the person who did the work if other than the person who approved the work).

Getting Your Aircraft Back in Service

As a general matter, the FARs permit the following persons to perform maintenance and authorize an aircraft for return to service. Here's a list of the authorized persons:

1. Holders of FAA mechanic certificates
2. Repairman certificate holders
3. Holders of inspection authorizations
4. Aircraft owners who are also FAA certificated pilots (at least a private pilot certificate) and perform preventive maintenance on an aircraft they own

FAA eligibility and certification rules for mechanics, inspection authorizations, and repairman are detailed in FAR Part 65. A summary of what each FAA certified maintenance provider can or can't do is provided below.

As a general rule, certificated mechanics can perform maintenance, preventive maintenance, or alteration of an aircraft or appliance (or parts of aircraft or appliances) for which he or she is rated. However, mechanics cannot perform major repairs or alterations to propellers or instruments (instruments are defined in FAR Part 1 as devices used to determine altitude, attitude, autopilots, or operation of an aircraft or aircraft part). Mechanics ratings include airframe and power plant. A mechanic with an airframe rating may approve an airframe (and related parts or appliances) after he or she has performed, supervised, or inspected its maintenance or alteration. He or she can also perform the 100-hour inspection on an airframe and approve it for return to service. A mechanic with a power plant rating may approve and return to service a power plant or propeller (and related parts or appliances) after performing, supervising, or inspecting its maintenance or alteration. He or she can also perform the 100-hour inspection on a power plant or propeller and return it to service. The big limitation on a mechanic is that he or she may not return an aircraft to service after major repairs or alterations.

A repairman certificate authorizes a certificated repairman to perform or supervise maintenance, preventive maintenance, or alteration of aircraft or aircraft components appropriate to the job for which the repairman was certificated. Often repairman certificates might be issued to persons with experience in a particular area of aircraft repair or alteration. Someone performing work under a repairman certificate does not have the ability to return your aircraft to service. Depending on the type of work done by the repairman, your aircraft will have to be returned to service by a mechanic or someone with an inspection authorization.

If your aircraft has undergone major repairs or alterations (as generally defined in FAR Part 1 and detailed in FAR Part 43, Appendix A), it will often require someone with an inspection authorization to return your aircraft to service. An inspection authorization is a special authorization for mechanics (holding both airframe and power plant ratings) who meet more extensive FAA requirements. In addition to being authorized to return an aircraft to service after major repairs or alteration, the IA is also authorized to perform annual or progressive inspections.

There are others who may return your aircraft to service after major repairs or alterations including FAA inspectors, FAA designees, manufacturers, certain persons authorized by Transport Canada Airworthiness Group, and FAA certificated repair stations. Of this group of "others" who can return an aircraft to service, you are most likely to run across the services of a repair station. Repair stations are certificated to maintain or alter any airframe, power plant, propeller, instrument, radio, or accessory (or parts of these items) for which they are rated. They may also approve for return to service any of the items listed above for which they are rated.

As an aircraft owner, you may also qualify to do some limited forms of maintenance on your aircraft. The big limitation is that you can perform only preventive maintenance. The other qualification(s) are that you can perform only work on your own aircraft and that you must hold at least a private pilot certificate. A list of items qualifying as preventive maintenance is found in FAR Part 43, Appendix A, Section (c). A reproduction of that list is found in Fig. 7-5.

Once you have completed your preventive maintenance, you may approve your aircraft for return to service. A sample entry for approving your aircraft for return to service is found in Fig. 7-6.

After an authorized person has returned your aircraft to service, your next step is to ensure that any required maintenance entries have been posted in the necessary logbook(s). Remember that it is your responsibility to ensure that the maintenance log book for your aircraft is up-to-date and accurate.

Maintenance Contract Issues

One of the more frequent complaints by general aviation aircraft owners is dissatisfaction with maintenance providers. Often the problems and resulting dissatisfaction could have been avoided with better communication between the parties. Of course, one of the best ways to communicate your expectations in any significant transaction is through a written agreement. There is often too much at stake to leave expensive maintenance work to a shop without a written agreement spelling out the essentials of the transaction. The following checklist of items should be helpful in getting you to think about what you expect from your next visit to a maintenance facility:

- Identify the legal names and addresses of the aircraft owner and maintenance facility contracted.

- Identify the aircraft to be delivered for service and the approximate date of delivery to the maintenance facility.

- Specify a date that the work will be completed by the maintenance facility.

- If necessary or advisable, identify damages if the work is not completed in a timely manner.

- If applicable, specify that the maintenance facility will prepare all documentation necessary to obtain FAA approval for return of your aircraft to service.

- Detail the work to be performed (this can be done on a separate schedule or exhibit to your agreement).

- Confirm the rate per hour for labor if the contract price is not a fixed fee.

- Get the maintenance facility to acknowledge its estimate in writing.

- Provide for what happens if the work necessary will go beyond the estimate (when you should be notified, what options you will have, etc.).

- Detail how and when your payments will be made (up front, after work completed, 30 days after invoice).

- Spell out any warranties on work, parts, and labor that you have negotiated with your maintenance provider.

(c) Preventive maintenance. Preventive maintenance is limited to the following work, provided it does not involve complex assembly operations:

(1) Removal, installation, and repair of landing gear tires.

(2) Replacing elastic shock absorber cords on landing gear.

(3) Servicing landing gear shock struts by adding oil, air, or both.

(4) Servicing landing gear wheel bearings, such as cleaning and greasing.

(5) Replacing defective safety wiring or cotter keys.

(6) Lubrication not requiring disassembly other than removal of nonstructural items such as cover plates, cowlings, and fairings.

(7) Making simple fabric patches not requiring rib stitching or the removal of structural parts or control surfaces. In the case of balloons, the making of small fabric repairs to envelopes (as defined in, and in accordance with, the balloon manufacturers' instructions) not requiring load tape repair or replacement.

(8) Replenishing hydraulic fluid in the hydraulic reservoir.

(9) Refinishing decorative coating of fuselage, balloon baskets, wings tail group surfaces (excluding balanced control surfaces), fairings, cowlings, landing gear, cabin, or cockpit interior when removal or disassembly of any primary structure or operating system is not required.

(10) Applying preservative or protective material to components where no disassembly of any primary structure or operating system is involved and where such coating is not prohibited or is not contrary to good practices.

(11) Repairing upholstery and decorative furnishings of the cabin, cockpit, or balloon basket interior when the repairing does not require disassembly of any primary structure or operating system or interfere with an operating system or affect the primary structure of the aircraft.

(12) Making small simple repairs to fairings, nonstructural cover plates, cowlings, and small patches and reinforcements not changing the contour so as to interfere with proper air flow.

(13) Replacing side windows where that work does not interfere with the structure or any operating system such as controls, electrical equipment, etc.

(14) Replacing safety belts.

(15) Replacing seats or seat parts with replacement parts approved for the aircraft, not involving disassembly of any primary structure or operating system.

Figure 7-5 List of items considered preventive maintenance.

(16) Trouble shooting and repairing broken circuits in landing light wiring circuits.

(17) Replacing bulbs, reflectors, and lenses of position and landing lights.

(18) Replacing wheels and skis where no weight and balance computation is involved.

(19) Replacing any cowling not requiring removal of the propeller or disconnection of flight controls.

(20) Replacing or cleaning spark plugs and setting of spark plug gap clearance.

(21) Replacing any hose connection except hydraulic connections.

(22) Replacing prefabricated fuel lines.

(23) Cleaning or replacing fuel and oil strainers or filter elements.

(24) Replacing and servicing batteries.

(25) Cleaning of balloon burner pilot and main nozzles in accordance with the balloon manufacturer's instructions.

(26) Replacement or adjustment of nonstructural standard fasteners incidental to operations.

(27) The interchange of balloon baskets and burners on envelopes when the basket or burner is designated as interchangeable in the balloon type certificate data and the baskets and burners are specifically designed for quick removal and installation.

(28) The installations of anti-misfueling devices to reduce the diameter of fuel tank filler openings provided the specific device has been made a part of the aircraft type certificate data by the aircraft manufacturer, the aircraft manufacturer has provided FAA-approved instructions for installation of the specific device, and installation does not involve the disassembly of the existing tank filler opening.

(29) Removing, checking, and replacing magnetic chip detectors.

(30) The inspection and maintenance tasks prescribed and specifically identified as preventive maintenance in a primary category aircraft type certificate or supplemental type certificate holder's approved special inspection and preventive maintenance program when accomplished on a primary category aircraft provided:

(i) They are performed by the holder of at least a private pilot certificate issued under part 61 who is the registered owner (including co-owners) of the affected aircraft and who holds a certificate of competency for the affected aircraft (1) issued by a school approved under § 147.21(e) of this chapter; (2) issued by the holder of the production certificate for that primary category aircraft that has a special training program approved under § 21.24 of this subchapter;

Figure 7-5   *(Continued)*

or (3) issued by another entity that has a course approved by the Administrator; and

(ii) The inspections and maintenance tasks are performed in accordance with instructions contained by the special inspection and preventive maintenance program approved as part of the aircraft's type design or supplemental type design.

(31) Removing and replacing self-contained, front-instrument-panel-mounted navigation and communication devices that employ tray-mounted connectors that connect the unit when the unit is installed into the instrument panel (excluding automatic flight control systems, transponders, and microwave frequency distance measuring equipment (DME)). The approved unit must be designed to be readily and repeatedly removed and replaced, and pertinent instructions must be provided. Prior to the unit's intended use, an operational check must be performed in accordance with the applicable sections of part 91 of this chapter.

(32) Updating self-contained, front-instrument-panel-mounted Air Traffic Control (ATC) navigational software data bases (excluding those of automatic flight control systems, transponders, and microwave frequency distance measuring equipment (DME)) provided no disassembly of the unit is required and pertinent instructions are provided. Prior to the unit's intended use, an operational check must be performed in accordance with applicable sections of part 91 of this chapter.

Figure 7-5 *(Continued)*

[Date] Total time: _____ hours. Replaced landing light bulb in accordance with manufac-turer's maintenance manual, chapter _____, page _____.

Joe Pilot Signature   Rating: Private Pilot   Certificate No. 000-00-0000

Figure 7-6   Sample preventive maintenance entry.

- You should require that your maintenance provider keep complete and accurate records of all materials, parts, and labor supplied under your agreement.
- You should be given the right to inspect the work accomplished and reject it on reasonable grounds.
- Will your maintenance provider indemnify you if there is damage due to its failure to perform? This can be a particularly difficult, yet important point to negotiate.
- Do you want your maintenance provider to be able to subcontract work related to your airplane? This issue should be addressed.
- Specify the state laws and courts that you wish to govern the agreement.
- Indicate whether attorney fees will be payable by the unsuccessful party to a lawsuit that is related to the agreement.
- Detail how modifications to the agreement may be made.

These are just some of the bigger items that you should consider when you are contracting for major repairs, maintenance, overhauls, prepurchase inspections, or other inspections of your aircraft. A specimen agreement with a maintenance provider can be found in the CD in the back of the book.

8

Airworthiness

In the last two chapters of this section, we've explored some of the common legal issues encountered in an aircraft owner's efforts to safely store and maintain an aircraft. Of course, the primary purpose of quality aircraft storage and maintenance is to make sure that your aircraft is fit to fly. In the aviation world we commonly refer to an aircraft's fitness to fly as "airworthiness." In this chapter we'll explore the legal standards you will have to meet to ensure your aircraft is fit to fly. Perhaps the best way to do this is to start with the FAA's regulations and interpretations first. After that, we can take a look at cases where the FAA's regulations and interpretations have been tested in real-life situations.

Is My Aircraft Airworthy?

FAR Section 91.7(a) states, "No person shall operate a civil aircraft unless it is in an airworthy condition." FAR Section 91.403 (a) states, "The owner or operator of an aircraft is primarily responsible for maintaining that aircraft in an airworthy condition...." The two critical words in these regulations are "operate" and "airworthy." Let's examine each.

The FAA makes it pretty easy to capture the meaning of "operate." In FAR Part 1, the FAA states, "Operate, with respect to aircraft, means use, cause to use or authorize to use aircraft, for the purpose...of air navigation including the piloting of aircraft, with or without the right of legal control (as owner, lessee, or otherwise)." The FAA's definition pretty clearly communicates that while piloting an aircraft clearly qualifies as "operating" an aircraft, simply owning an aircraft and authorizing its use also qualifies you as "operating" an aircraft.

Although the FAA is straightforward enough about what it means by the word "operate," the FARs are not so clear about what is meant by "airworthy." The FAA's definitions in Part 1 of the FARs do not define the term "airworthy." If you look in your dictionary, you will probably find that "airworthy" means fit for air navigation or operation in the air. If you dig further and research some of the FAA's literature on the subject, you will run across the FAA interpretation of the term "airworthy." In FAA Order 8130.2D CHG 2, the FAA states this as indicated in Fig. 8-1.

In the FAA's interpretation, the FAA's position is that in order for your aircraft to be considered airworthy, it must conform to its type certificate. If you give this some thought, you can easily conclude that this is a tall order. After all, your aircraft may only be "perfect" once it rolls off the assembly line at the manufacturing plant. After that, wear and tear and normal use will take a toll.

---

9. INTERPRETATION OF THE TERM "AIRWORTHY" FOR U.S. TYPE CERTIFICATED AIRCRAFT. The term "airworthy" is not defined in Title 49 or the regulations; however, a clear understanding of its meaning is essential for use in the agency's Airworthiness Certification program. Below is an analogy of the conditions necessary for the issuance of an airworthiness certificate. A review of case law relating to airworthiness reveals two conditions that must be met for an aircraft to be considered "airworthy." Title 49 Section 44704 (c) and 14 CFR part 21, Certification Procedures for Products and Parts (part 21), sections 21.183 (a), (b), and (c), all relate to the two conditions necessary for issuance of an airworthiness certificate. The statutory language establishes the two conditions as:

a. The aircraft must conform to its TC (type certificate). Conformity to type design is considered attained when the aircraft configuration and the components installed are consistent with the drawings, specifications, and other data that are part of the TC, and would include any STC and field approved alterations incorporated into the aircraft.

b. The aircraft must be in a condition for safe operation. This refers to the condition of the aircraft relative to wear and deterioration, e.g., skin corrosion, window delamination/crazing, fluid leaks, tire wear, etc.

NOTE: If one or both of these conditions were not met, the aircraft would be considered unairworthy. Aircraft which have not been issued a TC must meet the requirements of paragraph 9b above.

Figure 8-1   FAA interpretation of "airworthy."

Does this mean that all or most of the aircraft out on the ramp at your airport are unairworthy?

More than likely, the answer to that question is no. However, coming to that conclusion requires that you sift through some cases where other pilots and aircraft owners have been faced with FAA charges of operating unairworthy aircraft. A review of selected cases is presented below.

Review of Selected Airworthiness Cases

Sometimes, the best way to get a feel for what is meant by the term "airworthy" is to take a look at past cases decided by the National Transportation Safety Board (NTSB). The NTSB reviews cases involving alleged infractions of the FARs. After reviewing the selected cases below, we hope you will have a better feel for the challenges and issues related to determining whether an aircraft is "airworthy."

Administrator v. Pierce

In *Administrator v. Pierce,* NTSB Order EA-4965 (April 11, 2002), the FAA charged Pierce (in relevant part) with operating an unairworthy aircraft. Specifically, the FAA charged that Pierce violated FAR Section 91.7(a).

The FAA specifically alleged that the aircraft had a mixture control cable that was defective. Pierce argued that the cable was not completely broken and that he had been informed by the aircraft's owner (who also happened to be a certificated A&P mechanic) that a ferry permit was not required. However, Pierce did testify during the proceedings that the cable was "very sticky." The mechanic who repaired the mixture control cable found that it was seized and while trying to free it the cable was broken.

The NTSB found that the mixture control cable was inoperable and therefore rendered the aircraft unairworthy. Pierce's argument that it was not "completely broken" and that he was entitled to rely on the aircraft's owner/mechanic fell short with the board. In essence, the board found that if a mixture control cable is inoperable, it renders an aircraft unairworthy. The NTSB further found that Pierce "knew or should have known that he should not operate an aircraft with a binding cable that was already malfunctioning in a significant way. This is not so complicated that we would permit him to rely on the advice of a mechanic when he should know better." The board also noted that the fact that the flight safely arrived at its destination was immaterial.

Administrator v. Werve

In *Administrator v. Werve,* NTSB Order No. EA-4213 (1994), the FAA charged Werve with violating FAR Section 91.7 (a). In this case, the FAA argued that an aircraft was rendered unairworthy because a previous crew "wrote up" a cabin door that appeared to them to be stuck. The next crew tested the door and determined that it opened smoothly and consistently.

In this case, an NTSB law judge determined that the aircraft was unairworthy. The law judge concluded that the aircraft was unairworthy because the door in question did "not open each and every time smoothly."

However, the full NTSB Board reversed the law judge. It determined that "[a]irworthiness, and compliance with...91.7(a)...required the door to be 'reliably' operable from the inside." In applying this standard, the board found that the door did not create an unairworthy condition.

Administrator v. Calavaero

The case of *Administrator v. Calavaero*, 5 NTSB 1099 (1986), may be the most instructive case when it comes to close calls on airworthiness. In this case, an FAA inspector noted the following discrepancies during an aircraft inspection:

1. The right flap bearings/bushing were excessively worn.
2. The right aileron had a hole corroded through the skin at the right outboard hinge.
3. The right exhaust pipe on the number 2 engine had broken loose, and had worn a hole in the cowl flap.
4. The left exhaust stack (pipe) had worn a hole approximately 3″ by 3″ in the cowl flap.
5. The right engine cowl was cracked in two places, and the forward right side screws were missing.
6. Both right and left oil coolers were covered with white paints.
7. The left wing tip was damaged and filled with Bondo-type putty.

After a hearing on the matter, the law judge held that all the discrepancies listed by the inspector did not appear on this aircraft. Perhaps more important, the law judge stated that in any event, none of the FAA–listed discrepancies rendered the aircraft unairworthy.

The FAA appealed the law judge's decision. The position taken by the FAA was that any "structural defects at all" meant that an aircraft is not in conformity with its type certificate and therefore unairworthy regardless of whether the aircraft could be safely operated.

The Board disagreed with the FAA and presented the following practical advice on this question:

> We do not take issue with the Administrator's position that an aircraft, in addition to being in a safe position to operate, must be in conformity with its type certificate in order to be considered airworthy. However, we do not agree that every scratch, dent, "pinhole" of corrosion, missing screw or other defect, no matter how minor or where located on the aircraft, dictates the conclusion that the aircraft's design, construction, or performance has been impaired by the defect to a degree that the aircraft no longer conforms to its type certificate.

Analysis

What can we take away from these cases? It seems that there is no "bright-line" test of whether your aircraft will be considered airworthy or not. Certainly, there is no question that violating any regulation calling for periodic inspections or maintenance will cause your aircraft to be labeled "unairworthy." However, in less obvious cases, a decision on airworthiness comes down to a judgment call.

In the Supreme Court case of *Jacobellis v. Ohio,* Justice Potter Stewart was faced with the task of determining whether a film should be labeled "obscene." Justice Stewart stated: "I shall not today attempt to further define the kinds of material...embraced within that shorthand description [of obscenity]; and perhaps I could never succeed in intelligibly doing so. But I know it when I see it...." Perhaps this is the only reasonable way to look at the issue of airworthiness. It is perhaps impossible to lay out all the parameters that will cause an aircraft to be deemed "unairworthy." The trick may simply be to exercise good judgment so you will know it when you see it and to take prompt corrective action.

Liability Issues for Aircraft Owners

9

Insurance

The potential of liability is a big issue for most aircraft owners. Your first line of defense against liability exposure is safe aircraft operating practices. However, no matter how carefully you maintain and operate your aircraft, things can go wrong. The tried and true method for reducing your exposure to loss of property or liability to others is insurance coverage. Insurance permits the aviation community to spread the risk of property and personal injury loss. As an aircraft owner, you owe it to yourself, your family, and your community to have a fundamental understanding of aviation insurance coverage. This will allow you to determine the best kind of insurance coverage to protect yourself and others from the unexpected.

General Description of Policy

Inasmuch as the focus of this book is general aviation aircraft owners, this chapter will explore the basics of a noncommercial owner's policy. Sometimes you will hear this type of policy referred to as a "business and pleasure" policy. Usually, you can dissect these policies into the following key components:

- Data page
- Definitions
- Liability coverage
- Aircraft damage coverage
- General exclusions
- Coverage for medical expenses
- Coverage for use of nonowned aircraft
- General provisions

To assist you in a better understanding of general aviation insurance coverage, we'll walk through each of these key components. Within each provision you'll get a closer look at the types of things you should consider as you review your current or prospective aviation insurance policy.

Data Page

The data page is usually custom designed to fit your situation. It will include information identifying the policy period, insurance policy holder, any persons holding liens (such as financing banks), and information identifying your aircraft. Other significant sections of the data page will detail the coverage and limits of liability provided by your policy and a description of the pilots authorized to operate your aircraft for insurance purposes. Each of these sections is addressed below.

Aircraft damage coverage

This section of the data page will detail the type and amount of your insurance coverage for damage to your aircraft. This is often referred to in the aviation community as "hull" or "physical damage" insurance. You should understand what hull insurance is and how it works before you commit to an insurance policy.

You may have heard that hull insurance comes in different versions. One version is often referred to as "in-flight" or "in-motion." The other version is usually referred to as "not-in-flight" or "not-in-motion." You'll have to choose the version of hull insurance that will best suit your needs. As you consider this choice, you must also fully understand what is meant by "in-flight" or "in-motion." Some insurance policies define "in-motion" as being anytime the aircraft is moving under the power of its engine. "In-flight" often refers to the time between the start of your aircraft's takeoff run to the time it has landed and safely stopped or exited the landing runway. If you choose "in-motion" coverage and your parked aircraft is damaged by an uninsured vehicle on the airport ramp, you may have a difficult time recovering your loss. Some aircraft owners choose to ensure for both "in-flight" or "in-motion" and "not-in-flight" or "not-in-motion" (sometimes the combination of the two is known as "all-risk ground and flight") to ensure that they've covered all the possible bases. This makes for stronger coverage. However, the broader and deeper you make your coverage, the more it will cost you. If your aircraft is financed, the bank or finance company may very well require that you insure your aircraft with all-risk ground and flight coverage.

The next thing you will have to decide is the amount you wish to insure your aircraft for in case of damage. In order to make a prudent decision on how much value to insure your aircraft for, you should attempt to get a good idea of your aircraft's fair market value. Once you determine the fair market value, you can get a better idea of how much insurance you want to pay for (or can afford). You should keep tabs on any changes to the fair market value as you make modifications or improvements to your aircraft in the form of upgraded avionics or engine overhauls. Over- or underinsuring your aircraft may create problems. If your aircraft is underinsured, your insurance company may be more likely to treat your aircraft as a "total loss" after a mishap. On the other hand, overinsuring your aircraft carries the risk that your insurance company will prefer to repair a badly damaged aircraft rather than declaring it a "total loss."

Whatever you have chosen with respect to type of aircraft damage insurance and amount of coverage, you should carefully inspect your insurance policy data page to ensure that it properly indicates the insurance coverage you selected. If there are any discrepancies, you must bring them to the attention of your insurance carrier immediately.

Liability insurance

While hull insurance covers damage to your property (namely your aircraft), you also have to consider the possibility that a mishap may also cause injury to someone else's property or person. Insurance against harm to others is usually referred to as liability insurance. Your insurance policy data sheet will indicate the type and amount of coverage you have obtained for liability coverage.

As a general rule, you will have two choices when it comes to the type of liability coverage you can obtain. The choices are usually referred to as "sublimited coverage" versus "smooth limits" insurance.

Sublimited coverage sets two limits. The higher limit is often referred to as an "occurrence limit." This amount is the maximum amount that your insurance company will be liable for from any single mishap your aircraft may be involved in. The lower limits usually refer to the maximum coverage that may be available for each person injured or killed.

Here's how this works. Let's say you have a policy that covers you for $1,000,000 per occurrence with $200,000 sublimits per injury or death to each person involved in any single occurrence. This means that if you are involved in an accident, the absolute top dollar amount your insurer will cover you for is $1,000,000. However, and perhaps more important, your insurer will only cover you for up to $200,000 for each person injured or killed in that accident.

If you have sublimits that are listed as "per passenger," you may have a somewhat broader coverage than a "per person" sublimit. Sublimits listed "per passenger" mean that the sublimit applies only to aircraft passengers and not to persons who may be harmed on the ground.

Smooth limits provide you with the broadest coverage. This coverage combines your property damage and liability limits. Therefore, if you have insurance for $1,000,000 per occurrence, you are covered whether the damage or injury was to your passengers or persons or property on the ground. The entire $1,000,000 is available to cover the losses arising from the same accident. Smooth limits obviously provide you with greater protection. However, you will have to pay a higher premium for smooth limits.

Again, make sure that your insurance policy data page properly indicates the type and amount of liability insurance you are carrying. Any mistakes or misunderstandings can be costly.

Approved pilots

Information relating to approved pilots is another very important item on your insurance policy data sheet. It should not surprise you that your insurance company will limit its liability to situations where qualified pilots are at the controls of your aircraft. In just about every case, the insurance company will require that approved pilots hold certain FAA certificates or ratings. You'll also see frequent situations where your policy requires certain minimum total hours or hours in a particular aircraft.

Beyond the list of specifically approved pilots, your data page may include what is known as "open" pilot requirements. These are requirements for pilots other than specifically approved pilots. Often these qualifications are expressed in terms of hours and ratings and time in make and model. In all circumstances, in may be advisable to inspect logbooks to ensure that any pilot operating your aircraft meets the approval qualifications for your insurance. One other thing to keep in mind is that it is often not enough that the pilot has actually "flown" the hours required for approval. You should be prepared to provide evidence that the pilot actually logged the hours required for approved pilot status.

Definitions

Although it might not make for exciting reading, it is a good idea for you to become familiar with the definitions provided in your aircraft insurance policy. Most policies will use "defined terms" regularly throughout your policy. You can't get a good feel for the meaning of the policy unless you understand the underlying definitions. As far as the law is concerned, the definitions in your policy are agreed-upon terms. Therefore, the definitions in the policy are what guide your insurance carrier's actions. Dictionary definitions of the terms used will not be persuasive. To help you get a better feel for your insurance policy, some of the more commonly found definitions in noncommercial policies are discussed below.

Accident

Many aviation insurance policies refer to the term "accident" as a condition triggering your coverage. Thankfully, the definition of accident used by most policies is not very different from the dictionary definition of the word. Here are some of the common threads you will find in the definitions of the insurance policy term "accident":

1. A sudden event (or continued exposure to the same harmful conditions);

2. Resulting in bodily injury, death, or property damage;

3. That you did not expect.

If you are a pilot, you should note that this definition of the term "accident" is different from the definition of the same term found in the National Transportation Safety Board (NTSB) Regulation Section 830.2. The NTSB definition of an accident requires death, serious injury, or substantial damage to aircraft.

Commercial operations or purpose

Most insurance policies have a special definition for "commercial operations" or "commercial purpose." Read your policy's definition carefully. This term sometimes becomes a bone of contention when it comes time for your insurance carrier to pay on a claim. Operating your aircraft commercially does not necessarily mean that you must be transporting persons or property in order to make a profit. Nor does it necessarily mean that you must be accepting compensation on a regular basis.

Insured persons

You might think that you are the only insured person on your policy. Actually, many policies will also include persons who are using the aircraft with your authorization as long as the pilots meet the approval of your insurance company.

Territory covered

For most general aviation aircraft owners, the territorial limits of coverage will not pose much of a constraint. However, you should be aware of the noted limits especially if you intend to operate your aircraft outside the limits of the continental United States.

Liability Coverage

The liability coverage section of aircraft insurance policies is often the most significant to you as an aircraft owner. Although the typical general aviation aircraft is not inexpensive, there may be far more exposure to loss incurred by third persons or their property if your aircraft is involved in an accident. As you look through your insurance policy's liability coverage, you should note language regarding your insurance carrier's duty to defend. You should also carefully review the exceptions to liability coverage. Both of these issues are addressed below.

Duty to defend

Many aircraft owners believe that all they get with their insurance policy is protection against having to pay out of their own pocket for damage to their aircraft and/or someone else's property and/or person. Actually, one of the most valuable things you obtain with an aircraft insurance policy is the obligation of your insurance carrier to defend you if someone charges you with wrongdoing while operating your aircraft.

Most insurance policies indicate that they will provide you (or an "insured person") with legal counsel for a defense against covered claims for property damage or personal injury. That is true. However, it is also important to note that in most states, the duty of an insurer to defend is broader than its duty to cover. Essentially that means that in certain circumstances, your insurer may have a duty to defend you even though coverage for a claim may be questionable. Again, because legal costs are so high, this is a very important coverage that all aviation insurance policies provide you as an aircraft owner.

If your insurance company defends you against a claim, it will generally reserve the right to settle the claim against you as it deems appropriate. Even if your insurance company settles for the full amount of your policy coverage, that alone will not usually extinguish its duty to defend you if the claim exceeds the amount of your coverage.

Typical exclusions to liability coverage

Each aircraft insurance policy will present different exclusions to liability coverage. However, there are some common threads among the more significant exclusions. Here's a list of some of the more typical exclusions for liability coverage:

- *Assumption of liability.*   Your policy may exclude you from liability coverage if you contractually agree to assume liability for another person. An example of this would be your agreement to hold harmless and indemnify your hangar landlord for damage done to your aircraft in the hangar. In many cases this might negate your insurance coverage under this type of provision. In recent years, some aviation insurance carriers have recognized that many hangar contracts drafted in recent years contain the sort of "take it or leave it" language we discussed in Chap. 6. In view of this trend, a few policies have recognized this type of liability assumption as an exception to the exclusion and will not negate your coverage if you agree to a hold harmless or indemnify your hangar landlord.

- *Worker's compensation.*   Many policies exclude liability coverage if worker's compensation is available or required for an injured employee.

- *Damage to your property.*   Another common exclusion will block coverage if property that you own or rent is damaged. Some exceptions to this exclusion include personal effects of aircraft passengers and certain dollar limits for property you might be renting.

- *Injury to family members.*   Certain policies will either bar recovery to your family members (usually defined as sons, daughters, spouses, or parents). Sometimes this exclusion creates a maximum dollar amount a family member can recover that is well below the limits of your policy.

Aircraft Damage Coverage

As discussed earlier, this type of coverage is often referred to as hull insurance. Your aircraft damage insurance will protect you from loss if your aircraft is damaged in an accident. Remember that the applicability of this coverage may depend on whether you purchased insurance that will cover your aircraft on the ground and/or in-flight (or in-motion).

Scope of coverage

You can expect that most aviation insurance policies will cover the following costs in addition to paying for your aircraft loss:

- Transporting your aircraft to and from the place it will be repaired.
- Storage costs for your aircraft.
- Runway foaming and transporting your aircraft to a place where you can take off with the airplane.

Exclusions

As you might expect, the most prominent exclusion related to aircraft damage coverage is exclusion of coverage if you chose not-in-flight or not-in-motion insurance only and your aircraft is involved in an accident while in-flight or in-motion. Other typical exclusions you may run across are as follows:

- Damage incurred if your aircraft was seized (even if wrongfully) by a government agency or authority.
- Loss or damage if your aircraft is repossessed or taken by someone else claiming a right in your aircraft.
- Damages confined to "wear and tear" including mechanical or electrical failures, deterioration, or breakdown of parts of your aircraft.

Of all of these exclusions, the one that most frequently becomes an issue is the "wear and tear" exclusion. Here's how this common exclusion may come into play. Let's say you are unfortunate enough to land gear up. Typically, this type of mishap causes no personal injury (except pride). Just as typically, damage is limited to airframe dents and nicks and a bent propeller. If the propeller struck the surface, your insurance company may require a teardown of your aircraft engine(s) to ensure that no damage was sustained during the sudden stoppage of the propeller. Although the insurance company will pay for the teardown and any damage to the engine deemed to have resulted from the underlying accident, it will not pay for repair or replacement of engine components found to be in need of service because of ordinary wear and tear.

Limits to coverage

Limits to coverage for aircraft damage will depend largely on the insured value you determined for your aircraft. One of the more difficult calls that your insurance carrier may have to make is whether to repair an aircraft or simply declare it a "total loss" and pay out the limits of the policy. This sometimes becomes a contentious issue. That's why you should regularly assess your aircraft's insured value to make sure that it is reasonably close to the fair market value of your aircraft. In order to make the determination, some insurance policies provide that if estimates to repair or replace parts of your aircraft exceed a certain percentage of your aircraft's insured value, the insurance company will have the right to "total" your aircraft.

General Exclusions

Beyond the special exclusions for liability and aircraft damage coverage, most aircraft insurance policies also contain some general exclusions that pertain to all coverage provided under the policy. A few of the more significant general exclusions you may find are discussed below.

Commercial operations

Because we are discussing a noncommercial insurance policy, it is not surprising that your policy will exclude commercial operations. As suggested earlier, it is very important that you carefully review your policy's definition of "commercial operations" or "commercial purpose." As illustrated in Case 9-1, you may be surprised to learn what qualifies as a commercial operation in the context of an aviation insurance policy.

Case 9-1

Avemco Insurance Company v. Auburn Flying Service, Inc.
United States Court of Appeals for the Eighth Circuit
242 F.3d 819 (2001)

OPINION BY: Beam, Circuit Judge
I. BACKGROUND
The relevant facts of this case are as straightforward as they are tragic. On October 5, 1997, several organizations conducted a "fly-in" at the Auburn airport, near Auburn, Nebraska. As part of the event, attendees could pay $10 for a ten-to fifteen-minute airplane ride around the Auburn area in a plane piloted by Fred Farington. The money was collected at a table near the runway, which had a sign near it advertising the plane rides.

On the ninth trip of the day, while attempting to land, the plane struck a passing semi tractor-trailer and crashed. The three airplane passengers died in the crash. Farington died from his injuries four months later. The parties have stipulated that the money collected by Farington was not sufficient to cover the operating expenses of the flights.

Appellee, Avemco, had issued a noncommercial liability insurance policy to AFS, covering the plane in question. Farington was president of AFS, and was covered by name under the policy when piloting the plane.

The policy contained the following exclusion of coverage: "This policy does not cover bodily injury, property damage, or loss...when your insured aircraft is...used for a commercial purpose." The policy contained the following definition:

"Commercial purpose" means any use of your insured aircraft for which an insured person receives, or intends to receive, money or other benefits. It does not include:

a. the equal sharing among occupants of the operating costs of a flight.

After receiving evidence the district court granted Avemco's motion for summary judgment. Defendants and intervenors filed this appeal.

II. ANALYSIS

* * *

We hold that in interpreting an insurance exclusion clause, like the one in this case, the relevant inquiry is whether a reasonable person, viewing the totality of the circumstances surrounding the arrangement, would conclude that the arrangement was one of "shared expenses" or a flight made for the receipt of money. See *Meridian Mut. Ins. Co. v. Auto-Owners Ins. Co.,* 698 N.E.2d 770, 775-76 (Ind. 1998) (relying on a similar standard in interpreting an automobile policy with a "for charge" exclusion and a "shared car pool expense" exception); cf. *Mendenhall,* 546 N.W.2d at 775 (holding that in order to determine if a newly purchased vehicle was covered under a policy's automatic coverage clause, the inquiry is whether a *reasonable person* would have thought the vehicles already owned by the insured were in such a condition of inoperability that the insured would not have included them in a policy of liability insurance) (emphasis added).

The factors a reasonable person would examine would include those mentioned in the previous aviation policy cases: relationship of the amount paid to the expenses of a flight, existence of a community interest in taking the flight other than the flight itself, voluntariness of the payment and any indication of a quid pro quo. In automobile cases interpreting similar insurance contracts, courts have clearly followed a totality of the circumstances standard and taken a more comprehensive approach than the cited aviation insurance cases in considering all of the factors relevant to distinguishing "for charge" from "shared expense" arrangements. See, e.g., *Meridian Mut.,* 698 N.E.2d at 775.

For example, in *Meridian Mutual* the court considered whether the driver held herself out as providing transportation to the general public or just to a small group of people whose participation in the car pool generally did not change; whether there was significant formality in the arrangement between the driver and the passengers; whether the driver had some purpose in making the trip other than carrying the passengers (such as going to work herself); and whether the driver used the vehicle for her own personal use at other times.; see also *Johnson v. Allstate Ins. Co.,* 505 So. 2d 362, 367 (Ala. 1987) (stating that the significant factors are whether the insured charged a definite amount, whether the amount was proportionate to actual expenses of the trip, whether payment was voluntary (or paid as consideration for the trip) and whether the driver and passengers were engaged in some common enterprise).

* * *

III. CONCLUSION

By offering flights to anyone willing to pay the fee, Farington increased the number of flights he would make with the plane, and increased the risk to the insurer beyond that contemplated in the insurance contract. Consequently, the flight in question was for receipt of money and was not an equal sharing of expenses.

Accordingly, we affirm the decision of the district court.

The lesson learned from this case is that you must be meticulous about ensuring that you are not testing the boundaries of your insurance policy. The first step in avoiding problems is for you to gain a basic working knowledge of your policy and its limits.

Operations outside policy territory

This is a pretty easy exclusion to avoid. Again, just make sure you read the definitions and review the charts/maps your insurance company may provide you with containing the policy's operational boundaries.

Unlawful acts

The exclusion in your policy for unlawful acts should not come as a surprise. If your aircraft is being used for illegal smuggling of contraband or drugs and it is involved in a mishap, don't expect any help from your insurance carrier.

War and nuclear events

One other general exclusion that is found in most policies is the exclusion for war and nuclear detonation or contamination. As you might anticipate, insurance companies are not willing to pay claims in these "doomsday" situations.

Operation by unapproved pilots

As indicated above, your insurance carrier is only going to provide you with protection when you authorize approved pilots to operate your aircraft. In order to make sure that your authorized users qualify under your policy, you may have to review logs. Case 9-2 illustrates this very important point.

Case 9-2

Ideal Mutual Insurance Co., v. Last Days Evangelical Association, Inc. and Junk Air, Inc. and William David Jenkins d/b/a Junk Air Co.
United States Court of Appeals for the Fifth Circuit
783 F.2d 1234 (1986)

OPINION BY: Goldberg, Circuit Judge
OPINION:
Last Days Evangelical Association decided to lease a Cessna 414 from William Jenkins' Junk Air….the relationship between Junk Air and Last Days ended tragically on the afternoon of July 28, 1982, when the airplane crashed in the vicinity of Lindale, Texas, killing the pilot, Don Burmeister, and his eleven passengers.

Last Days promptly filed a claim under the insurance policy. Soon thereafter, Ideal filed a declaratory judgment action against Last Days, Junk Air, and William Jenkins, and pleaded that several exclusionary clauses in the policy precluded coverage. Ideal alleged that coverage did not exist under the policy for the following reasons: (1) the aircraft was overloaded at the time of the crash and was operated in a manner requiring Federal Aviation Administration authorization by special permit or waiver; and (2) Burmeister did not meet the requirements of the pilot clause endorsement, in that he did not have 1,045 logged hours and had not attended the manufacturer's flight training school. Pursuant to the policy's Lienholder's Interest Endorsement, Ideal also sought to recover from Jenkins the balance of principal and interest due on a promissory note and reasonable attorneys' fees incurred in bringing suit on the note.

Rejecting all but one of Ideal's arguments, the district court found that coverage did not exist because appellants failed to prove that Burmeister had the 1,045 "total logged hours" required by the policy. The district court also held Jenkins liable on the note, found that Jenkins was Last Days' agent for the purpose of procuring insurance, and awarded Ideal $5,000 in attorneys' fees against Jenkins.

Since the Texas Supreme Court's decision in *Puckett v. United States Fire Insurance Co.,* 678 S.W.2d 936 (1984), which was handed down shortly after the district court's decision, a finding of no coverage must be accompanied by a finding that the breach of the policy contributed to the loss. Although we find no error in the district court's finding that appellants breached the policy, we must nonetheless reverse the judgment below and remand to allow appellants an opportunity to prove that the breach of the policy did not contribute to the loss.

FACTUAL BACKGROUND

* * *

Jenkins leased the ill-fated Cessna 414 to Last Days in February, 1982. Under the terms of the lease, Jenkins agreed to provide hull and liability insurance on the plane with both Jenkins and Last Days named as insureds. Jenkins contacted Aviation Assurance Agency by Dickens ("Assurance"), an insurance broker. On February 22, 1982, Assurance issued to Jenkins and Last Days an insurance binder on the plane through Global Aviation Insurance Managers, Inc. with an effective period of one year, beginning on February 24, 1982. On March 23, 1982, Ideal issued its own binder through Global Aviation. At the time of the crash, all premiums had been fully paid. This insurance policy contains the conditions and terms of coverage at issue in this suit.

Exclusion No. 2 of the policy provides as follows:

This Policy does not apply:

* * *

2. to any occurrence or to any loss or damage occurring while the aircraft is operated in flight by other than the pilot or pilots set forth under Item 7 of the Declaration [which provides]:

PILOT CLAUSE. Only the following pilot or pilots holding valid and effective pilot and medical certificates with ratings as required by the Federal Aviation Administration for the flight involved will operate the aircraft in flight:

SEE ENDORSEMENT #1 [which provides:]

IT IS HEREBY UNDERSTOOD AND AGREED THAT ITEM 7 OF THE POLICY DECLARATIONS SHALL BE COMPLETED TO READ AS FOLLOWS: TOM KRAUS HAVING A COMMERCIAL PILOT CERTIFICATE WITH A MULTI ENGINE LAND AND INSTRUMENT RATING AND HAVING A MINIMUM OF 950 TOTAL LOGGED FLYING HOURS OF WHICH NOT LESS THAN 10 HOURS SHALL HAVE BEEN IN MULTI ENGINE AIRCRAFT INCLUDING NOT LESS THAN 10 HOURS IN THE SAME MAKE AND MODEL AIRCRAFT INSURED BY THIS POLICY.

DON BURMESER [SIC] HAVING A COMMERCIAL PILOT CERTIFICATE WITH A MULTI ENGINE LAND AND INSTRUMENT RATING AND *HAVING A MINIMUM OF 1,045 TOTAL LOGGED FLYING HOURS* OF WHICH NOT LESS THAN 50 HOURS SHALL HAVE BEEN IN MULTI ENGINE AIRCRAFT INCLUDING NOT LESS THAN 15 HOURS IN THE SAME MAKE AND MODEL AIRCRAFT INSURED BY THIS POLICY.

PILOTS TOM KRAUS AND DON BURMESER [SIC] MUST ATTEND THE MANUFACTURER'S GROUND AND FLIGHT SCHOOL.

ANY PERSON HAVING A PRIVATE PILOT CERTIFICATE WITH A MULTI ENGINE LAND AND INSTRUMENT RATING AND HAVING A MINIMUM OF 1500 TOTAL LOGGED FLYING HOURS OF WHICH NOT LESS THAN 500 HOURS SHALL HAVE BEEN IN MULTI ENGINE AIRCRAFT INCLUDING NOT LESS THAN 25 HOURS IN THE SAME MAKE AND MODEL AIRCRAFT INSURED BY THIS POLICY. [emphasis added].

The clause "having a minimum of 1,045 total logged flying hours" is now the sole source of controversy and found its way into the policy in the following manner. John Wallace, the underwriter involved at Ideal, asked Sharon Lewalling, the employee at Assurance handling the matter of Jenkins, for each pilots' logged flying hours. The application binder completed by Assurance and sent to Jenkins to be signed required that pilot experience be listed as "Total Logged Hours." In the space provided for the specific numbers, however, "Will fill out pilot forms" was written. Lewalling had sent Burmeister a form that requested "pilot experience."

The form required that Burmeister "list each aircraft by Make and Model and Hours as Pilot-in-Command in each." It did not, however, specifically request "logged" hours; rather, it simply requested "total hours." Burmeister filled out the form and returned it to Assurance. Many of the hours listed had been acquired during Burmeister's military training, and Lewalling assumed that military time was logged time. She then forwarded the form along with Assurance's binder to Wallace at Ideal, who incorporated the information provided by Burmeister in Endorsement #1.

The district court found that both Lewalling and Wallace "clearly intended to include a provision in the policy that Burmeister must have 1,045 logged flying hours." However, as is clear from the pilot experience form's request for "total hours," Assurance perhaps failed to convey this intent to Burmeister. Thus did Ideal innocently transform Burmeister's statement that he had 1,045 total hours into a statement that he had 1,045 total "logged" hours.

DISCUSSION

Special rules govern the construction of insurance contracts under Texas law. As a general rule, exceptions and limitations found in an insurance policy are construed against the insurer. *Ranger Insurance Co. v. Bowie*, 574 S.W.2d 540, 542 (Tex. 1978). "When the language of a policy is susceptible of more than one reasonable construction, the courts will apply the construction which favors the insured and permits recovery." *Ramsay v. Maryland American General Insurance Co.*, 533 S.W.2d 344, 346 (Tex. 1976). "The insurer may not escape liability merely because his or its interpretation should appear to us a more likely reflection of the intent of the parties than the interpretation urged by the insured. The latter has to be no more than one which is not itself unreasonable." *Continental Casualty Co. v. Warren*, 152 Tex. 164, 254 S.W.2d 762, 763 (1953). Of course, when the language of the policy permits only one reasonable construction and that construction favors the insurance company, recovery is denied. *Puckett, supra,* at 938. Moreover, we attempt to construe a contract so as to avoid rendering any of its terms meaningless. *Blaylock v. American Guarantee Bank Liability Insurance Company*, 632 S.W.2d 719, 722 (Tex. 1982).

As necessity is the mother of invention, so is ambiguity the father of multiple reasonable constructions, and where lawyers are involved, one never lacks an eager parent of either gender. Ambiguity in a contract is a question of law, *Airmark, Inc. v. Advanced Systems, Inc.*, 715 F.2d 229, 230 n. 1 (5th Cir. 1983) (applying Texas law), and a court may look beyond the plain language of the contract for evidence of surrounding circumstances to determine whether a writing is ambiguous. *Id.* at 230. "When terms of an insurance policy are unambiguous, they are to be given their plain, ordinary and generally accepted meaning unless the instrument itself shows that the terms have been used in a technical or different sense." *Ramsey, supra,* at 346. Unlike the deconstructionists at the forefront of modern literary criticism, the courts still recognize the possibility of an unambiguous text.

Appellants argue that the endorsement's use of the phrase "having a minimum of 1,045 total logged flying hours" gives birth to two ambiguities. They claim first that the entire phrase merely describes Burmeister and does not purport to act as a condition precedent to coverage. Second, they contend that even if the phrase prescribes a condition precedent, the word "logged" should be construed to describe the general writing submitted by Burmeister on his pilot form, rather than a written record of each one of his flights. We reject both arguments and find, as did the district court, that the policy unambiguously required that Burmeister have a minimum of 1,045 hours of logged time as a condition precedent to coverage.

The argument that the phrase beginning with "having" is not a condition precedent to coverage rests on Ideal's alleged failure explicitly to make it such. Had this been Ideal's desire, so goes the argument, Ideal could have written "*provided* he has a minimum of 1,045 hours of total logged time." See, e.g., *Ideal Mutual Insurance Co. v. C.D.I. Construction, Inc.,* 640 F.2d 654, 657 (5th Cir. 1981); *Schepps Grocer Supply, Inc. v. Ranger Insurance Co.,* 545 S.W.2d 13, 14 (Tex. Civ. App.—Dallas 1976, writ ref'd n.r.e.). Appellants argued that without some such language, the reference to pilot hours is merely descriptive.

This construction of the policy urged by appellants is unreasonable because it effectively works an Amelia Earhart disappearing act on the offending phrase. While this result is from appellants' perspective advantageous, it is certainly not reasonable, as it fails to give any effect at all to the phrase. A reasonable construction is one that gives meaning to the disputed language in the context of the writing, not one that strips the language of meaning altogether.

To give effect to this language, one need not look farther than the insurance company's business need to assess a risk prior to insuring it and setting the premium. An obvious element of that risk is the experience of those who will pilot the plane, and we can safely assume that both the risk and the premium will be higher if the pilot is Wrong Way Corrigan instead of Chuck Yeager. Thus, it does not require any speculation to conclude that Ideal included Burmeister's experience as a warranty that the risk it insured initially would be the same risk it paid out on.

That this construction is the only reasonable one is evidenced from the policy itself. Exclusion No. 2 provides that the policy does not apply if the aircraft is operated by other than the pilots set forth in the Pilot Clause, and the Pilot Clause reemphasizes that only the pilots listed in Endorsement #1 are allowed to operate the aircraft. Endorsement #1 then states that one of these pilots is "Don [Burmeister]...having a minimum of 1,045 total logged flying hours." Clearly, if Burmeister flew the plane and did not have 1,045 logged hours, the plane was being operated by other than the pilot named in the pilot clause and the policy would not apply. That is, even if we read the disputed language as a "description," then it does not correctly describe its intended referent, Burmeister, unless he in fact has the requisite logged flying hours. Thus, if the "Burmeister" described in the policy is not the real Burmeister, then it does not apply to him, and he is not a covered pilot. To read this language any other way is to read it completely out of the policy, leaving the insurer vulnerable to a risk it did not insure.

Appellants also seek to squeeze some ambiguity from the use of the word "logged." They argue that "logged" merely means written down in some manner, and that

Burmeister's pilot experience form, in which Burmeister listed in less than half a page the time he had spent flying over a period of many years, constitutes a "log."

"Logged," however, means more than this. In *Stewart v. Vanguard Insurance Company*, 603 S.W.2d 761 (Tex. 1980), the Court stated that the "common-sense meaning of the term is that a record, however informal, is made of the event....Human memory is so frail that a record needs to be made of the time, duration, point of departure, and destination of the flights...."

Not all time spent in flight is logged time, nor need it be; but logged time is given greater credence, and the FAA counts only that time which is logged. In *Republic Aero, Inc. v. North American Underwriters*, 462 S.W.2d 635 (Tex. Civ. App.—Waco 1970, no writ), the court counted all flight time even though it was not logged and held that a pilot had met the policy requirement that he have 500 hours of total time. The court stated that if the defendant insurance company had wished to recognize for insurance purposes only those hours also recognized by the FAA for certificate or rating purposes, it could have done so by placing a provision to this effect in the policy. *In this case, the insurer did include that provision in the policy. It did so by requiring that "logged" time be included in the number of hours to be counted. Republic National Life Ins. Co. v. Spillars*, 368 S.W.2d 92 (Tex. 1963).

Id. at 763 (emphasis added).

If there was any ambiguity concerning the type of hours requested, it arose only in Burmeister's mind when he confronted the pilot form's request for "total hours." The policy itself is unambiguous, and there is no evidence from the surrounding circumstances that any of the parties to this suit intended "logged" to have any but its ordinary meaning.

CONCLUSION

We affirm the district court's finding that appellants breached the policy. However, we reverse the district court's judgment that coverage does not exist and remand so that appellants may prove, if possible, that Burmeister's lack of the requisite number of logged hours did not contribute to the crash. In practical terms, appellants must prove that pilot error was not a cause of the crash.

The lesson from this case is that flight experience does not equal logged flight time when it comes to insurance matters. Make sure you and any other pilot authorized to operate your aircraft present evidence of pilot's qualifications. It would be most prudent for you to keep copies of the evidence presented to you.

Coverage for Medical Expenses

Many aircraft insurance policies offer medical payments coverage. This coverage will pay for medical expenses up to stated limits arising from injuries in any single occurrence. Depending on the policy you are looking at, medical payments coverage may or may not cover the pilot of the aircraft. Sometimes limits are expressed in terms of total dollars per person or dollars per day in a hospital. This type of coverage may be helpful because it may allow you to avoid dipping into your liability insurance for minor injuries. Often you can confirm that you've got this coverage by reviewing your insurance policy's data page. The data page should also indicate the limits to your coverage for medical expenses.

Nonowned Aircraft

You might be surprised to find out that your aircraft insurance policy also covers you (under certain circumstances) when you rent an aircraft. Many policies provide for liability, medical, and aircraft damage coverage for a nonowned aircraft if you also purchased these types of coverage for your owned aircraft. The usual conditions for this coverage to apply are:

1. You are using a nonowned aircraft while your insured aircraft is being repaired or serviced.
2. You or you and your spouse are the policyholders.

Typical exclusions for this type of coverage are as follows:

- Use of aircraft that doesn't have "standard" airworthiness category certification.
- Aircraft used without the owner's consent.
- Rotorcraft, unless your insured aircraft is a rotorcraft.
- Turboprop or turbojet aircraft, unless your insured aircraft is the same.
- Aircraft with certain seating capacities, unless your aircraft matches or exceeds the seating capacity limits.
- Amounts above the limits in your "in-flight," "not-in-flight," "in-motion," and/or "not-in-motion" policy limits.

General Provisions

Most policies have a set of general provisions to cover important, but miscellaneous, issues. Here's a discussion of a few of the more significant items you are likely to run across as you review your policy.

Modifications

You should clearly understand that the complete agreement with your insurance carrier is embodied in the policy. Any statements made to you by agents or understandings that you may have may not be worth much if you have to make a claim and the policy does not clearly cover your claim. If you wish to make any changes to your policy, they must be accomplished in writing. They must also be evidenced by written modifications (usually referred to as "endorsements") issued by your insurance carrier. With your insurance policy you have to remember that "what you see is what you get." If you want changes, get them in the form of written endorsements. Most policies will specifically state that no changes are valid unless they have issued such an endorsement.

Suing your insurance company

You should take a good look to see if your insurance policy places any conditions or restrictions on your ability to sue your insurance company. Some policies will indicate that you cannot sue for coverage under liability coverage unless you have a judgment against you. You might not see an immediate problem with that, but it could put you in a very disadvantageous situation. For instance, if a person injured in an accident involving your aircraft sues you, one of the first things you will do is to contact your insurance company. If your insurance company denies coverage, you are now on your own. That means you will also have to pick up the cost of your defense. Now is the time you'd want to determine whether your insurance carrier has at least a duty to defend you in court. Having to wait until a judgment has been rendered against you can place you in a precarious financial position with legal costs. If you see a provision like this in your policy, discuss it with your carrier or agent to make sure you fully understand the risks.

Subrogation rights

Your insurance company may be willing to pay on a claim for liability coverage or aircraft damage coverage. However, if someone else caused the damage, your insurance carrier should have the legal right to step into your shoes and pursue a claim against the wrongdoer. It will not be unusual for you to find a provision in your policy stating that you may not do anything to impair your insurance company's ability to pursue a claim against a third party. Some policies will also indicate that you must also provide positive assistance to your carrier in recovering damages that you might have been able to recover against a third-party wrongdoer.

Making your claim

Many policies will have express provisions containing instructions and guidelines for making a claim against your policy. One issue that comes up from time to time is the issue of your compliance with the insurance policy at the time that you applied for your insurance and thereafter. You have to remember that your insurance carrier is obligated to pay, but only if you have complied with the terms of the policy.

One of the more difficult issues that has cropped up over the years is the issue of "causal connection." This problem surfaces from time to time when an insurance company denies a claim on the grounds that the insured failed to comply with the provisions of the insurance policy. But what if the insured merely failed to comply with a provision that was totally unrelated to the accident causing the loss? The answer to this question may depend entirely on which state law you are using to decide the issue.

In a majority of states, there does not have to be any connection between your breach and the accident. If you breached the agreement in any way, your insurance company has the legal right to back out of your claim. In a minority of states, the courts have taken the position that your breach should only trigger a denial of coverage if it somehow caused the accident and the ensuing loss. To demonstrate just how vexing an issue this is, take a look at Cases 9-3 and 9-4.

Case 9-3

Western Food Products Company, Inc., Appellant, v. United States Fire Insurance
Company, Appellee
Court of Appeals of Kansas
699 P.2d 579 (1985)

OPINION BY: Parks, J.

OPINION: Plaintiff, Western Food Products Co., Inc., was the owner of an airplane
insured by defendant, United States Fire Insurance Company. The plane was totally
destroyed in a crash on January 5, 1981, and plaintiff filed a claim for its loss with
the insurer. Defendant denied coverage for the loss and plaintiff filed this action.
Both parties requested summary judgment based on stipulated facts and the trial
court granted judgment to defendant insurer.

Plaintiff appeals.

The parties stipulated that at the time of the crash, the airplane was being operated by Charles Newton Benscheidt, who had been issued a private pilot's certificate
on January 14, 1976, with ratings for single engine land aircraft. Mr. Benscheidt was
issued a third-class medical certificate with no limitations on October 3, 1977, which
expired on October 31, 1979. The records of the Federal Aviation Administration fail
to indicate that any subsequent medical certificate was issued to Mr. Benscheidt.

The district court held that because defendant's insurance policy requires the pilot
to have valid medical certification as a condition of coverage, defendant was not
liable for the loss of the aircraft. Plaintiff contends that this holding is erroneous
because the policy provision requiring a medical certificate did not apply to the person piloting the plane when it crashed. Arguing that this clause is ambiguous, plaintiff contends that it must be construed in its favor and held inapplicable under the
circumstances. The clause at issue states as follows:

"THE PILOT FLYING THE AIRCRAFT: The aircraft must be operated in flight
only by a person shown below who must have a current and proper (1) medical certificate and (2) pilot's certificate with necessary ratings, as required by the FAA for
each flight. There is no coverage under the policy if the pilot does not meet these
requirements."

Immediately after the above clause was typed the following language:

"STEVE BENSCHEIDT, JAMES DRISCOLL, III; OTHERWISE, PILOTS WHO
HAVE A CURRENT PRIVATE OR COMMERCIAL CERTIFICATE AND A MINI-
MUM OF 750 LOGGED PILOT HOURS OF WHICH AT LEAST 25 HOURS HAVE
BEEN IN THE SAME MAKE AND MODEL AIRCRAFT WE COVER IN ITEM 5."

It is undisputed that at the time the insured airplane crashed, it was being piloted
by an individual who is not named in the above paragraph. In addition, plaintiff concedes that the pilot of the plane did not have a current medical certificate. Thus,
plaintiff seeks to avoid the exclusionary force of the insurance provision by initially
contending that it is ambiguous.

If this insurance provision can be said to be ambiguous, the insured is entitled
to the benefit of various rules of construction. *American Media, Inc. v. Home
Indemnity Co.,* 232 Kan. 737, Syl. paras. 1-6, 658 P.2d 1015 (1983). However, if the
language of the policy is clear and unambiguous, the words are to be taken and
understood in their plain, ordinary and popular sense, and there is no need for
judicial interpretation or the application of rules of liberal construction; the court's

function is to enforce the contract according to its terms. *American Media, Inc.,* 232 Kan. at 740.

The provision of defendant's policy appears clear. The printed language indicates that the plane must be piloted by one of the persons enumerated in the space provided on the form and that this pilot must have both a valid medical certificate and a pilot's certificate. The contract specifically states that there is no coverage if the pilot does not meet these requirements. The typed-in list of approved pilots includes two individuals and any others who meet specific standards of experience. Thus, three categories of approved pilots are named—the category consisting of Steve Benscheidt, that naming James Driscoll and a third category including pilots who have a certain amount of experience.

Plaintiff contends that there is ambiguity in whether the requirements for a current medical and pilot's certificate apply to pilots falling in the third category. It argues that since these "other pilots" must meet flying-time standards more stringent than those met by either Steve Benscheidt or James Driscoll, this experience criteria supersedes the restrictions contained in the printed portion of the clause.

This argument defies the plain structure of the insurance provision. The policy provides the insured with the prerogative to choose its pilots within certain limits. Two specific individuals are listed as approved (so long as they have current medical and pilot's certificates) but other pilots will be covered only if they have the requisite experience and certification. The structure of the paragraphs indicates that the entire typed paragraph is incorporated by reference into the printed portion of the provision. The "person shown below" could include any person falling within the three described categories. In sum, the insurer has defined the scope of its coverage in clear terms. Any pilot flying the plane at the time of an accident must have both the medical and pilot's certificate in order for the loss to be protected. We find no ambiguity in this requirement.

Plaintiff contends that even if the provision of the policy excluding coverage of a flight piloted by a person without a medical certificate is unambiguous, it should not be enforced absent proof of a causal connection between the crash and the health of the pilot. Although such a holding would certainly depart from the ordinary rule mandating enforcement of an unambiguous provision according to its terms, the argument is not without some support in other jurisdictions. See, e.g., *Bayers v. Omni Aviation Managers, Inc.,* 510 F. Supp. 1204, 1207 (D. Mont. 1981); *Avemco Insurance Company v. Chung,* 388 F. Supp. 142 (D. Hawaii 1975); and *S. C. Insurance Company v. Collins,* 269 S.C. 282, 237 S.E.2d 358 (1977). These opinions generally stress the social purpose of insurance and the unfairness to the insured if he loses his expected protection because of a technical breach of the policy which is unrelated to the risks insured. However, in each of these cases the court was able to say that the provision relied on by the insurer failed to include language of express exclusion. The opinions could fairly conclude that the provision relied upon to exclude coverage under the aviation policy was a forfeiture clause or condition subsequent which is not ordinarily deserving of enforcement. See, e.g., *Am. States Ins. Co. v. Byerly Aviation, Inc.,* 456 F. Supp. 967 (S.D. Ill. 1978). Thus, in these cases the court was generally concerned with a provision viewed as attempting to eliminate extended coverage rather than one which defines the intended scope of coverage in the first instance.

In this case, the provision states unequivocally, "there is no coverage under the policy if the pilot does not meet these requirements." It is not a question of the insured being forced to forego coverage for which he has already paid; no protection was ever extended to cover the circumstances of the loss. Thus, since the policy specifically excludes coverage, its application cannot be categorized as a forfeiture.

A second distinguishing point of the cases holding that a causal connection must be shown to exclude coverage in an aviation insurance policy is that in each of the cases relied upon by plaintiff, there is state law either strongly favoring coverage in the absence of causation or requiring liberal construction of exclusionary clauses even without a specific finding of ambiguity. For example, in *S. C. Insurance Company,* 269 S.C. at 288, the court relied on a line of cases involving automobile liability and life insurance. These precedents held in varying situations that the violation of a provision limiting coverage would effectively operate to exclude liability only if a causal connection were demonstrated between the risk sought to be excluded and the actual loss. See also *Avemco Insurance Company,* 388 F. Supp. at 147 (applying Hawaiian law, the court noted that exclusionary clauses must be liberally construed).

By contrast, Kansas decisions have considered the purpose and reasonableness of an exclusionary provision in other types of insurance policies to determine its scope, but they have not demanded a relationship between the failure to satisfy the limitation and the actual loss sustained. For example, in *Mah v. United States Fire Ins. Co.,* 218 Kan. 583, 588, 545 P.2d 366 (1976), the Court was concerned with whether a geographical limitation in an insurance policy covering a mobile home was unambiguous and enforceable. The insured argued that there was no relevant connection between the geographical limitation and the extended coverages and that, thus, the parties did not intend this limitation to apply. The Court pointed out that there was indeed good reason for the insurer to seek to geographically limit its extended coverage and held the provision doing so to be unambiguous. The Court then refused to further consider the relationship between the limitation imposed and the actual loss suffered, stating as follows: "Further discussion of the relevance, materiality or importance of location of insured property is not warranted here, for the issue here is not whether the geographic limitation *should be* applicable, but whether it is applicable under the policy terms. We hold that it is." *Mah,* 218 Kan. at 588.

Thus, the Court refused to discuss an argument similar to that made here—that there must be some relevant connection between the actual loss sustained and the limiting provision sought to be enforced. The connection between the limitation and the abstract risks insured may be examined to determine the scope of the limitation but there is no need for the loss to be caused by the condition named in the exclusion in order for the exclusion to be effective.

Similarly, in *Heshion Motors, Inc. v. Trinity Universal Ins. Co.,* 229 Kan. 412, 414, 625 P.2d 437 (1981), the Court discussed the purpose of a provision conditioning coverage for a theft by deception on the possession of legal title to the car stolen. However, the purpose of this discussion was confined to determining the intended meaning of the policy provision requiring full and valid title. The Court did not consider, once finding the provision unambiguous and applicable, whether there was a causal connection between the failure to have legal title and the theft of the car. The limitation on coverage was to be enforced so long as the insured failed to prove it had title to the car. See also *Adventure Line Mfg. Co. Inc. v. Western Casualty & Surety Co.,* 214 Kan. 820, 522 P.2d 359 (1974).

Plaintiff cites *Leiker v. State Farm Mutual Automobile Ins. Co.,* 193 Kan. 630, 396 P.2d 264 (1964), in support of its argument that causation must be proven by the insurer before coverage may be excluded. The issue in *Leiker* was whether the particular facts of an automobile accident fell within the terms of an exclusionary clause. The clause provided that the insurance would not apply to "bodily injury sustained in the course of his occupation by any person while engaged (1) in duties incident to the operation, loading or unloading of...a...commercial automobile." The Court was initially concerned with whether the decedent was engaged in duties incident to the operation of a commercial auto. To determine this question, the Court had to decide whether the provision required proof of a connection between the accident and the operation of the truck. The Court's concern with causation resulted from its need to determine the scope of the exclusionary clause. This was not a case where the court found the exclusion applied on its face but demanded proof of causation before liability could be avoided. Therefore, *Leiker* is not authority for the argument put forth by plaintiff. Indeed, our research discloses no Kansas insurance cases suggesting that a causal connection must be established before an insurer may take advantage of a provision which by its terms excludes coverage.

Finally, we note that despite the existence of cases in which plaintiff's causation argument has been adopted, the majority of jurisdictions have held to the contrary. See, e.g., *Ochs v. Avemco Ins. Co.,* 54 Or. App. 768, 636 P.2d 421 (1981); *Security Mut. Cas. Co. v. O'Brien,* 99 N.M. 638, 662 P.2d 639 (1983); *Hollywood Flying Service v. Compass Ins. Co.,* 597 F.2d 507 (5th Cir. 1979); *Glades Flying Club v. Americas Aviation & M. Ins. Co.,* 235 So.2d 18 (Fla. Dist. App. 1970). The rationale behind these decisions is summarized as follows:

"We hold that a causal connection between an exclusion clause and an accident is not necessarily essential before coverage can be denied. In *Glades Flying Club v. Americas Aviation & M. Ins. Co.,* 235 So.2d 18 (Fla. App. 1970), a Florida court, faced with a similar set of facts and an identical policy exclusion, held that no causal connection was needed. The court stated:

'An aircraft insurance policy may validly condition liability coverage on compliance with a governmental regulation and, while non-compliance with such a regulation continues, the insurance is suspended as if it had never been in force. There need be no causal connection between the non-compliance and the loss or injury. (Citations omitted.)'

Id. at 20. To hold otherwise would allow courts to ignore the plain language of insurance policy exclusions whenever they feel an insurer should not be allowed to avoid liability for an accident unrelated to a policy exclusion. This rationale is contrary to substantial legal precedent as well as long-standing public policy. Insurance coverage must not be afforded aircraft owners who ignore or refuse to comply with established certification requirements commonly part of policy exclusions." *Security Mut. Cas. Co. v. O'Brien,* 99 N.M. at 640.

In conclusion, we remain unconvinced by plaintiff's argument that the ordinary rules of construction should be abandoned in this case. The exclusionary provision of the insurance contract is unambiguous and we refuse to alter this plain language to forestall its effect. Therefore, like the majority of jurisdictions, we must conclude that a causal connection between the accident causing the loss and the purpose of an exclusionary clause need not be proven before coverage can be denied by the aircraft insurer on the basis of the exclusion.

Plaintiff's final argument is that since the insurer should have known from the information submitted to it that the pilot did not have a current medical certificate, it is estopped from denying coverage. It is a general rule, acknowledged in this jurisdiction, that waiver and estoppel may be invoked to forestall the forfeiture of an insurance contract but they cannot be used to expand its coverage. *Ron Henry Ford, Lincoln, Mercury, Inc. v. Nat'l Union Fire Ins. Co.,* 8 Kan. App. 2d 766, 769, 667 P.2d

907 (1983). We have concluded that the defendant's policy unambiguously excludes coverage of the factual situation in this case There is no forfeiture of coverage being effected; the insured was never protected for the circumstances which took place. Therefore, the equitable relief of estoppel claimed by plaintiff is inappropriate because it would operate to expand the plain scope of the insurance policy.

Affirmed.

Case 9-4

South Carolina Insurance Company, Appellant, v. Lois S. Collins, Co-Administratrix of the Estate of Metz W. Collins, and Evelyn C. Lee, Co-Administrartix of the Estate of Metz W. Collins, and Wesley B. Nesbitt, Respondents
Supreme Court of South Carolina
237 S.E. 2d 358 (1977)

OPINION BY: Rhodes, J.

OPINION: This appeal presents the question of whether, in order to avoid liability under an aircraft insurance policy, the insurer is required to demonstrate a causal connection between the crash of the aircraft and the insured pilot's failure to have a valid and effective medical certificate as provided by the terms of the policy. For the reasons set forth herein, we affirm the relief granted by the special circuit judge and hold that such causal connection must be shown.

The plaintiff-appellant, South Carolina Insurance Company (hereinafter appellant), issued to Metz W. Collins, the named insured, its aircraft liability insurance policy, number AC-801297, effective for the period April 27, 1975 to April 27, 1976. The contract of insurance covered a Piper Colt aircraft, Federal Aviation Agency registration number N5723Z, owned by Collins. On May 23, 1975, Collins, while piloting the airplane described in the policy, crashed, resulting in Collins' death and injuries to one Wesley B. Nesbitt, a passenger in the airplane. Subsequently, Lois S. Collins and Evelyn C. Lee were appointed administratrices of the estate of the deceased, Metz W. Collins, and Nesbitt commenced an action against the estate seeking damages for injuries he sustained in the crash. Nesbitt's action is pending in the Court of Common Pleas of Horry County.

The appellant refused the demand of the administratrices to defend the insured's estate against the lawsuit instituted by Nesbitt and subsequently commenced this action for declaratory judgment pursuant to S.C. Code § 15-53-10 (1976) *et seq.* The appellant sought an order declaring that the policy issued to the insured was not in effect during the flight of May 23, 1975, and that it did not afford the estate of the deceased any coverage. Named as defendants in the appellant's action were both administratrices and the injured passenger, Nesbitt. (All defendants are hereafter referred to as "respondents.")

The respondents answered and sought affirmative relief, demanding that the appellant's complaint be dismissed and seeking an order declaring that the aforesaid policy was in full force and effect at the time of the crash. The trial judge, after hearing arguments and considering briefs, issued an order which granted the relief prayed for by the respondents.

In the trial judge's order it is stated that the parties, through responses to requests for admissions and stipulations made before him, agreed that the following facts are not in dispute: The Federal Aviation Regulations promulgated by the Federal Aviation Agency require a medical examination of pilots under the supervision of the

Federal Air Surgeon or his authorized representative. The insured held a third-class medical certificate which was valid for a twenty-four (24) month period. The last medical certificate issued to the insured was in February, 1973, and it expired on the last day of February, 1975, or nearly three (3) months before the date of the crash. The Federal Aviation Regulations also require the insured to obtain and to have in his possession a valid and effective pilot certificate. The insured had his last required flight review on November 3, 1974, or a period of six (6) months and twenty (20) days before the date of the accident. It was stipulated by the parties that at the time of the accident the insured possessed a valid and effective pilot certificate but that he did not have a valid and effective medical certificate. Moreover, for the purposes of this declaratory judgment action only, it was stipulated that the insured, to the best of the parties' knowledge, had no physical or mental defects at the time of or immediately prior to the accident and that there was no causal connection between the accident and the failure of the insured to have a valid and effective medical certificate. The above statement by the trial judge of the undisputed or stipulated facts has not been challenged on this appeal.

The appellant argues vigorously that the failure of the insured to have a valid and effective medical certificate on the date of the accident amounted to a breach of a condition subsequent or promissory warranty under the terms of the policy, thereby suspending coverage and permitting the appellant to avoid liability. The appellant contends that the trial judge's classification of the pertinent policy provisions as being merely an "exclusion" of the insured's liability was erroneous. Additionally, the appellant maintains that the case law does not support the court's holding that the insurer, in order to avoid liability on a policy such as that involved here, must show that there exists a causal connection between the resulting loss and the insured's failure to have the required effective medical certificate.

The appellant relies upon the following provisions of the policy as supportive of its contention that the policy stated a condition subsequent or promissory warranty which the insured breached:

EXCLUSIONS

This Policy does not apply:

...

2. to any occurrence or to any loss or damage occurring while the aircraft is operated in flight by other than the pilot or pilots set forth under Item 7 of the Declarations;

...

4. to any Insured:

(a) who operates or permits the aircraft to be operated in any manner which requires a special permit or waiver from the Federal Aviation Administration, whether granted or not, unless this Policy is specifically endorsed to include such operation;

...

DECLARATIONS

...

7. PILOT CLAUSE. Only the following pilot or pilots holding valid and effective pilot and medical certificates with ratings as required by the Federal Aviation Administration for the flight involved will operate the aircraft in flight:

METZ W. COLLINS.

...

CONDITIONS

...

25. Declarations. By acceptance of this Policy the Named Insured agrees that the statements in the Declarations are his agreements and representations, that this Policy is issued in reliance upon the truth of such representations, and that this Policy embodies all agreements existing between himself and the Company or any of its agents relating to this insurance.

The appellant maintains that the above-quoted provisions, when viewed together, established a condition with which the insured was bound to comply.

While provisions of aircraft liability insurance policies similar to the provisions quoted above have been passed upon in other jurisdictions, we have not found any case in this state which has dealt with such a policy. However, a line of South Carolina cases involving contracts of automobile liability and life insurance are, in our judgment, dispositive of the issues raised on this appeal.

The first in this series of cases was decided by this Court in 1932. In *Reynolds v. Life & Casualty Ins. Co. of Tennessee,* 166 S.C. 214, 164 S.E. 602, the mother of the insured brought suit in her capacity as beneficiary under two (2) policies of life insurance issued by the defendant company to her son. The contract of insurance involved in the first cause of action contained the following provision: "This policy does not cover...loss sustained by the insured...while committing some act in violation of law." The contract involved in the second cause of action provided that "[w]ithin two years from date of issuance of this policy, the liability of the company under same shall be limited, under the following conditions, to the return of the premium paid thereon:...(2) If the insured shall die...as a result of acts committed by him while in the commission of...some act in violation of law."

During the effective period of these contracts of insurance, an ordinance of the City of Greenville, the place of residence of the insured, provided that "[n]o person shall be allowed to ride upon the running board of any motor vehicle...." The defendant company moved for a directed verdict at the close of the testimony in the trial of the case upon the ground that at the time of the accident, in which fatal injuries were sustained, the insured was riding on the running board of a truck in violation of the ordinance. The trial judge overruled the defendant's motion and the jury found for the plaintiff in the entire policy amounts sued for. The defendant appealed from the denial of its motion and also from the court's charge to the jury that the insurer must establish a causal connection between the insured's loss or death and his violation of the ordinance.

In *Reynolds,* this Court stated that under the evidence and the particular facts adduced at the trial, it could not be held as a matter of law, as defendant contended, that the insured's death was proximately caused by his riding upon the running board of the truck or that his death was the necessary or natural consequence of his act. The Court ruled that these questions had been properly submitted to the jury for determination.

In discussing the issue of causal connection, The Court stated the following:

[I]t is clear that the court properly refused defendant's motion, for, even if it should be admitted that the insured was violating the ordinance at the time he was injured, this alone would not be sufficient ground for direction of a verdict. In order to defeat recovery under policies excluding or limiting liability where death or injury results from an unlawful act on the part of the insured, there must be shown, in addition to the violation of the law, some causative connection between such act and the death or injury....

In *Insurance Co. v. Bennett,* 90 Tenn. [256], 267, 16 S.W. 723, 725, 25 Am. St. Rep. 685, the Tennessee court, construing a provision of this kind in an insurance policy, said: "In order to defeat a recovery because of such provision, there must appear a connecting link between the unlawful act and the death. It is not sufficient that there was an unlawful act committed by the insured, and that death occurred during the time he was engaged in its commission. There must be some causative connection between the act which constituted a violation of the law and the death of the insured....The provision of the policy excluding liability for injury received by the insured while committing an unlawful act refers to such injuries as may happen as the necessary or natural consequence of the act." *Reynolds, supra,* 166 S.C. at 217, 218, 164 S.E. at 603.

Approximately two years after the decision in *Reynolds* this Court again had occasion to rule with respect to the necessity on the part of an insurer to show causal connection. In the case of *McGee v. Globe Indemnity Co.,* 173 S.C. 380, 175 S.E. 849 (1934), the insured instituted an action on a policy of insurance whereby the insurer had agreed to indemnify the plaintiff against loss or damage to his automobile occasioned by accidental means. The insurer sought to escape liability by alleging that the policy contained a provision that it should not apply "in respect of any automobile...(2) while used or maintained by any person in violation of law as to age or by any person under the age of sixteen (16) years." The undisputed facts showed that at the time the automobile was damaged it was being operated by the insured's son who was under 16 years of age. In overruling a demurrer to the insured's reply which alleged that there existed no causal connection between the age of the driver and the collision, the trial judge relied upon *Reynolds v. Life & Casualty Ins. Co. of Tennessee, supra.*

On appeal, the insurer attempted to distinguish the *Reynolds* case from the *McGee* case, contending that in *Reynolds* the court was dealing with a provision limiting liability of the insurance company on account of some act of the insured (in other words, an exclusion), while in *McGee* the provision of the policy relied upon to defeat recovery related to a condition. The appellant urged that while it was logical to hold that it is necessary to show some causative connection between the act or acts named in the limiting provision of a policy and the accident involved, a policy which limits liability under a certain condition *absolutely excludes* liability under the named condition, even though the condition might have no causative connection with the accident and resulting loss.

In rejecting this theory advanced by the insurer, this Court held:

We are of the opinion that the distinction advanced between the excluded act and condition is without any logical basis....The rule established by the *Reynolds Case* is obviously founded upon the reasonable view that, *when the parties made the contract of insurance, they were not inserting a mere arbitrary provision, but that it was the purpose of the insurance company to relieve itself of liability from accidents caused by the excluded condition.* And there is no more reason that the parties to the contract of insurance would arbitrarily exclude liability under a certain condition than they would arbitrarily exclude liability in the commission of a certain act. This case is controlled by the *Reynolds* Case. *McGee, supra,* 173 S.C. at 384, 175 S.E. at 850. (emphasis added)

Subsequent decisions of this Court in the *Reynolds-McGee* line affirm the requirement on the part of the insurer to show casual connection. *Bailey v. United States Fidelity & Guaranty Co.,* 185 S.C. 169, 193 S.E. 638 (1937); *Smith v. Sovereign Camp, W.O.W.,* 204 S.C. 193, 28 S.E. (2d) 808 (1944); *Young v. Life &*

Casualty Ins. Co. of Tennessee, 204 S.C. 386, 29 S.E. (2d) 482 (1944); *Outlaw v. Calhoun Life Ins. Co.,* 238 S.C. 199, 119 S.E. (2d) 685 (1961).

The appellant cites cases from other jurisdictions holding that under circumstances virtually identical, or closely related, to those in the instant case, the insurer need not show a causal connection between damages and injuries sustained in the crash of the aircraft and the insured's failure to comply with certain terms of the policy. These decisions hold essentially that the purpose of the exclusionary language of such policies is not that the risk is excluded if damage to the aircraft is *caused* by the failure of the pilot to be properly certificated, but that the risk is excluded absolutely if loss *occurs while* the aircraft is being flown by a pilot not properly certificated. *See Grigsby v. Houston Fire & Casualty Ins. Co.,* 113 Ga. App. 572, 148 S.E. (2d) 925 (1966); *Glades Flying Club v. Americas Aviation & Marine Ins. Co.,* 235 So. (2d) 18 (Fla. App. 1970); *Baker v. Insurance Company of North America,* 10 N.C. App. 605, 179 S.E. (2d) 892 (1971); *Aetna Casualty and Surety Co. v. Urner,* 264 Md. 660, 287 A. (2d) 764 (1972); *Omaha Sky Divers Parachute Club, Inc. v. Ranger Insurance Co.,* 189 Neb. 610, 204 N.W. (2d) 162 (1973); *Bruce v. Lumbermens Mutual Casualty Co.,* 222 F. (2d) 642 (4th Cir. 1955); *Roberts v. Underwriters at Lloyds London,* 195 F. Supp. 168 (S.D. Idaho 1961); *National Insurance Underwriters, Inc. v. Bequette,* 280 F. Supp. 842 (D. Alaska 1968); *Arnold v. Globe Indemnity Co.,* 416 F. (2d) 119 (6th Cir. 1969).

In some of these cited cases the courts have referred to the certification provisions sometimes as "conditions" and, at other times, as "exclusions." However, the language quoted above from the *McGee* case resolves adversely to the appellant its contention that the trial judge in the instant case erred in characterizing the policy provisions as an exclusion rather than a condition subsequent or a promissory warranty. While there is a recognized distinction among these various terms, such a difference is immaterial in this case. We rely upon *McGee* in so holding.

After examination of the decisions cited by the appellant, we find them unpersuasive. Only their number, not their reasoning, lends support to a reversal here. We find that the reasoning used in the *Reynolds-McGee* line of cases no less compelling when applied to an aircraft liability policy.

In view of the stipulation by the parties that there was no causal connection between the loss and injuries resulting from the crash and the failure of the insured to have a valid and effective medical certificate at the time of the accident, we hold that the lower court acted correctly in awarding the relief sought by the respondents.

Affirmed.

Notice that both cases were decided in the same year and decided the same issue. In one case the court held that the lack of a medical certificate negated coverage. That court felt that it was not necessary to establish a connection (causation) between the policyholder's breach and the damage or loss caused by the accident. In the other case, the court held that the lack of a medical certificate should not negate coverage. That court concluded that it was necessary to establish a connection (causation) between the policyholder's breach and the damage or loss caused by the accident. The only difference in these two cases was the location of the court where the case was decided.

Policy cancellation

Your insurance policy should also spell out the details for cancellation of the policy by either party. Some of the questions that should be addressed include the following:

- How much notice do you have to give to cancel your policy?
- How much notice does your insurance carrier have to give to cancel your policy?
- How must the notice be transmitted?
- Are you entitled to a refund of premiums paid?
- Is the refund calculated differently if you are the canceling party?

Policy cancellation is a serious step by either party. Make sure you understand the process if you are faced with a cancellation issue.

10

Liability Releases

In an attempt to supplement the protection supplied by insurance, a few air-craft owners have turned to liability release forms in an attempt to reduce their exposure to aircraft-related lawsuits. On the face of it, the concept seems simple enough—if someone wants to fly in your airplane, they will have to promise to release any legal claims against you if they are injured or killed during any flights in your aircraft.

Releases may or may not have value to you as an aircraft owner. This chapter will focus on the legal issues surrounding liability releases and the different circumstances affecting the enforceability of such agreements. First, we'll take a look at the general state of the law relating to liability releases, including a look at some recent court decisions and what trends may be developing in this area of the law. Next, we'll review some basics that you should consider with your attorney if you decide to use a liability release form.

Will a Release Work?

The law imposes duties on you as an aircraft owner. One of those duties is to carry out your duties as an aircraft owner/pilot in a reasonably careful manner. If you are careless in tending to your duties as an aircraft owner/pilot, you may be guilty of negligence, which exposes you to paying money damages to any-one harmed by your negligence.

Liability releases may work to reduce your exposure, but there is no guar-antee. Often courts disfavor these clauses because they believe they are contrary to public policy. The public policy question is a bit tricky. Cases seem to indicate that in certain situations, the courts will be more likely to refuse enforcement of a liability release. Here's a rundown of the most common circumstances where liability releases have been shot down on public policy grounds:

- Where the party utilizing the clause has vastly superior bargaining power over the other.

- Where the goods or services referred to in the liability release are necessities.

- Where a party attempts to escape liability for gross negligence, recklessness, or intentional misconduct.

A careful examination of these three circumstances tends to indicate that they are typically inapplicable to general aviation aircraft owners. Usually, your aircraft is being used for recreational or business purposes. Typically, that means that the use of your aircraft is not a necessity. The person agreeing to fly in your aircraft has many choices, one of which is not to fly at all. Therefore, it is likely that you are also able to avoid the appearance of having superior bargaining power over the passengers in your aircraft. Again, they do not have to fly in your aircraft—there are lots of other choices they may make. Finally, most aircraft mishaps are caused by simple negligence where you might be found guilty of being careless. It is less frequent that a mishap is caused by gross negligence (intentional failure to perform your duties). Therefore, in most cases, the issue of gross negligence or intentional misconduct does not pertain.

Even if the public policy hurdle is cleared, you next have to ensure that your liability release is (1) effectively called to your passenger's attention and (2) clearly written. You want to avoid burying your liability release in the middle of a lengthy form contract. If it is part of another agreement, it should at least be highlighted in bold letters. Further, even if you are careful in bringing your release to the attention of the reader, your release will only work to avoid the types of liability that are explicitly and clearly spelled out. If your release is ambiguous, the courts will construe any ambiguities against you. For a vivid example of just how explicit you will have to be, read Case 10-1.

Case 10-1

Holly F. Kissick, as Personal Representative of the Estate of Michael W. Kissick, Petitioner, v. Kay M. Schmierer, as Personal Representative of the Estate of Otto Schmierer, Judith Jonsen, as Personal Representative of the Estate of Alan Jonsen, and Linda Leblanc, as Personal Representative of the Estate of Ernest Leblanc, Respondents

Supreme Court of Alaska
816 P.2d 188 (1991)

OPINION BY: Rabinowitz, C. J. and Matthews, J.
FACTUAL AND PROCEDURAL BACKGROUND
On July 4, 1988, Michael Kissick invited Alan Jonsen, Ernest LeBlanc, and Otto Schmierer to fly with him to Coghill Lake for a fishing trip. Kissick and the three passengers died when the plane crashed into a mountain bordering Burns Glacier.

Kissick was a major in the United States Air Force and a member of the Air Force Elmendorf Aero Club. The Aero Club is an instrumentality of the Air Force, established and managed according to Air Force Regulation 215-12. The plane that crashed was owned by the Aero Club and rented by Kissick. The plane was kept at a lake on the Air Force base. The three passengers were civilians. Aero Club members are authorized to rent the Club's planes and may fly with non-member civilian passengers if

Air Force regulations are satisfied. Of primary concern in this case is the requirement that passengers sign AF Form 1585 agreeing not to bring a claim against "the US Government and/or its officers, agents, or employees, or Aero Club members...for any loss, damage, or injury to my person or my property which may occur from any cause whatsoever. . . ."

Prior to departure, Supervisor of Flying Steven Wright directed Jonsen, LeBlanc, and Schmierer to complete and sign Air Force Form 1585, Covenant Not to Sue and Indemnity Agreement, and the data in the emergency notification section of an Aero Club membership application. The passenger only needs to print his name near the top and sign and date the bottom to complete the Covenant Not to Sue. Wright did not explain the forms in any detail to Jonsen, LeBlanc and Schmierer, but he testified that he "usually tell(s) people who go flying: 'This is a covenant not to sue in the event of an accident. It indicates that you won't sue the Air Force or the Aero Club.'" Wright also checked Kissick's qualifications, and reviewed the flight plan and weather with him. Although Wright did not specifically discuss the risks of flying in a small aircraft with the passengers, he testified that they nonetheless "knew about the limitations regarding the weather and the aircraft, the size, the weight and all that [because t]hey were all present and, I assumed, listened to this conversation that was going on."

Following the accident, the widows of Jonsen, LeBlanc and Schmierer filed wrongful death claims against the Kissick estate. Kissick asserted as an affirmative defense that AF Form 1585 barred all claims. The parties made cross-motions for summary judgment regarding the effect to be given the covenant. In addition, Kissick argued that Air Force regulations preempted plaintiffs' state tort claims. The trial court ruled that federal preemption was not an issue, and strictly interpreted the form according to state law. It concluded that the agreement did not bar wrongful death actions because "the covenant doesn't even talk about death. . . . It is ambiguous. The ambiguity must be construed against the government, against the parties seeking to rely on the exculpatory provision." Kissick sought review.

* * *

Kissick contends that respondents' claims are barred by operation of the preemption doctrine since signing the covenant not to sue was required by an Air Force regulation. Kissick also asserts that inclusion of the agreement in a regulation alters the court's review authority and requires the Aero Club's form to be interpreted like a statute or regulation instead of like other exculpatory contracts. If the covenant is interpreted as a statute or regulation, the purposes of the enacting body and the plain language of the regulation will be given primary consideration, and the covenant will not be subjected to the strict scrutiny that exculpatory clauses customarily must survive to be upheld. In Kissick's view, in enacting the regulation the Air Force intended to shield itself from all liability, and the omission of the word "death" is a technicality that should not act to interfere with the regulation's clear purpose.

Cases have evaluated covenants not to sue promulgated by the federal government in accordance with state law as though the parties thereto were private individuals or entities. *Rogow v. United States,* 173 F. Supp. 547 (S.D.N.Y. 1959), was an action by the widow of a free lance writer who died when the Air Force plane on which he was a passenger crashed. Prior to the flight, Mr. Rogow signed a covenant not to sue that was similar to the Aero Club's agreement except that it expressly released claims arising "on account of my death." *Id.* at 550-51 n.7. The court applied New

York law to interpret the covenant and found that it did not bar the claim. Further, even though one can infer that the covenant was the product of administrative regulations, the court did not interpret it differently from similar private covenants. See cases cited *id.* at 551-52.

A similar situation was considered in *Montellier v. United States,* 202 F. Supp. 384 (E.D.N.Y. 1962), aff'd 315 F.2d 180 (2nd Cir. 1963). Plaintiff's husband, a civilian reporter, died in the crash of an Air Force plane after signing a covenant not to sue. The covenant resembled the Aero Club's except that it included "death." The court interpreted the covenant in light of Massachusetts's Death Act and held that the covenant did not bar plaintiff's action. *Id.* at 394. In deciding whether Mr. Montellier had assumed the risk of a crash, the court relied on cases in which the defendant was not a governmental entity. See also *Green v. United States,* 709 F.2d 1158, 1165 (7th Cir. 1983) (in a medical malpractice action against an Air Force doctor the effect of a release from liability must be determined according to the law of the state where the tort occurred); *1 S. Speiser & C. Krause, Aviation Tort Law* § 3:53, at 300 (1978) (government pre-flight covenants are governed by state rather than federal law).

In the absence of a direct conflict, state law is only preempted "when Congress intends that federal law occupy a given field." *California v. ARC America Corp.,* 490 U.S. 93, 100, 109 S. Ct. 1661, 104 L. Ed. 2d 86 (1989). There has been no showing that Congress intended to occupy the state tort field when it authorized the Secretary of the Air Force to promulgate regulations regarding Aero Clubs. The existence of regulations governing the operation of Aero Clubs is not enough to find that state tort actions are preempted. The applicable regulation merely requires passengers to execute the form covenant not to sue. The regulation does not suggest that the covenant should not be construed according to state law. If federal law was intended to govern the meaning of the covenant, this intention could have been directly stated. Similarly, if the regulation itself had been meant to bestow immunity, that intention could readily have been expressed.

Based on the foregoing, we conclude that the covenant is to be construed like a covenant between private parties in accordance with Alaska law.

B. The Exculpatory Agreement as Interpreted by State Law

As noted, the trial court ruled that the covenant not to sue was not a bar to respondents' claims under state law. The court noted that exculpatory agreements are strictly construed against the party seeking immunity from suit and found that since the word "death" was missing from the covenant, the term "injury" was ambiguous and must be construed to exclude death. We agree.

Although there are no Alaska cases directly on point, it is well settled that ambiguities in a pre-recreational activity exculpatory clause will be resolved against the party seeking exculpation, and that to be enforced the intent to release a party from liability for future negligence must be conspicuously and unequivocally expressed. See, e.g., *Gross v. Sweet,* 49 N.Y.2d 102, 400 N.E.2d 306, 309, 424 N.Y.S.2d 365 (N.Y. 1979). ("It has been repeatedly emphasized that unless the intention of the parties is expressed in unmistakable language, an exculpatory clause will not be deemed to insulate a party from liability for his own negligent acts.") *Ferrell v. Southern Nevada Off-Road Enthusiasts Ltd.,* 147 Cal. App. 3d 309, 195 Cal. Rptr. 90, 95 (Cal. App. 1983), is representative.

To be effective, an agreement which purports to release, indemnify or exculpate the party who prepared it from liability for that party's own negligence or tortious conduct must be clear, explicit and comprehensible in each of its essential details.

Such an agreement, read as a whole, must clearly notify the prospective releasor or indemnitor of the effect of signing the agreement.

The efficacy of this salutary rule is especially applicable in the case before us in which the adhesive agreement was prepared by one of the purported releasee-indemnitees, and was presented on a take-it-or-leave-it basis as a condition of being allowed to enter the race.

Professors Williston and Prosser have also expressed the view that exculpatory provisions are disfavored.

Generally, an indemnity agreement will not be construed to cover losses to the indemnitee caused by his own negligence unless such effect is clearly and unequivocally expressed in the agreement.

A promise not to sue for future damage caused by simple negligence may be valid. Such bargains are not favored, however, and, if possible, bargains are construed not to confer this immunity. S. Williston, *A Treatise on the Law of Contracts* § 1750A, at 143-45 (3d ed. 1972) (footnotes omitted); see also W. Keeton, D. Dobbs, R. Keeton & D. Owen, *Prosser and Keeton on the Law of Torts* § 68, at 483-84 (5th ed. 1984) (footnotes omitted):

> If an express agreement exempting the defendant from liability for his negligence is to be sustained, it must appear that its terms were brought home to the plaintiff; and if he did not know of the provision in his contract, and a reasonable person in his position would not have known of it, it is not binding upon him, and the agreement fails. . . . It is also necessary that the expressed terms of the agreement be applicable to the particular misconduct of the defendant, and the courts have strictly construed the terms of exculpatory clauses against the defendant. . . .
>
> If the defendant seeks to use the agreement to escape responsibility for the consequences of his negligence, then it must so provide, clearly and unequivocally, as by using the word "negligence" itself.

Courts in a number of contexts have recognized that an ambiguity exists as to whether the term "injury" includes death. *Tobin v. Beneficial Standard Life Ins. Co.,* 675 F.2d 606 (4th Cir. 1982) (clause in insurance policy which excluded coverage for injury found to be ambiguous with respect to whether injury included death; this ambiguity was resolved against the insurer); *Ziolkowski v. Continental Casualty Co.,* 365 Ill. 594, 7 N.E.2d 451 (Ill. 1937) ("injury" in insurance policy exclusion did not include death); *Cal-Farm Ins. Co. v. TAC Exterminators,* 172 Cal. App. 3d 564, 218 Cal. Rptr. 407, 411 (Cal. App. 1985) (whether "injury" includes death held to be ambiguous). In view of this ambiguity, the rule of construction disfavoring exculpatory agreements applies. The covenant therefore does not bar respondents' claims.

The decision of the superior court is AFFIRMED.

DISSENT BY: COMPTON

DISSENT: COMPTON, Justice, dissenting.

For the purpose of this case, I will assume that the court's decision on federal preemption is correct. I will assume also that the Covenant Not to Sue and Indemnity Agreement should be construed according to state law. However, I am unpersuaded that the covenant is ambiguous, and therefore dissent.

The covenant, signed by Otto Schmierer, Alan Jonsen and Ernest LeBlanc, provides: I, for myself, my heirs, administrators, executors, and assignees, hereby covenant and agree that I will never institute . . . any demand, claim, or suit against . . . Aero Club members, participants, [or] users, . . . for any loss, damage, or injury to my person or my property which may occur from any cause whatsoever as a result of my participation in the activities of the Aero Club.

The statute under which the personal representatives of the estates of each of the above signators are suing the estate of Aero Club member Michael W. Kissick, AS 09.55.580, provides for recovery of an amount exclusively for the benefit of the decedent's spouse, children or other dependents, if there are any, or the decedent's estate, if there are not. If there is a spouse, children or other dependents, the award should fairly "compensate for the injury resulting from the death" considering at least the following factors: "(1) deprivation of the expectation of pecuniary benefits to the beneficiary or beneficiaries . . . ; (2) loss of contributions for support; (3) loss of assistance or services . . . ; (4) loss of consortium; (5) loss of prospective training and education; (6) medical and funeral expenses." As is clear from the statute and cases construing it, see, e.g., *Tommy's Elbow Room, Inc. v. Kavorkian,* 727 P.2d 1038 (Alaska 1986), it is injury to beneficiaries for which compensation is being awarded, the compensation representing losses of various kinds reduced to dollars and cents.

In construing the covenant, the court declines to apply the "reasonable construction" rule articulated in *Manson-Osberg Co. v. State,* 552 P.2d 654, 659 (Alaska 1976), and similar cases. However, reasonable construction is intended to resolve ambiguities ("unambiguous language . . . as 'reasonably construed'"). Here the court strictly construes "injury" to create an ambiguity, implicitly conceding that the language in the covenant, reasonably construed, is not ambiguous. Otherwise, Manson-Osberg adds nothing to the debate. Furthermore, "injury" is construed without reference to the language preceding and following it, and without reference to the statute under which the suit has been brought.

I am incredulous that the phrase "any loss, damage, or injury to my person or my property" can be construed, whether "reasonably" or "strictly," to exclude death. This is particularly so in the context of AS 09.55.580. For this reason alone I do not agree that the covenant does not state a viable defense.

Even if the language in this covenant is ambiguous, it is to be strictly construed only against the party who prepared it. However, it must be remembered that Michael W. Kissick did not prepare the covenant at issue; the United States Air Force did. It is Michael W. Kissick's estate, not the United States Air Force, that is seeking to enforce this covenant. Thus I question application of the strict construction rule at all. If the strict construction rule is inapplicable, the covenant should be upheld as stating a viable defense to the wrongful death actions.

Notice that this case appears to have turned on the fact that the liability release included the word "injury" but not "death." Most reasonable persons would assume that this release intended a release in the case of death (see the dissenting opinion by Judge Compton). However, reasonableness sometimes takes a back seat when it comes to judicial interpretation of liability releases.

Besides the need for precision in the language of releases, you should also be sensitive to the fact that even if a liability release is legally enforceable, it is only legally enforceable against certain persons.

For instance, your liability release form will not be enforceable against persons or their property that might be injured on the ground. Obviously, it would be inappropriate to hold this class of potential plaintiffs to a liability release they have never seen and certainly have not agreed to.

Another question comes up with respect to wrongful death actions. Wrongful death actions are allowed by the laws of every state (the *Kissick* case is an example of a wrongful death action). It is a type of lawsuit brought on behalf of a deceased person's beneficiaries. Typically, a wrongful death action will include claims that the plaintiff's death was due to the negligence or willful misconduct of another person. The interesting question here is whether a liability release signed by your passenger will also legally bind the dead passenger's family or personal representative.

In the majority of states, the release will bar a subsequent wrongful death action. In these states, the view is that the deceased person's beneficiaries merely step into the shoes of the deceased. Therefore, if the deceased waived his claims, the beneficiaries are also deemed to have waived their claims.

In a minority of states, the release does not bar a subsequent wrongful death action. In these states, it appears the view is that the wrongful death action is a new cause of action (a new case) with potential benefits going to the decedent's beneficiaries, not to the decedent's estate.

These nuances with regard to wrongful death actions will require you and your counsel to be sensitive to the state law controlling your liability release. It is essential that you consult with counsel familiar with state law before your liability release is drafted.

Although it will probably be difficult for you and your counsel to place absolute confidence in a liability release, it is possible to gauge trends in various states. On the basis of more recent cases involving liability releases, it seems that the trend is to support the enforcement of such agreements in the context of aircraft use. The key seems to be clearly worded agreements that the courts will not consider "patently offensive."

In any case, although the use of liability releases may be an effective protection from liability exposure, you should not be foolish enough to think of them as a substitute for adequate insurance coverage. Whatever you do, make sure that your first line of defense against liability is a solid insurance policy.

Drafting Considerations

As indicated earlier, the first thing you should do is hire an attorney if you want to use a liability release form for your aircraft activities. The nuances of drafting such an agreement will vary from state to state and simply "filling in the blanks" from a form agreement can be dangerous.

The first thing to consider is the overall structure of the release. One approach that may best serve you is to incorporate (1) a release, (2) a covenant not to sue, (3) an indemnification, and (4) an assumption of risk clause into your agreement.

The release portion of your agreement is the part where you are asking passengers on your aircraft to give up any rights or claims they may have against you if your negligence harms them in any way. Your agreement's covenant not to sue requires that your passengers promise not to sue you (with respect to the operation of your aircraft). The indemnification clause states that if your passenger violates the agreement, he or she may be held liable for any damages that you or your estate may incur, including attorney fees. The threat of this indemnification clause should force any potential plaintiffs to carefully consider their claim before they start a lawsuit. The assumption of risk provision requires your passengers to acknowledge their understanding of the risks involved and that they are voluntarily exposing themselves to the risk. A sample liability release is found in Fig. 10-1.

LIABILITY RELEASE AND COVENANT NOT TO SUE

THIS LIABILITY RELEASE AND COVENANT NOT TO SUE is executed as

of this _____ day of _____, 2003, by and between Airplane Owner ("Owner"), an

individual whose address is _____, and Pete Passenger ("Passenger"),

an individual whose address is _____.

RELEASE AND COVENANT NOT TO SUE

In anticipation of Passenger's flight(s) on Owner's aircraft, Passenger does, for

himself, his heirs, successors, personal representative and assigns, release, acquit,

covenant not sue and forever discharge Owner, his heirs, successors, personal

representative and assigns from any and all actions, causes of action, claims, demands,

damages, costs, loss of services, expenses, compensation and all consequential damage

on account of, or in any way arising out of, any damages, including but not limited to

injury or death, occurring any time during the time Passenger is in, entering or leaving

Owner's aircraft whether caused by the negligence of Owner or otherwise.

INDEMNIFICATION

Passenger agrees to indemnify and save and hold harmless Owner, his heirs,

successors, personal representative and assigns from any and all actions, causes of action,

claims, demands, damages, costs (including reasonable attorney fees), loss of services,

expenses, compensation and all consequential damages caused by Passenger's breach of

this agreement.

Figure 10-1 Liability release and covenant not to sue.

ASSUMPTION OF RISK

Passenger assumes full responsibility for and risk of bodily injury, death or property damage due to Owner's negligence or otherwise, while entering, in or exiting Owner's aircraft.

SCOPE OR RELEASE AND INDEMNITY

Passenger agrees that this RELEASE AND COVENANT NOT TO SUE is intended to be as broad and inclusive as permitted by the law of _____, and that if any portion of it is held invalid, it is agreed that the balance, notwithstanding, continue in full legal force and effect. Passenger has read and voluntarily signs this RELEASE AND COVENANT NOT TO SUE and further agrees and acknowledges that no oral representations, statements or inducements apart from the foregoing written agreement have been made.

IF PASSENGER IS UNDER THE AGE OF 18 YEARS, NOTARIZED SIGNATURE OF PARENT OR GUARDIAN IS REQUIRED.

Figure 10-1   *(Continued)*

IN WITNESS THEREOF, I have hereunto set my hand and seal on the date written below:

Date: _____

PETE PASSENGER:

In the presence of:

_____

_____
Witness 1

_____
Witness 2

Figure 10-1 (*Continued*)

Liability for Federal and State Laws

When aircraft owners think about liability, they often think first about liability to other persons or private entities if their aircraft is involved in an accident. However, aircraft owners should also become familiar with the types of liability they are exposed to from federal and state laws and regulations. This chapter will first explore your responsibilities to comply with federal laws and regulations enforced by the FAA and U.S. Customs Service. Next we'll take a look at state laws affecting aircraft owners.

The FAA and Aircraft Ownership

Many aircraft owners tend to think that pilots are the party ultimately responsible for compliance with the FARs. Although it might be true that pilots are indeed the target of the vast majority of FAA enforcement cases, it does not follow that aircraft owners are free from exposure to this type of regulatory liability. Some of the more common issues that arise for aircraft owners include responsibility for compliance with aircraft operation rules, registration rules, and FAA requests for documentation resulting from ramp checks. We'll take a closer look at these three issues below.

Aircraft operation

Can the FAA hold you liable for the illegal operation of your aircraft when someone else was piloting the aircraft? Many aircraft owners immediately answer this question with a "no." After all, how can an aircraft owner be held liable when he or she had no legal control over the person operating the aircraft?

The FAA responds to this question by directing aircraft owners' attention to the United States Code Article 49 U.S.C. Section 40102(a)(32) which states:

> "operate aircraft" and "operation of aircraft" mean using aircraft for the purpose of air navigation, including—
> (A) the navigation of aircraft; and

(B) causing or authorizing the operation of aircraft with or without the right of legal control of the aircraft.

FAR Section 1.1 similarly defines operate as follows:

Operate, with respect to aircraft, means use, cause to use, or authorize to use aircraft, for the purpose of . . . air navigation including the piloting of aircraft with or without the right of legal control (as owner, lessee, or otherwise).

As indicated in the law and regulations, the keyword for an aircraft owner is "authorize." The various definitions of "operate" clearly encompass you as an aircraft owner if you have authorized someone to use your aircraft. A close look through Part 91 of the FARs will quickly lead you to see that the general operating and flight rules of the FARs consistently refer to the prohibitions against operating an aircraft in a certain manner ("no person may operate an aircraft . . .").

How often does this become a problem for aircraft owners who were not piloting their own aircraft when things went wrong? Thankfully, this is not a frequently encountered issue. Can it become a problem for you? Let's take a look at one case where the FAA went after the owner of an aircraft for operating an aircraft in a careless or reckless manner when the owner was not on board the aircraft. See Case 11-1.

Case 11-1

In the Matter of Ramon C. Fenner
FAA Civil Penalty Decision 1996-17
Decision and Order

This case involves an aircraft owner, Ramon C. Fenner, whose airplane was involved in two near mid-air collisions. In the ensuing Federal Aviation Administration (FAA) investigation, Mr. Fenner refused to identify the person who was at the controls of his airplane when the near mid-air collisions occurred.

The central issue in this case is whether Mr. Fenner can be held responsible for the unsafe actions of the pilot, who had permission to fly the airplane. Chief Administrative Law Judge John J. Mathias held that Mr. Fenner is indeed responsible for the safety violations at issue and assessed a $4,000 civil penalty against Mr. Fenner. Mr. Fenner has appealed. This decision denies Mr. Fenner's appeal and affirms the law judge's assessment of a $4,000 civil penalty for the violations alleged in the complaint.

A Georgia National Guard UH1 helicopter conducting a marijuana search-and-eradication mission was flying in the vicinity of Mr. Fenner's residence one day in August 1992. Mr. Fenner's residence was not a particular object of surveillance; it was simply part of a larger search area.

Mr. Fenner and his wife have a private grass airstrip adjacent to their home. The pilot of the National Guard helicopter saw an airplane taxi down the grass strip. To avoid interfering with the airplane's takeoff, the pilot asked his copilot, who was at the flight controls, to move the National Guard helicopter away from the airstrip. (Tr. 16-17.)

Shortly thereafter, the co-pilot heard the crew chief yell, "Aircraft on the left." (Tr. 59.) Then a Cessna 182 airplane bearing the identification number N8531T, which was later identified as belonging to Mr. Fenner, passed closely underneath the

National Guard helicopter from left rear to right front. Immediately after the Cessna passed under the National Guard helicopter, it pulled its nose up abruptly to the right, and as a result, it nearly collided with the helicopter. The Cessna was so close to the National Guard helicopter that one of the Cessna's wings took up the entire windshield of the helicopter. A Georgia state trooper aboard the National Guard helicopter testified that the Cessna was less than 150 feet away from the helicopter. (Tr. 79.)

The first reaction of the co-pilot of the National Guard helicopter, who was at the controls, was to move the helicopter to the right, but at the pilot's instruction, he quickly turned the nose up and back to the left, narrowly avoiding a collision. The co-pilot testified that if he had not turned to the left, the two aircraft would have collided. (Tr. 59.) The co-pilot further testified that if he had not slowed down the National Guard helicopter, its rotary blades would have hit the Cessna's tail as the Cessna popped up in front of the helicopter. (Tr. 60.)

The Cessna was close enough so that the crew of the National Guard helicopter could see that the pilot of the Cessna was a male between the age of 35 and 40. They could also see that the Cessna's pilot had short brown hair, and that he was wearing a short-sleeved shirt and sunglasses. (Tr. 60, 83.)

A few minutes later, the crew chief yelled, "Oh my God, he is coming back again." The National Guard helicopter stayed straight and level, avoiding any abrupt movements that might create more danger. (Tr. 84.) Again the Cessna passed underneath the National Guard helicopter, though a little further away this time, and again the Cessna pulled up and to the right after passing underneath the National Guard helicopter. (Tr. 18.) The Georgia state trooper took a photograph of the Cessna as it passed underneath the National Guard helicopter. (Tr. 19). This photograph was later admitted into evidence at the hearing. (Complainant's Exhibit 1.) The trooper estimated that the Cessna was approximately 250 feet away from the helicopter at the time he took the photograph. (Tr. 82.)

The crew of the National Guard helicopter then saw the Cessna coming back a third time. (Tr. 81.) At this point, the pilot of the National Guard helicopter decided to take his helicopter and crew out of the situation as quickly as he could. (Tr. 19.) He took over the controls and flew the helicopter to a nearby pasture area, where he descended to a hover below treetop level so that the airplane could not pass under his helicopter or otherwise disturb it again. (*Id.*) The Cessna then left the area. Later, the crew of the National Guard helicopter wondered if the Cessna had tried to "scare them off" because marijuana was in the area. (Tr. 43.)

When the FAA investigated the incident, the Cessna turned out to be owned by Mr. Fenner, although Mr. Fenner was in Fort Lauderdale, Florida at the time of the incident. (Tr. 99; Complainant's Exhibit 5.) Accordingly, the FAA inspector determined that Mr. Fenner was not the pilot of the aircraft. Both Mr. Fenner and his wife, who was the sole witness for Mr. Fenner at the hearing, knew the pilot's identity, but refused to disclose it. (Tr. 99, 136.) When asked about her reasons for declining to disclose the pilot's identity, Mrs. Fenner said that she did not want someone else to go through what she and Mr. Fenner had been through since the time of the incident. (Tr. 136.) Mrs. Fenner testified that the individual piloting the airplane had permission to do so. (Tr. 145-146.)

FAA inspector David Dees testified that if the agency had known the identity of the pilot of the airplane, the agency would have sought to suspend his pilot certificate, rather than seeking to impose a civil penalty on Mr. Fenner. (Tr. 100-101.) The pilot of Mr. Fenner's aircraft has never come forward to identify himself or to explain his actions.

Mr. Fenner filed an appeal brief in which he states that he does not take issue with the law judge's findings concerning whether the alleged near mid-air collision took place. (Appeal Brief at 4.) Instead, Mr. Fenner challenges the law judge's holding that he, Mr. Fenner, is legally responsible for the "suicidal" actions of an unidentified person in an airplane simply because he owned the airplane on the date in question. (*Id.*) According to Mr. Fenner, the law judge's holding violated his right to due process of law under the Fifth Amendment to the U.S. Constitution. Mr. Fenner also argues that the law judge erred, both in stating that the pilot was an employee of Mr. Fenner, and also in determining that Mr. Fenner was liable under agency theory.

It is true, as Mr. Fenner asserts on appeal, that the record does not support the law judge's statement that the pilot of Mr. Fenner's aircraft was in Mr. Fenner's employ. However, it makes no difference. Regardless of whether the pilot was employed by Mr. Fenner or whether agency principles permit a finding of liability, the statutory definition of the term "operate" indicates that Mr. Fenner "operated" the aircraft, because the pilot had permission to use the aircraft. The Federal Aviation Act, as amended, defines "operate aircraft" as follows:

"operate aircraft" and "operation of aircraft" mean using aircraft for the purpose of air navigation, including—

(A) the navigation of aircraft; and

(B) causing or *authorizing the operation of* aircraft with or without the right of legal control of the aircraft.

49 U.S.C. § 40102(a)(32) (emphasis added).

In addition, the Federal Aviation Regulations provide that:

Operate, with respect to aircraft, means use, cause to use, or *authorize to use* aircraft, for the purpose (except as provided in § 91.13 of this chapter) of air navigation including the piloting of aircraft with or without the right of legal control (as owner, lessee, or otherwise).

14 C.F.R. 11.1 (emphasis added).

Neither the statutory nor the regulatory definition of the term "operate" contains any language indicating that in order to have "operated" an aircraft, a person must have authorized the pilot's *particular use* or *manner of use* of the aircraft. Granting Mr. Fenner's appeal would require reading into the statutory and regulatory definitions language that is simply not there.

Mr. Fenner's unsworn assertions, made for the first time on appeal, that the record contains no evidence that he authorized or approved the pilot to fly his airplane, are without merit. Contrary to Mr. Fenner's argument, a preponderance of the reliable, probative, and substantial evidence indicates that the pilot had permission to use the airplane. Complainant is correct that the presumption is that the owner authorized use of the aircraft, absent some evidence to the contrary. Here, Mrs. Fenner testified that the pilot had permission to use the aircraft. Moreover, Mr. Fenner did not take the stand to testify that he had not authorized use of the aircraft, though he was present at the hearing and had the opportunity to do so.

Although Mr. Fenner argues that "it cannot be seriously argued that owners are liable for all infractions committed in their aircraft, a doctrine equivalent to strict liability" (Appeal Brief at 13), that is not what this case is about. While aircraft owners may not be liable for all infractions committed in their aircraft, they can be held liable for infractions committed by a pilot who had permission to use their aircraft. The FAA has a statutory duty to protect the public from dangerous actions.

Moreover, holding aircraft owners responsible in cases like this may help ensure that aircraft owners grant permission to use their aircraft only to persons they know to be responsible.

Mr. Fenner's argument that his right to due process was violated also lacks merit. Although Mr. Fenner asserts that "the adjudicator should not be allowed to make up the rules as the case unfolds," that did not occur here. The statutory and regulatory definitions of the term "operate" are not new. Indeed, the definition of "operate" was part of the Federal Aviation Act when it was first enacted in 1958.

For the foregoing reasons, the law judge's decision, finding that Mr. Fenner violated the regulations alleged in the complaint and imposing a $4,000 civil penalty, is affirmed.

The Fenner case is a clear signal that you may be held liable for the operational violations of pilots who use your aircraft with your permission. The key here is to know who is using your airplane. You may be liable to the FAA for their mistakes.

Registration

As indicated in FAR Section 47.3 (b), you cannot operate an aircraft that you own unless your aircraft is registered to your name. FAR Section 47.5(b) further states that "[a]n aircraft may be registered only by and in the legal name of its owner." Although many aircraft owners sense instinctively that this is the case, they may not be aware of the hazards that await them if they fail to comply with this rule and the law supporting the rule.

The hazard lies in laws that were designed to fight drug trafficking. As you might guess, people involved in transporting illegal drugs might tend to shy away from registering aircraft they owned in their own names. Law enforcement agencies quickly became aware of this problem. Laws were then drafted to make it easier for law enforcement authorities to seize aircraft that were not registered in the names of their owners. Here's how the law is laid out.

Article 49 of the United States Code, Section 1401, lays out the law with respect to aircraft registration. 49 USC Section 1472(b)(1)(C) indicates that it is against the law for anyone "who is the owner of an aircraft eligible for registration, to knowingly and willfully operate, attempt to operate, or permit any other person to operate such aircraft if such aircraft is not registered. . . ." This section is followed by 49 USC Section 1472(b)(3)(A) which warns that "aircraft used in connection with, or in aiding or facilitating a violation of [49 USC section 1472(b)(1)] whether or not a person is charged in connection with such violation may be seized and forfeited by the Drug Enforcement Administration. . . . ."

In plain English, this means that if you know your aircraft is not properly registered to its owner (whether to you or another entity that legally owns the aircraft), you are risking the forfeiture of the aircraft. It does not matter whether you had any involvement in illegal drug trafficking. Case 11-2 is one example of the application of these laws.

Case 11-2

United States of America, Plaintiff, v. One Helicopter, Defendant. No. 90 C 6777
United States District Court for the Northern District of Illinois, Eastern Division
1993 U.S. Dist. LEXIS 7749

OPINION BY: Lundberg, J.
OPINION: MEMORANDUM OPINION AND ORDER

The government has filed an in rem action for forfeiture of a Hughes helicopter,
registration no. N277ST. On February 9, 1993, the court ordered the parties to rebrief
the government's motion for summary judgment on count I of the complaint, which
alleges that the defendant aircraft was operated without being registered to the owner.
The parties were to address the issue of whether claimant Daniel Bonnetts was the
"owner" of the defendant helicopter in August 1990 at the time of the alleged violations
of 49 USC app § 1472(b)(1)(C). Section 1472(b)(1)(C) provides that it is unlawful for
any person "who is the owner of an aircraft eligible for registration under [49 USC app
§ 1401], to knowingly and willfully operate, attempt to operate, or permit any other person to operate such aircraft if such aircraft is not registered under section 1401. . . ."
Section 1472(b)(3)(A) provides that an "aircraft used in connection with, or in aiding
or facilitating, a violation of paragraph (1) [49 USC app § 1472(b)(1)] whether or not
a person is charged in connection with such violation, may be seized and forfeited by
the Drug Enforcement Administration. . . ." 49 USC app § 1472(b)(3)(A).

* * *

Panoff sold the helicopter to . . . Bonnetts on December 15, 1989. Bonnetts paid in
full for the helicopter, received an executed bill of sale from Panoff and accepted
delivery of the helicopter at Midway Airport. Bonnetts claims that shortly after the
sale, he discovered the existence of the Maryland National Bank lien and complained
to Panoff. Panoff does not recall any complaints by Bonnetts concerning the lien until
after Bonnetts ordered a title search from the Aircraft Owners and Pilots Association
("AOPA") on June 8, 1990. Whether or not Bonnetts complained in December of 1989,
the parties are in agreement that Bonnetts complained following the title search in
June of 1990. The title search showed the security agreement but no release.

Panoff contacted the Maryland National Bank and requested the release of the
chattel mortgage/security agreement. On June 12, 1990, Panoff faxed a copy of
the release to Bonnetts. On June 13, 1990, Ann Lennon of the AOPA further advised
Bonnetts via fax that she was filing a copy of the release from the Maryland National
Bank with the FAA. A copy of the release, dated June 12, 1990, was recorded by the
FAA on July 10, 1990. Although the recording problem was resolved, Bonnetts failed
to register the helicopter with the FAA.

Bonnetts subsequently gave Richard Randazzo permission to operate the helicopter on August 3, 1990 and August 6, 1990. On October 16, 1990, DEA agents
seized the helicopter following investigation of complaints of unauthorized landings
and operation of the helicopter on August 3 and 6, 1990 in Lombard, Elmhurst and
Orland Park.

The FAA has prescribed regulations for registering aircraft under 49 USC § 1401.
14 CFR § 47.5(b) provides that "an aircraft may be registered only by and in the legal
name of its owner." The regulations define an owner as "a buyer in possession, a bailee,
or a lessee of an aircraft under a contract of conditional sale, and the assignee of that
person." 14 CFR § 47.5(d). There is no factual dispute that Bonnetts was a "buyer in
possession" of the defendant helicopter. Bonnetts paid in full before taking possession
of the helicopter and received a valid bill of sale from Panoff. Bonnetts further

possessed the helicopter from December of 1989 until it was seized by the DEA in November of 1990. As a "buyer in possession," Bonnetts was required to register the helicopter before permitting the helicopter to be operated. 49 USC app § 1472(b)(1)(C).

Bonnetts implies he no longer was "in possession" of the helicopter once it was placed at a hanger pending resolution of the lien dispute. Bonnetts claims that he and Panoff entered into an "agreement" to keep the helicopter at the hangar while Panoff cured the alleged defect in title—the failure to properly record the release of the Maryland National Bank lien with the FAA.

Bonnetts' deposition testimony belies his contention that any such agreement existed with Panoff. Bonnetts initially testified that he held the helicopter as security while the lien was being resolved. (Bonnetts 3/6/92 Dep at 71-73.) When asked the nature of his "agreement" with Panoff at a subsequent deposition, Bonnetts testified: "I don't remember." (Bonnetts 2/1/93 Dep at 58, 59.) When asked whether he needed Panoff's permission to keep the helicopter at the hangar, Bonnetts testified: "I didn't really feel I had his permission. I didn't think so. I got the helicopter. . . . I don't know if he gave me permission or not." (Bonnetts 2/1/93 Dep at 60.) In any event, Bonnetts' testimony removes any doubt as to the issue of possession: "I got the helicopter."

Even if an agreement did exist between Bonnetts and Panoff, the duration of the "agreement" was limited to curing the alleged defect in title. Once the release of lien was recorded on July 10, 1990, any agreement terminated. The parties do not dispute that Bonnetts was in possession of the helicopter on August 3 and 6, 1990 when he gave Richard Randazzo permission to operate the helicopter.

Claimant's arguments concerning title are irrelevant to this forfeiture proceeding. Congress has proscribed aircraft owners from operating unregistered aircraft. FAA regulations require aircraft to be registered in the owner's name and define "owner" to include a "buyer in possession." An owner can even be a bailee or a lessee under a conditional sales contract. The definition of owner does not include any requirement of title. Because it does not appear that Congress intended to restrict the application of the statute to those with "perfect title," this court will not read state law concepts of title into the definition of ownership.

As the "buyer in possession" or "owner" of the helicopter, Bonnetts was required under 49 USC app § 1472(b)(1)(C) to register the helicopter with the FAA before operating or allowing another person to operate it. Because there is no genuine issue of material fact as to ownership or the other elements of the statute, summary judgment is granted for the government and the defendant helicopter is forfeited to the government pursuant to 49 USC app § 1472(b)(3)(A).

ORDERED: The government's motion for summary judgment on count I of the complaint is granted. The defendant helicopter is ordered to be forfeited to the government pursuant to 49 USC app § 472(b)(3)(A). Judgment is to be entered for the government and against the defendant helicopter on a separate document pursuant to FRCP 58.

Notice that the case never addresses the issue of whether Bonnetts was involved in any form of illegal drug activity. All that was needed to support the forfeiture was the court's opinion that Bonnetts knew the aircraft was not properly registered to him. As indicated in Chaps. 3 and 4, timely filing and recording of your bill of sale and registration application are critical to avoiding legal problems.

Record inspections

Aircraft owners and pilots frequently voice concern over their duties when faced with an FAA examination of their paperwork. This type of examination is commonly referred to as a "ramp check." However, the FAA's request for records does not necessarily have to take place as a random check on an airport ramp. An investigation following an incident or accident, a complaint filed with the FAA, or any number of other things may precipitate a request from the FAA for required documents.

For aircraft owners, the critical regulation on this issue is FAR Section 91.417(c). This regulation states that:

> [t]he owner or operator shall make all maintenance records required to be kept by this section available for inspection by the Administrator or any authorized representative of the National Transportation Safety Board (NTSB). In addition, the owner or operator shall present Form 337 [for special fuel tank installation in the passenger or baggage compartment] . . . for inspection upon request of any law enforcement officer.

As a general rule, a request from the FAA or NTSB to review maintenance records must be reasonable. If you are on the ramp of an airport that is hundreds of miles from your home base, you will not be expected to fly back to your home base and return with the requested records. In any case where maintenance records have been requested, it is perfectly appropriate to make copies of your aircraft maintenance records and mail them to the FAA or NTSB. If originals are requested, you may want to arrange the inspection so that you are present. It is not a good idea to send your originals to the FAA or NTSB. They will usually treat them with caution. However, if they are lost, you will have no recourse for the lost value of your aircraft without original maintenance records.

Since FAR Section 91.417(d) requires that you keep Form 337s for fuel tanks installed in the passenger or baggage compartment of your aircraft, it is expected that you would be able to present this form to a law enforcement officer for inspection "on the ramp." In most cases, your Form 337 will be inspected if you are entering the United States and are checking in with Customs. Make sure you have your Form 337 ready and in good order for this type of spot inspection.

As an aircraft owner, you must also ensure that your aircraft is never operated without a current airworthiness certificate and registration certificate [FAR Section 91.203 (a) and (b)]. The airworthiness certificate should be displayed at or near your aircraft's cabin or cockpit entrance so that it is visible to passengers and/or crew. The only exception to this rule is when you operate your aircraft under the authority of a special flight authorization or "ferry permit." Your special flight authorization must be displayed in a similar manner to the airworthiness certificate.

Beyond maintenance records and required certificates, you may be asked for additional documentation if you are also a pilot. Under FAR Section 61.3(l), the FAA, NTSB, or federal, state, or local law enforcement officers may request an inspection of your pilot or medical certificates. Remember that your pilot and medical certificates are required to be in your physical possession or readily available in your aircraft if you are exercising the privileges of those certificates.

Finally, you may also be asked to present your pilot logbook to the FAA, NTSB, or federal, state, or local law enforcement officers in accordance with FAR Section 61.51(i)(1). Unlike your pilot and medical certificates, you do not have to make your logbook readily available when you are flying your airplane. However, upon reasonable request, you must make it available for inspection. Similar to your aircraft maintenance records, you may wish to consider presenting only copies of your pilot logbooks if the presentation is made by mail. Your pilot logbooks are important documents and you would not want to risk loss or damage.

To protect yourself, it is always prudent to request that the government official or officer requesting your documents present credentials. After all, if they are asking to see your documents, they should at least verify who they are and what authority they have to make their request. It is also important for you to note that the request for inspection of your documents does not have to be for any specific reason. FAA inspectors have a perfect legal right to randomly inspect your aircraft and airman documents without cause. However, they do not have the legal right to touch or enter your aircraft. You are certainly within your legal rights to limit any inspections to documents. If an inspector wants to touch or board your aircraft, you are advised to courteously request that he or she refrain. Perhaps most important, you have no legal obligation to respond to any questions posed by the inspector other than appropriate requests for documents. You should limit the inspection to an inspection of documents. Any information you volunteer beyond your required documents may put you at risk.

Customs

If your aircraft will ever be crossing international borders, there are a myriad of laws and regulations that you must become familiar with. Most of these rules relate directly to the pilot of your aircraft. However, there are some important rules that apply to aircraft owners whether or not they are piloting their own aircraft.

As you might guess, U.S. Customs may seize an aircraft if it is found to have illegal contraband on board. However, an aircraft may be subject to seizure if its owner or the aircraft itself is in any way connected with illegal transportation of goods. United States Code Article 19, Section 1595a is one of the statutes controlling this type of case. The threshold for determining the connection between the aircraft and illegal activity is sometimes a bit fuzzy.

For instance, in *United States v. One Twin Beech Airplane,* 533 F2d 1106 (1976), the U.S. Court of Appeals held that the government seizure of an aircraft was appropriate when the government reasonably believed that the aircraft was being used in marijuana smuggling. The court gave great weight to credible evidence provided by a Mexican informant and the apparent evasive behavior of the aircraft owner when asked about his aircraft.

In other cases, where informants were not found to be so credible and/or traces of drugs found on an aircraft were minuscule, federal courts have not upheld the seizure of aircraft. See *United States v. One Gates Learjet,* 861 F2d 868 (1988) and *United States v. One 1967 Cessna Aircraft,* 454 F Supp 1352 (1978).

In addition to the law cited above, 19 USC Section 1590 refers directly to aviation smuggling. This law is notable because it references circumstances in which an aircraft will be presumed to have been involved in illegal activities. Some of the circumstances prompting the presumption are as follows:

- Operation without lights when operating lights would be required by law or regulation.

- Auxiliary fuel tanks not installed or documented in accordance with applicable law (this presumption has placed several "innocent" aircraft owners in jeopardy of losing their aircraft when they failed to keep proper documentation of fuel tank installation in the form of an FAA Form 337).

- Failure to correctly identify an aircraft by registration number or country of registration.

- Displaying false registration numbers.

- Presence of compartments or equipment designed for smuggling illegal contraband.

- Presence of illegal contraband.

Note that in many of these circumstances, illegal contraband is not required for an aircraft to be seized. It is up to you as an aircraft owner to ensure that you do not find yourself placed in a situation where your aircraft is presumed to have been involved in illegal smuggling activities.

State Laws

As an aircraft owner you are probably very sensitive to the need to comply with pervasive federal laws and regulations relating to aircraft and aircraft operations. However, you should also be aware of state laws that might affect you as an aircraft owner. A discussion of relevant state law issues is found below.

Civil liability

Many states have adopted (in part) the FARs. This means that a violation of the adopted FARs may be a violation of both federal and state law. As a practical matter, aircraft owners may be most affected by laws that track the FAA's definition of "operate." As explained above in the discussion regarding the Fenner case, the federal law and regulations state that you will be deemed to be the operator of an aircraft if you cause or authorize the use of your aircraft. You do not have to be physically present on your aircraft to gain exposure to liability under this type of definition.

The intriguing legal question that follows is whether the liability an aircraft owner may have to the FAA can be expanded to private third persons or entities if a state adopts statutes tracking federal law and regulations. There are not many cases addressing this issue. Moreover, available cases appear to leave the issue unsettled.

In the following older, but often cited case, the Fifth Circuit Court of Appeals was applying Mississippi law. Mississippi law prohibited the operation of an aircraft in a careless or reckless manner that endangered the life or property of another. The law also defined the term operation of aircraft to include causing or authorizing the operation of an aircraft. On the basis of these laws, the court found that the aircraft owner was liable for injuries inflicted on a third person by the owner's aircraft—even though the owner was not physically present at the time of the mishap. For a closer look at the court's reasoning, see Case 11-3.

Case 11-3

Carlton C. Hays, W. Nathan Hays and Blanche Hays Hall v. Ernest Morgan
United States Court of Appeals
221 F.2d 481 (1955)

OPINION BY: Holmes, C. J.
OPINION: This appeal is from a judgment for the appellee, plaintiff below, for personal injuries sustained by him on September 15, 1951, when he was struck and seriously injured by an airplane, owned by appellants and operated by one Fortney, a duly licensed pilot, who was spraying cotton on a farm in Mississippi. The appellee's complaint alleged that his injuries were proximately caused by the negligence of Fortney, who was a partner of appellants or joint adventurer with them at the time of the accident; and that the appellants, as holders of the legal title to the plane, were responsible for Fortney's negligence, regardless of whether or not they and the pilot were partners or joint adventurers. Fortney was not made a party.

It is without dispute that the appellants owned the plane; that Fortney was flying it; that Ernest Morgan was employed by the Holly Grove Farm Company, whose cotton was being sprayed; and that Morgan was acting as flagman for the plane when he was injured. The case, by consent of the parties, proceeded to a trial on the merits before the court, without a jury. The court found from the evidence that Fortney was guilty of negligence in the operation of the plane, which caused the accident; that the appellee was not guilty of any contributory negligence; that the appellee was entitled to recover the sum of $18,806.53; and that judgment should be entered accordingly, which was done.

The Mississippi Code of 1942 contains a chapter on Aeronautics, Section 7533 through Section 7544-17, which with amendments was effective after March 21, 1950. One of the purposes of the act, as set out in Section 7536-01, was to promote aeronautics so that those engaged therein might do so with the least possible restriction consistent with the safety and rights of other persons. Section 7536-26(9) and (10), of the Mississippi Code, as amended, is as follows:

(9) "Operation of aircraft" or "operate aircraft" means the use of aircraft for the purpose of air navigation, and includes the navigation or piloting of aircraft. Any person who causes or authorizes the operation of aircraft, whether with or without the right of legal control (in the capacity of owner, lessee, or otherwise) of the aircraft, shall be deemed to be engaged in the operation of aircraft within the meaning of the statutes of this state. "Person" means any individual, firm, partnership, corporation, company, association, joint stock association, or body politic; and includes any trustee, receiver, assignee, or other similar representative thereof.

It is the evident intent of the statute to protect the public from any negligence and financial irresponsibility of pilots. It does not say that one on the ground must assume the risk of being hit by an airplane, but that the owner who authorizes the use of his airplane shall be deemed to be engaged in the operation thereof within the meaning of the statutes of Mississippi. The law may be compared to the statutes of some states that make the insurer directly liable for the negligence of the driver of an automobile when driving with the owner's consent. See also 49 U.S.C.A. § 401(26), from which said Section 7336-26(9) was copied.

Section 7536-12 of said Code of Mississippi makes it unlawful for any person to operate an aircraft in the air or on the ground or water, in a careless or reckless manner, so as to endanger the life or property of another; and the court in determining whether the operation was careless or reckless must consider also the standards of safe operation prescribed by federal statutes or regulations governing aeronautics. The primary standard for airplanes prescribed by federal statutes is careful operation so as not to endanger the life or property of another, which standard was found by the court below to have been violated by the pilot in this case.

No federal statute or constitutional provision, and no state constitutional provision, is contravened by the state statutes above quoted. Appellants cite 49 U.S.C.A. § 524, which exempts mortgagees, security holders, and lessors under a bona fide lease of thirty days or more, but their arrangement with these pilots does not bring them within the exemption of the act, which shows a federal legislative intention to hold the owner liable as an operator, and to relieve lienors and lessors only under conditions that are not present here. There is no showing or allegation in the present case of a lease for thirty days or more, and the agreed facts are to the contrary.

The appellants reserved the right to direct when and how their crops should be sprayed or dusted. There was no lease of any kind, and the bailment was terminable at the will of the bailors. The statute clearly means that the owner who authorizes the operation of any aircraft, whether with or without the right of legal control thereof, shall be deemed to be engaged in the operation thereof within the meaning of the statutes of Mississippi. The liability of the owner is there just as much as if he were the operator of the aircraft. The owner who authorizes a pilot to use his plane becomes liable for the negligence of the pilot in the operation of the plane. Under the statute, the liability arises out of the facts as a matter of public policy, the essential facts being the defendant's ownership of the aircraft, his authorization of the pilot to operate it, the pilot's negligence in operating it, and the consequent injury to the plaintiff.

Mere disagreement with the judge's findings of fact is not ground for reversal. *Wald v. Eagle Indemnity Co.,* 5 Cir., 178 F.2d 91, 93; *Grace Bros. v. Commissioner of Internal Revenue,* 9 Cir., 173 F.2d 170, 173. Rule 52 of the Federal Rules of Civil Procedure, 28 U.S.C.A. provides that findings of fact shall not be set aside unless clearly erroneous, and that due regard shall be given to the opportunity of the trial court to judge of the credibility of the witnesses. *Graver Tank & Mfg. Co., Inc., v. Linde Air Products Co.,* 336 U.S. 271, 69 S.Ct. 535, 537, 93 L.Ed. 672.

We find no reversible error in the record, and the judgment appealed from is affirmed. Affirmed.

However, not all courts agree with the Court of Appeals' opinion in this matter. An Illinois appellate court faced with similar facts and law comes to a different conclusion as detailed in Case 11-4.

Case 11-4

Bonnie Ferrari, Admr. of the Estate of John Ferrari, Deceased, Plaintiff-Appellant, v. Byerly Aviation, Inc., Defendant-Appellee
  Appellate Court of Illinois, Third District, Second Division
  268 N.E.2d 558 (1971)

OPINION BY: Stouder, J.
OPINION: This is a wrongful death action brought by Bonnie Ferrari, as administrator of the estate of John Ferrari, plaintiff-appellant against Byerly Aviation, Inc., defendant-appellee. Leave to appeal was granted from an interlocutory order denying plaintiff's motion to amend her complaint in accord with Supreme Court Rule 308.

As alleged in the complaint, defendant Byerly Aviation, Inc. as owner, rented an airplane to Charles Burress on January 31, 1965. John Ferrari was a passenger in the plane piloted by Burress when the plane crashed in Peoria County killing both Ferrari and Burress. This action was commenced against Byerly Aviation, Inc. only. The complaint as it stood prior to the proposed amendment alleged that Byerly was negligent in permitting an unskilled pilot to fly its airplane, renting a defective airplane and failing to warn of the defective airplane and failing to inspect its airplane.

In seeking to amend the complaint plaintiff proposed to allege that Byerly Aviation, Inc., "E. Negligently operated an aircraft which it owned through an unskilled pilot." Defendant objected to the proposed amendment because it failed to state a cause of action and subsequently the court denied the motion for leave to make the amendment.

The court made the requisite finding concerning "substantial difference of opinion" as required by Supreme Court Rule 308 and indicated the question to be "does the language of this section impute the actions of the operator of an aircraft to the person who authorizes the operation of the aircraft, for purposes of civil liability to a passenger therein, contrary to the common law of this state?"

So far as this appeal is concerned there is no issue relating to any negligence of Byerly and there is no claim that Burress was an agent or employee of Byerly. As quoted above the proposed amendment (Subsection E) to the complaint is somewhat vague in concept but both parties have assumed and concede it a sufficient basis for the question of law certified by the trial court.

It is plaintiff's theory that Byerly as owner of the plane, is liable because it rented the plane to Burress who negligently operated the plane causing Ferrari's death. This claim is based primarily on ch. 15 1/2, par. 22.11, Ill. Rev. Stat. 1965 (the statute referred to in the legal question certified) which provides, "'Operation of aircraft' or 'operate aircraft' means the use of aircraft for the purpose of air navigation, and includes the navigation or piloting of aircraft. Any person who causes or authorizes the operation of aircraft, whether with or without the right of legal control (in the capacity of owner, lessee, or otherwise) of the aircraft, shall be deemed to be engaged in the operation of aircraft within the meaning of the statutes of this State." Par. 22.1 *et seq.* of ch. 15 1/2, commonly referred to as the Aeronautics Act, was adopted in Illinois in 1945 and had no historical antecedent. The Act was substantially copied from the Civil Aeronautics Act of 1938, now known as the Federal Aviation Act, USCA, Title 49, Section 1301.

There is general agreement by both parties to this appeal that the bailor of a chattel is at common law not liable for the negligence of his bailee merely because of the relationship. 4 I.L.P., Bailments, Section 16, see also 2 C.J.S., Aerial Navigation, Section 26, discussing such rule as applied to the bailment of airplanes. (See also *Johnson v. Central Aviation Corp.*, Cal.App.2d 229 P.2d 114.) From this mutually accepted premise the arguments of the parties diverge with plaintiff arguing that par. 22.11 quoted above changes the common law and the defendant insisting that it does not. The gist of the controversy then is the meaning of par. 22.11.

Both parties have directed our attention to the several rules relating to construction and interpretation of statutes. Furthermore since there are no Illinois cases related to the problem at hand, each party has referred to cases from other jurisdictions where the problem has been involved to some extent.

Aviation and aerial navigation by their nature suggest problems involving Federal-State relations. Congress early recognized the need for a Federal policy regarding control of the airspace and provided for a regulatory system. Such regulatory system has not pre-empted the field of aeronautical regulations (*Southwest Aviation Inc. v. Hurd,* Tenn.355 S.W.2d 436) but the paramount Federal authority contemplates that state action may not be inconsistent with Federal standards. To promote consistency at least in the basic statutory authorization, some states, including Illinois, substantially copied the Federal statutes. In adopting substantially similar statutes the State legislature could well have believed that inconsistency could thus be avoided and the tensions between the dual sovereignties in an area where uniformity was desirable could best be accommodated by relying on the Federal enactment. At first glance the adoption of substantially similar statutory provisions would be an eminently satisfactory method of accomplishing the purpose sought. The laudable purpose has however, as applied to this type of case, tended to obscure or make more difficult the determination of State legislative intent.

We are unable to agree with plaintiff that the plain meaning of par. 22.11 establishes or creates civil liability in the owner of an airplane for the negligent operation thereof by another. Par. 22.11 quoted above, is one of the definition sections of the Illinois Aeronautics Act as is the provision from which it is copied in the Federal Act. The language of the definition does not in our opinion by its plain meaning, indicate any concern with the civil liability of any party and in particular does not purport to establish some special or different relationship between an airplane owner and a pilot. Liability for acts or conduct of another is neither a new nor novel concept. Language to describe such a well recognized concept is readily available and could have been easily employed if such concept was intended. (*Haskin v. Northeast Airways, Inc.,* Minn. Supreme, 123 N.W.2d 81). We believe that par. 22.11 describes in general terms those parties included within or affected by the other provisions of the Aeronautics Act or the regulations promulgated under the authority thereof. See par. 22.1, ch. 15 1/2, Ill. Rev. Stat. 1965 (the initial Section of the Illinois Aeronautics Act).

Nor do we believe that the statutory provision can or ought to be given the meaning ascribed by plaintiff because of the general policy of the Act, "to promote safety." The general purpose of an Act of the legislature is appropriately resorted to where specific provisions thereof may have some uncertainty as applied to the facts. It is still the primary function of the legislature to implement its specific purpose. In the case at bar we find neither any uncertainty in the provisions under consideration nor any support for the assertion that the purpose of the Act will be defeated if liability is not imposed.

Plaintiff argues that its interpretation of par. 22.11 is supported by the weight of authority of cases from other jurisdictions and cites *Hoebee v. Howe,* 97 A.2d 223, *Hayes v. Morgan,* 221 F.2d 481 and *LaMaster v. Snodgrass,* 85 N.W.2d 622, as persuasive on the issue. Such authorities are according to defendant, of little persuasive authority and have been because of inadequate reasoning, distinguished or not followed in the recent cases of *Rosdail v. Western Aviation, Inc.* 297 F. Supp. 681 and *Rogers v. Ray Gardner Flying Service, Inc.* U.S. Court of Appeal, Fifth Circuit, No. 28666, November, 1970.

The three cases cited by plaintiff (*Hoebee, Hayes* and *LaMaster*) each involve states which like Illinois substantially copied the Federal Act including the definition which is the subject of this controversy. There is language in the *Hoebee* and *Hayes* cases indicating that the state enactment of the legislation modified the common law bailment rule. As pointed out in *Rosdail, supra*, an examination of the cases reveals that in each case application of common law rules justified liability and consequently the admixture of reasoning relating to both common law liability and the meaning of the statute reduced the persuasiveness of the holdings. For example in *Hoebee* the Court concludes that an airplane was a dangerous instrumentaility and in *Hayes* that the owner maintained control and direction of the pilot both of which circumstances would have been a proper basis for liability at common law. *LaMaster v. Snodgrass*, 85 N.W.2d 622 (Iowa), quotes from and applies *Hoebee* and *Hayes* without considering that such conclusions went substantially beyond the facts and necessary principles of the cases. Furthermore each of the three cases assumes to interpret the meaning of the Federal law based on erroneous interpretation of an amendment thereto and without any regard for the basic dual Federal-State problem or the preemptive effect of the interpretation so reached.

As observed in *Rogers v. Ray Gardner Flying Serv., Inc., supra* (approving *Rosdail v. Western Aviation, Inc., supra*) no intent can be discerned from the language of the Federal law to deal with the civil liability of an airplane owner as bailor. The court dismisses the authority of the *Hoebee, Hayes* and *LaMaster's* cases and suggests that to the extent that such cases relied on the supposed meaning of the Federal Statute such interpretation was at least by implication erroneous.

Because of the pre-emption it well might be that the certainty in ascertaining legislative intention regarding Federal acts may be different from the same certainty when it is a State statute which is questioned. It is our conviction that such provisions ought to be accorded the same meaning. As is pointed out in the *Rogers* case if civil liability may be imported from the definition then such a rule of liability would be in effect in each of the states contrary to the generally accepted premise that the states were free in this regard to exercise discretion in this area. We are impelled to believe that the Illinois state legislature had no particular intent to deal with civil liability of an owner by adopting the language of the Federal Act. Consequently there is no legislative basis for plaintiff's theory of recovery.

For the foregoing reasons the judgment of the Circuit Court of Peoria County is affirmed.

Judgment affirmed.

The Illinois court decision seems to turn, at least in part, on a reading of the state legislature's intent. The Illinois court does not believe that the state legislature intended the language of the statutes to impart civil liability on aircraft owners.

Regardless of which side of this legal debate may prevail, the lesson is still clear and bears repeating. Know who you are authorizing to use your aircraft. Their negligence may cost you.

Criminal liability and aircraft forfeiture

From time to time states have also attempted to seize aircraft in furtherance of their battle against illegal drug trafficking. Obviously, a state should have the ability to exercise its police power by seizing an aircraft if its owner has knowingly used it in furtherance of illegal activity. But what happens when the state law permits the law enforcement agencies to seize an aircraft simply because it fits the profile of an aircraft that could be used in drug trafficking? This was the question facing the Supreme Court of Florida after county law enforcement officers seized a Piper Navajo on the grounds that it had extra fuel tanks that allegedly did not conform to FAA regulations. The court's analysis of this issue follows in Case 11-5.

Case 11-5

In Re: Forfeiture of: 1969 Piper Navajo, Model PA-31-310, S/N-31-395, U.S. Registration N1717G
Supreme Court of Florida
592 So. 2d 233 (1992)

OPINION BY: Barrett, J.

OPINION: we have for review In re Forfeiture of 1969 Piper Navajo, 570 So.2d 1357 (Fla. 4th DCA 1990), in which the district court declared section 330.40 of the Florida Statutes (1987) unconstitutional. We affirm the decision of the district court.

On February 8, 1988, Broward County sheriff's deputies seized a 1969 Piper Navajo aircraft parked at a private field. The sheriff alleged that the aircraft was equipped with extra fuel tanks which did not conform to Federal Aviation Administration (FAA) regulations, and which were not approved by the FAA, and therefore the aircraft was subject to forfeiture pursuant to section 330.40 of the Florida Statutes (1987). A petition for forfeiture of the aircraft was timely filed pursuant to sections 330.40 and 923.703(1) of the Florida Statutes (1987). Appellee, Anacaola Trading (Anacaola), the owner of the aircraft, sought dismissal of the petition on the basis that, among other reasons, section 330.40 violated due process of law.

The trial court dismissed the petition for forfeiture and found the forfeiture provision contained in section 330.40 to be unconstitutional. The trial court assumed that the statute's main purpose was the seizure and forfeiture of aircraft employed in illegal drug trafficking. The trial court reasoned, however, that "it is perfectly plausible for an airplane to be equipped with extra fuel tanks for purposes other than smuggling. Therefore, the statute brings within its ambit otherwise innocent activities." In re Forfeiture of 1969 Piper Navajo, 570 So.2d at 1359. Accordingly, the court found that section 330.40, "as it relates to the 'Florida Contraband Forfeiture Act,'" was not rationally related to any legislative objective and thus violated substantive due process of law. The district court affirmed and adopted the trial judge's order in substantial part.

Section 330.40, Florida Statutes (1989), provides in full:

In the interests of the public welfare, it is unlawful for any person, firm, corporation, or association to *install, maintain,* or *possess* any aircraft which has been equipped with, or had installed in its wings or fuselage, fuel tanks, bladders, drums, or other containers which will hold fuel if such fuel tanks, bladders, drums, or other containers do not conform to federal aviation regulations or have not been approved by the Federal Aviation Administration by inspection or special permit. This provision also includes any pipes, hoses, or auxiliary pumps which when present in the aircraft could be used to introduce fuel into the primary fuel system of the aircraft from such tanks, bladders, drums, or containers. Any person who violates any provision of this section is guilty of a felony of the third degree, punishable as provided in s. 775.082, s. 775.083, or s. 775.084. Any aircraft in violation of this section shall be considered contraband, and said aircraft may be seized as contraband by a law enforcement agency and shall be subject to forfeiture pursuant to ss. 932.701-932.704. (Emphasis added.)

Thus, under section 330.40, originally enacted in 1983, the possession of Anacaola's aircraft, if it had unapproved fuel tanks, was a felony of the third degree. § 330.40, Fla. Stat. (1983). In 1987, the legislature amended section 330.40 to expressly authorize forfeiture of such nonconforming aircraft as contraband per se under the Florida Contraband Forfeiture Act. Ch. 87-243, § 22, Laws of Fla.

In considering whether a statute violates substantive due process, the basic test is whether the state can justify the infringement of its legislative activity upon personal rights and liberties. The general rule is that when the legislature enacts penal statutes, such as section 330.40, under the authority of the state's police power, the legislature's power is confined to those acts which reasonably may be construed as expedient for protection of the public health, safety, and welfare. Art. I, § 9, Fla. Const.; *State v. Saiez,* 489 So.2d 1125, 1127 (Fla. 1986); see *Nebbia v. New York,* 291 U.S. 502, 523, 78 L. Ed. 940 , 54 S. Ct. 505 (1934); *Hamilton v. State,* 366 So.2d 8, 10 (Fla. 1978); *Carroll v. State,* 361 So.2d 144, 146 (Fla. 1978); *Newman v. Carson,* 280 So.2d 426, 428 (Fla. 1973); *State v. Leone,* 118 So.2d 781, 784 (Fla. 1960).

In addition, due process requires that the law shall not be unreasonable, arbitrary, or capricious, and therefore courts must determine that the means selected by the legislature bear a reasonable and substantial relation to the purpose sought to be attained. Art. I, § 9, Fla. Const.; *Saiez,* 489 So.2d at 1128; see *Nebbia,* 291 U.S. at 510-11; *Hamilton,* 366 So.2d at 10; *Carroll,* 361 So.2d at 146; *Newman,* 280 So.2d at 429; *Leone,* 118 So.2d at 784-785.

Anacaola argues that the trial judge correctly found that there was no reasonable relation between drug smuggling and expanded fuel tanks. If the statute simply prohibited the possession of extra fuel capacity, Anacaola's position would have merit. However, the legislature has not prohibited merely the possession of extra fuel capacity. Section 330.40 makes it "unlawful for any person . . . to *install, maintain,* or *possess* any aircraft which has been equipped with . . . fuel tanks . . . which . . . do not conform to federal aviation regulations." § 330.40, Fla. Stat. (1989) (emphasis added). As noted above, the state, through the exercise of its police power, has the authority to pass laws to preserve the public safety. E.g. *Saiez.* It is primarily the responsibility of the FAA, as a division of the Department of Transportation, to promulgate rules, regulations, and standards to promote flight safety in air commerce. See 49 U.S.C. § 1421(a) (1988); *Landy v. Federal Aviation Admin.,* 635 F.2d 143, 148 (2d Cir. 1980), *cert. denied,* 464 U.S. 895, 78 L. Ed. 2d 232 , 104 S. Ct. 243 (1983); 14 C.F.R. §§ 1-199 (1991). Thus, we find that assuring conformity with FAA regulations for the purpose of public safety is within the legislative province.

Having decided that the state can infringe upon an individual's property rights by regulating for the public safety, we must then decide whether the means chosen by the legislature (forfeiture) are narrowly tailored to achieve the state's objective (aircraft safety) through the least restrictive alternative. See *Department of Law Enforcement v. Real Property,* 588 So. 2d 957, 1991 Fla. LEXIS 1256, 16 Fla. Law W. S 497, S499 (Fla. 1991).

* * *

In this case the method chosen by the legislature to prohibit operation of aircraft with nonconforming fuel tanks is not sufficiently narrowly tailored to the objective of flight safety in air commerce to survive constitutional scrutiny. This is particularly so because property rights are protected by a number of provisions in the Florida Constitution. Article I, section 2 provides that "all natural persons are equal before the law and have inalienable rights, among which are the right . . . to acquire, possess and protect property. . . ." Article I, section 9 provides that "no person shall be deprived of life, liberty or property without due process of law. . . ." Article I, section 23 provides that "every natural person has the right to be let alone and free from governmental intrusion into his private life. . . ." As we have previously noted, "these property rights are woven into the fabric of Florida history." *Shriners Hosp. v. Zrillic,* 563 So.2d 64, 67 (Fla. 1990). The main thrust of these protections is that, so long as the public welfare is protected, every person in Florida enjoys the right to possess property free from unreasonable government interference. See *id.*; *Town of Bay Harbor Islands v. Schlapik,* 57 So.2d 855, 857 (Fla. 1952).

While the state undoubtedly has a substantial interest in promoting air safety, the legislature does not have the authority to confiscate airplanes simply because they possess additional fuel capacity. The central concern of substantive due process is to limit the means employed by the state to the least restrictive way of achieving its permissible ends.

We note that in authorizing forfeiture under section 330.40, the legislature has even exceeded the awesome authority exercised in the Florida Contraband Forfeiture Act. On its face, section 330.40 automatically converts every aircraft with nonconforming fuel tanks, whether airworthy or not, and whether involved in criminal activity or not, into contraband subject to forfeiture under the forfeiture act. Here, as we have said, Anacaola's aircraft was parked; the sheriff did not allege that the airplane had been, was being, or was about to be used in the commission of a felony as required for forfeiture under the forfeiture act. As the trial court noted, the legislature has available many other less restrictive means to assure compliance with FAA regulations. And in the event that an aircraft is being used as a criminal instrumentality, the forfeiture act already provides for forfeiture of such aircraft.

Accordingly, we hold the forfeiture provision contained in section 330.40, Florida Statutes (1987), unconstitutional under due process of law as guaranteed by article I, section 9 of the Florida Constitution. The decision of the district court is affirmed.

It is so ordered.

As indicated by the court, while the legislature may have had good intentions when it passed the forfeiture law, the law was unconstitutional. To date, there have not been any other notable cases where similar laws were passed and tested. However, in the current climate of sensitivity to security issues, aircraft owners will have to be extra vigilant to ensure compliance with state laws that impact their aircraft.

Registration

Is your aircraft properly registered? Even if you have properly registered your aircraft with the FAA, you may still have to go one step further and register with appropriate state authorities. Many states have imposed a registration requirement. Often, one of the motivations behind these state registration laws is revenue enhancement. The states (at press time) requiring registration fees are listed along with a contact number for more detailed information in Table 11-1.

TABLE 11-1   States Requiring
Aircraft Registration

State	Contact number
Arizona	602/294-9144
Connecticut	—
Hawaii	808/838-8701
Idaho	208/334-8775
Illinois	217/785-8514
Indiana	—
Iowa	515/237-3138
Maine	207/287-2577
Massachusetts	617/973-8883
Michigan	517/335-9283
Minnesota	612/296-8046
Mississippi	601/923-7131
Missouri	—
Montana	406/444-2506
New Hampshire	603/271-1094
New Mexico	505/827-1525
North Dakota	701/328-9651
Ohio	614/793-1233
Oklahoma	405/521-3271
Oregon	503/378-4880
Rhode Island	401/737-4000
South Dakota	605/773-3862
Utah	800/662-4335
Virginia	804/236-3637
Washington	206/764-4131
Wisconsin	608/267-2030

There are a variety of formulas used by states to impose registration fees. In some cases the registration fee is a flat fee regardless of the type or size of your aircraft. In other cases, the fee is based on the gross weight or ramp weight of your aircraft. Sometimes registration and payment of registration fees will exempt you from personal property taxes on your aircraft. The best advice is to do some checking with the state that your aircraft is based in to ensure that you are properly complying with the law.

Insurance

Some states have passed laws requiring aircraft owners to have a liability insurance policy in force before an aircraft is operated. The laws generally pertain to an aircraft that is based or hangared within the state. The states requiring the coverage also mandate certain minimum levels of coverage. Currently, there are six states requiring liability insurance on noncommercial aircraft. The six states are listed below:

- California
- Maryland
- New Hampshire
- Minnesota
- South Carolina
- Virginia

This list is subject to change so you should check into this issue whenever you purchase an aircraft. Of course, whether the state you are in requires insurance coverage or not, it is critical that you obtain adequate liability and hull insurance to protect yourself, your family, and your community from the risk of a mishap involving your aircraft.

Sales, use, and personal property taxes

Perhaps the single most common liability imposed by states on aircraft owners is the tax liability for sales, use, and/or personal property taxes. These issues are discussed in detail in Chap. 17.

Aircraft Cost Sharing

12

Aircraft Leasing

One common approach to sharing aircraft expenses is to lease or rent your aircraft to other users. In the context of aircraft, a lease is a contract in which the aircraft owner grants another person the right to possession and use of his/her/its aircraft for a specified period of time in return for periodic payments (better known as "rent"). The obvious and most important difference between a sale and a lease is that a sale conveys title to an aircraft—a lease does not. For this reason, many aircraft owners prefer a lease to other cost-sharing devices because they will not have to share ownership rights with others.

You should also keep in mind that there might be tax consequences to your aircraft leasing activities. Chapter 16 provides a discussion of the tax issues that you are likely to encounter if you lease your aircraft.

Lease Agreements

The key to a successful aircraft leasing arrangement is a professionally drafted lease agreement. The accompanying CD materials related to this chapter include two sample leases. CD materials related to the tax issues addressed in Chap. 16 contain a tax-sensitive agreement that you and your tax counsel may consider if you are planning on making your aircraft available on an hourly basis through an FBO or flight school. Every agreement will be different, but these sample agreements include many of the points that are addressed in most aircraft leases. You may wish to refer to these sample agreements as we reference certain points in this chapter.

As far as the law is concerned, a lease is a lease. However, for purposes of our discussion, it may be most practical to break down leases of lighter, general aviation aircraft into three categories:

- Nonexclusive lease
- Exclusive lease
- "Leasebacks" with flight schools and FBOs

The nonexclusive lease is most commonly used when you wish to permit several other users (typically no more than one or two) to use your aircraft on an as-needed basis. Often, this type of lease also refers to rentals by the hour rather than by the week, month, or year. The duration of the lease may be short term or long term.

Some aircraft owners make the business decision that it would be best to lease their aircraft exclusively to one user for a specified period of time. As a general rule, we call these types of leases "exclusive" leases. If you decide to go this route, it means that you may not be able to make use of your aircraft during the term of the lease. In many cases, this type of arrangement is necessary where the lessee (the person renting the airplane) is a commercial or corporate aircraft operator. In those instances, the renter may need the use of your aircraft on a moment's notice. A lease giving the renter exclusive use of the aircraft will generally fit the bill. This type of lease is also quite practical when someone is considering a purchase of your aircraft, but they'd first like to try it out for a period of time. Sometimes the exclusive aircraft lease will also include an option to purchase.

The "leaseback" is the type of lease featured in Chap. 16's discussion of tax issues. In the typical leaseback, the owner is offering the aircraft to the qualified flying public for short-term rentals. Just as in the nonexclusive lease discussed above, this type of lease usually calls for rental rates based on hours of use rather than days, months, or years.

Regardless of the type of lease, the ingredients for a solid lease have common threads. Let's take a look at some of the important features that you will want to include in your aircraft lease.

Get it in writing

Does your aircraft lease have to be reduced to a written agreement? The answer to this question is generally "no." Should my aircraft lease be reduced to a written agreement? The answer to this is almost always a resounding "yes."

The questions that are left unanswered by the usual informal, verbal lease carry the seeds for potential disputes in the future. For instance, who is responsible for insurance on the aircraft? Who can use the aircraft and when can they use it? These are issues that have plagued many aircraft owners who cavalierly ignored the need for a professionally drafted (or reviewed) aircraft lease agreement.

Identification of parties

As always, you will want to start the agreement by identifying yourself and the aircraft lessee involved in the agreement. Remember to include permanent mailing addresses. It is also important to note if any of the parties are separate legal entities such as LLCs or corporations. The state where any LLC or corporation is chartered should be clearly indicated in your written aircraft lease agreement.

Identification of the aircraft

Next, you'll have to carefully identify the aircraft involved in your lease. The following items should be sufficient to make a tight enough identification:

- Make
- Model
- Year
- Serial number
- FAA registration number

Lease payments

Here's your opportunity to spell out how much rent you will be charging and when you expect it to be paid. In exclusive leases, you may prefer a monthly charge on the basis of hours of usage with a minimum dollar amount per month. Remember that if you entered into an exclusive lease, you could be stung if the lessee uses your aircraft only sporadically. A provision for minimum lease payments will help prevent any such problems.

For nonexclusive leases and agreements to rent through FBOs and flight schools, the usual approach is an hourly rate. Here you should be sensitive to the fact that fuel prices and other costs may rapidly change. Another problem may be lack of demand or a sudden increase in demand for your aircraft. You may have to include some flexibility in the agreement for the possibility of increases or decreases in rate to meet market and business needs.

How do you want to be paid? Most aircraft owners still prefer the old-fashioned check in the mail. However, you and/or your lessee may prefer electronic payments.

Another thing to consider is whether you should collect a security deposit. A security deposit will provide you with some protection in case your lessee fails to meet the obligations of the lease or damages your aircraft. Here you will need some guidance from local counsel to ensure that there are no state or local laws restricting the amount of the security deposit. Some states have such restrictions for security deposits on real and/or personal property. You'll also need to know whether the law requires security deposits to be held in separate trust accounts and whether these accounts must bear interest.

Lease term

This may be one of the most important provisions in your aircraft lease. Depending on your circumstances, you may want a longer or shorter lease period. Remember that locking the lessee into a longer-term lease also locks you into the same lease. If for some reason you decide to sell your aircraft during the course of a long lease term, you may have to wait it out until the end of a long-term lease unless you can get the lessee to agree to modify the lease—and that may cost you.

One way to gain maximum flexibility in this area is to select a reasonable length of time for the lease term—let's say 1 year. That will require both parties to stick to the terms of the written lease for that period of time. However, you can also insert a provision that permits either party to terminate the agreement with a certain number of days'/months' written notice to the other. This will allow either party an "out" in case circumstances change. It may also serve to avert unpleasant conflicts by allowing a prompt termination. The negative to this approach is that it can allow your lease to terminate before you really wanted it to be terminated. This is an issue where you should carefully consult with your counsel to determine what is best for you.

Another issue that often comes up is the option to renew. There is nothing wrong with giving your lessee the option to renew. However, you have to be very careful to spell out the new terms of the next lease term. For instance, will you still want the same monthly rental? It's likely you will have to hike the rate a bit just to keep up with inflation. Also, how much notice will you want for the renewal option to be exercised? Will you want the renewal option to continue term after term or only for the term immediately after the original term of your aircraft lease? These are all questions you should examine carefully before you agree to renewal options.

Title to aircraft

It is prudent to include a section in your lease agreement confirming that you are the only party who will hold title in the aircraft. If nothing else, this confirms that you are only transferring possession and use according to the lease. You are not transferring title. Pursuant to FAR Section 49.31, the lessee may have the ability to record the lease. However, recording a lease is similar to recording an encumbrance against an aircraft; it is not the same as acquiring title to the aircraft.

Another drafting consideration may be the need for a provision in which your lessee promises not to pledge, loan, mortgage, or permit any liens to attach to your aircraft. You don't want any unexpected surprises when you are preparing to sell your aircraft and discover that your lessee has engaged in transactions that encumbered your title.

Restrictions on aircraft use

You should consider drafting express restrictions on the lessee's use of your aircraft. Depending on the business circumstances, some of the restrictions addressed in your lease may include:

- *Geographic or location restrictions.* You may want to limit your lessee to flights within a particular region or within the confines of the region outlined in your insurance policy.
- *FAA restrictions.* In order to protect yourself, you may consider drafting language in your lease that adopts the FARs or certain applicable provisions

of the FARs as operating rules. Violations of these rules may be drafted so they will automatically trigger the termination of the lease at your discretion.

- *Purpose.* If you want your aircraft used for business and pleasure flying only, it should be specified in the lease. Often exclusive leases may allow for commercial operations (typically Part 135 operations). In any case, it is a good idea to lay out your understanding of what your aircraft can and cannot be used for during the lease term.

Pilot restrictions

Your lease should contain some reference to pilot restrictions. In circumstances where you are responsible for carrying aircraft insurance, you will want any pilots operating your aircraft to at least meet the requirements of the applicable insurance policy. In other situations, you may want to specify that the pilots must at least meet the lessee's insurance coverage requirements. Regardless of the circumstances, these restrictions can create important protections for your aircraft and you.

Insurance

Who will be responsible for insurance on your aircraft? As a general rule, if you are leasing your aircraft on an exclusive basis, the lessee will be responsible for providing for insurance coverage. In order to protect yourself, your lease agreement should provide that you'll be named as an additional insured on the policy purchased by the lessee. Additionally, you will want to require that you receive a certificate of insurance evidencing your coverage under the policy. The certificate of insurance should also promise that you will be notified a reasonable amount of time before the insurance policy is canceled. This may be your only protection if the lessee fails to make payments or simply lets the policy lapse and the policy is canceled.

If you are leasing your aircraft on a nonexclusive basis, you will probably be the party purchasing coverage for the aircraft. However, the lessees should be notified through the lease that in case of damage incurred during their operation of the aircraft, they will be liable for any deductibles. You may also wish to notify the lessees that if they caused damage to your aircraft, your insurance company may pursue them for damages to your aircraft. With a nonexclusive lease, you should also check with your insurance provider to make sure that the lease arrangement will not require you to purchase a commercial policy. Some insurers exclude nonexclusive leases to approved pilots from being categorized as commercial activities or operations, but others may not.

For a lease with a flight school or FBO, you may try to buy your own policy. However, this may not be easy or cheap. You'll probably need a commercial policy for the type of activities the aircraft will be engaged in during the lease. It may be most cost effective to work with the flight school or FBO in adding your aircraft to their policy. If so, you should require proof of insurance and notice of cancellation to ensure that insurance coverage is not terminated without your knowledge.

Maintenance and repair of aircraft

Your lease should clearly spell out each party's responsibility for aircraft maintenance during the term of the lease. Again, the question of who is responsible may be largely dependent on the type of lease you are entering into.

If your lease is an exclusive lease, it is likely that you will want to turn over all maintenance responsibilities to the lessee. In some cases, the exclusive lessee may have a preferred maintenance facility or its own facility. Even if you turn over responsibility for maintenance to your exclusive lessee, you may wish to restrict routine maintenance to a particular maintenance facility or facilities that you have had an opportunity to investigate.

In the nonexclusive lease situation, you will most likely be the person with full responsibility for maintenance. Obviously, maintenance costs will be built into your rental rates and should allow you to recoup your costs because of extra aircraft usage. Your lease may provide that the lessee is authorized to proceed with emergency repairs by FAA certified personnel or facilities when operated by the lessee away from your home base. Even in this situation, you may want to require that you be notified prior to the authorization of any such repairs. To be reasonable, you can set a dollar limit indicating that repairs above a certain dollar limit will require your consent.

Aircraft leases with flight schools and FBOs should also contain a provision relating to maintenance responsibilities. Sometimes the question of who is responsible will be tax driven (see Chap. 16). If you want to actively participate in your leasing activity, you should take on maintenance responsibilities and have any charges billed directly to you. If you prefer to let the FBO or flight school handle these matters, your lease should clearly turn over maintenance and repair responsibilities to your lessee. Again, even if you turn over authority, it might be prudent to specify facilities or personnel permitted to work on your aircraft. You may also want to retain authority to make decisions on maintenance or repairs exceeding a specified dollar ceiling.

Indemnification

It is always prudent to require that the lessee indemnify you against liability you may incur as a result of a breach of the lease by the lessee and/or lessee's operation or maintenance of your aircraft. Within this provision, you should also consider requiring the lessee to reimburse you for any reasonable attorney fees you incur defending yourself against liability caused by the lessee or successfully pursuing a breach of contract claim against the lessee.

Liquidated damages

A liquidated damages provision can simplify the issue of damages in case your lessee breaches the lease agreement. Liquidated damages are damages that the parties to the lease agree to ahead of time. In the case of a lease, the formula can be as simple as the number of months of unpaid rent multiplied by the rent per month plus penalties and interest. The parties can calculate the liquidated damage amount in any way they deem fit. However, the provision for liquidated damage will not be enforceable if it appears to be a penalty rather than a recovery of economic damages.

By way of example, let's say that your lease called for rent of $2,000.00 per month. A liquidated damage provision calling for $2,000.00 per month plus $200.00 for each month your lessee is more than 5 days late with rent will most likely be enforceable. However, if the liquidated damages provision called for $4,000 per month in damages for each month the rent is more than 5 days late, it might be considered a penalty and the courts may refuse to enforce the provision.

Rights in event of breach

In order to clarify your rights in case of a breach, you can expressly state in the agreement what actions you may be entitled to take against the lessee in the case of a breach by your lessee. As indicated above, you may wish to provide for liquidated damages. Additionally, it might be prudent to exclude the lessee from use of your aircraft if the lessee fails to pay rent or breaches the agreement in any other way.

If you are dealing with a nonexclusive aircraft lease or a lease to an FBO or flight school, you can require that keys be returned. If you are leasing your aircraft on an exclusive basis, you may have to spell out your rights to repossess your aircraft upon a breach by the lessee.

Although all of these remedies may be automatically triggered under the law in the case of lessee breach, it may be wise to spell out certain rights you can take immediately to protect your interest. It is important not to limit your rights, so language indicating that you are not limited to the remedies noted in your agreement is important.

Option to purchase

Provisions for purchase options are most likely applicable to exclusive leases only. Sometimes the exclusive lessee really wants a chance to use your aircraft on a longer-term trial basis before deciding whether he or she wishes to purchase the aircraft.

If you decide to include an option to purchase in your lease, you will have a few issues that should be carefully considered. The first thing you'll have to address is the option price. In some cases, you might prefer to leave the price open and merely give the lessee a "first right of refusal." In other cases, your negotiations with the lessee may lead you to a more specific pricing scheme. If you are getting specific with price, you may also negotiate an offset or application of all or part of the previous rent paid.

Another item you'll have to consider is when you want the lessee to provide you with notice of intent to exercise the option to purchase. This should be specified in your lease agreement in order to avoid any confusion. You don't want to expend a lot of time, money, and effort to line up potential purchasers for your aircraft only to learn that your lessee is seeking to exercise the option to purchase.

Assignment/subletting

For your protection, the assignment of your lease or subletting of your aircraft should be either prohibited or subject to your express written authorization. You do not want your lessee placing you in a position where someone you are not familiar with is operating your aircraft.

Remember that if you want to control the ability of your lessee to assign your lease or sublet your aircraft, you must include restrictions on both. There is a difference between an assignment and a sublease. An assignment would occur if your lessee transferred the entire remaining portion of your aircraft lease to a third person. A sublease would involve a transfer of only a portion of your remaining lease to a third party.

For instance, if 8 months remained on your aircraft lease and your lessee transferred the lease to a third party for the remaining 8 months, your lease is being assigned. On the other hand, if 8 months remained and your lessee transferred your lease for only 4 months, your aircraft is being sublet.

If your lease is assigned, the third party who is assigned the lease assumes your lessee's role. However, if the third party fails to pay rent, you can pursue both the third party and the original lessee. If your aircraft is sublet, the general rule is that your only recourse for unpaid rent is against the original lessee.

Choice of governing law

To avoid expensive confusion in case of disagreement, it is necessary that you and your lessee agree on the law to govern your lease agreement and the specific jurisdiction to submit disputes. Typically, it may be to your advantage to choose your home state and local courts. This will at least save you travel time and perhaps make it easier for you and your counsel to draft the lease agreement. Sometimes, you and the lessee may agree that it is better to submit any disputes to arbitration through a nationally recognized arbitration association. These are important decisions that should be made with the advice of your counsel.

Attorney fees

Who will pay for attorney fees in case a dispute arises in your leasing arrangement? If you are leasing your aircraft, you may wish to require that your lessee pay reasonable attorney fees if you are forced to hire counsel in efforts to collect delinquent rent due. Otherwise, you may want to require that the prevailing party in any dispute be entitled to reasonable attorney fees.

13

Multiple Ownership Arrangements

Another method of sharing the cost of aircraft ownership is through multiple ownership of the same aircraft. This approach can take on many legal forms. A group of more than one can form a co-ownership, partnership, corporation, or limited liability company (LLC) to own an aircraft. Chapter 2 discusses the legal basics of each of these ownership forms that are most commonly referred to by aircraft owners as "flying clubs." Once you have made your decision on which ownership approach might be the best fit for you and your co-owners, it is time to address the details of how you are going to make the flying club work.

This chapter is devoted to exploring the nuts and bolts legal issues for flying clubs. Perhaps the most important thing you can do to ensure the success of your flying club is to think through and detail in writing what each participant expects his or her duties and privileges to include. It may be impossible to foresee every conceivable situation that might arise during the term of your arrangement. However, a good faith effort to come to an agreement on how to handle the most realistic of possibilities is critical to ensuring harmony in your shared aircraft ownership.

The CD material related to this chapter includes the following specimen agreements and other documents that will acquaint you with some of the legal issues you will have to address when you and your counsel are preparing the necessary documentation to suit your situation:

- Co-Ownership Agreement
- Corporate Bylaws
- Corporate Articles of Organization
- Corporate Shareholder's Agreement
- Limited Liability Company Operating Agreement
- Partnership Agreement

It cannot be emphasized enough that these agreements are not meant to be "fill-in-the-blank" tools designed to transform you into a do-it-yourself lawyer. Every situation is different. The agreement for your situation will most likely require customized provisions tailored to the number of owners, the expectations of the owners, the amount of money at stake, whether your aircraft activities will be for profit or not for profit, and many other variables. These illustrations are simple documents and agreements that are intended to serve as a catalyst for discussion between you and your counsel.

The bulk of this chapter is devoted to outlining and discussing the common threads that will be necessary for most successful multiple ownership arrangements. The issues discussed and the drafting points noted can be integrated into the preparation of any type of flying club forms including a co-ownership, corporation, LLC, or partnership. Toward the end of the chapter, there is a brief section devoted to an update on FAA's proposed rules for fractional aircraft ownership arrangements.

Drafting Considerations

The following discussion is a review of the most commonly encountered drafting issues for flying clubs. The first part discusses the key provisions for any kind of multiple ownership situation. The second part deals with special provisions for aircraft operating rules.

Identification of the parties

The first thing you'll need to do in drafting any kind of flying club agreement is to properly identify the names and addresses of all the parties involved. In most cases, the parties involved will be individuals. However, it is possible that if you are forming a co-ownership, LLC, or corporation, you may have owners, members, or shareholders who are not individuals (remember that a partnership may not have corporate owners—see Chap. 2). Therefore, it is very important that your agreement carefully identifies the persons and/or entities involved. LLCs and corporations should be identified by their legal names. It is also a good idea for you to document the states where any LLCs or corporations were formed.

Formation and name, office, purpose, term

Your agreement should provide an opportunity to outline the background for your flying club. Some of the information you provide here might be available by a review of other documents. However, it still may be prudent to summarize some of these basics in one document so that future or potential owners can get a clearer picture of your organization by simply reviewing one document.

One place to start might be a section stating the legal origin of the entity you have formed. For instance, if your entity is an LLC, you might indicate that

"The parties to this agreement have formed an LLC pursuant to the laws of the State of _____, and for that purpose, they have caused Articles of Organization to be prepared, executed, and filed with the Secretary of State."

If your entity is an LLC, corporation, or partnership, you can also take this opportunity to formally introduce the name. If there are any thoughts to doing business using any other names, you can indicate something to the effect that "The company/corporation/partnership may do business under that name or any other name or names upon which the members/shareholders/partners may agree. If it is decided to operate under a different name than the name set forth in the articles of organization/incorporation/partnership, then the company will file a trade name certificate as required by law."

A brief statement of your entity's purpose is also appropriate in this introductory section to the agreement. You should include something that will capture the spirit and intent of your organization. If accurate, your statement can indicate that "This company/corporation/partnership is formed for the purpose of providing its members/shareholders/partners a convenient and cost-effective means for private flying." It might also be prudent to state that your organization will also have the authority to do anything permitted under applicable state laws.

You should also provide the address of the principal office and resident agent in your agreement. For corporations this information can often be found in the articles of incorporation and bylaws. LLCs usually include resident agent information in the articles of organization. For partnerships and co-ownerships, there may not be a need for a resident agent, but setting up an "official address" may be a good idea in order to allow correspondence, bills, and other documents to be properly delivered.

One final, but very important administrative detail will be the identification of owners and ownership interests. This can be done with a schedule listing each person's name and a percentage of ownership interest. Sometimes this kind of information is included in the body of the agreement. Other times, it may be best to include this in a separate schedule or exhibit that can be amended as needed to include updates of owners and ownership percentages.

Initial/additional financial contributions

Here's where you'll have to start making some decisions. In order for multiple owners to succeed in aircraft ownership, you and the rest of the owners will have to carefully consider the amount and timing of initial and possibly subsequent financial contributions to the entity.

Your legal documents must clearly indicate the amount and timing of the initial financial contribution that must be made by each member, shareholder, partner, or co-owner. You may also want to indicate whether contributions in the form of services or goods will be accepted in lieu of cash. Sometimes, one or several owners may not have cash to contribute. However, they may be able to contribute valuable aircraft repair or restoration services. If this is the case, how will such contributions be valued?

Will any interest accumulate on initial financial contributions? In most cases the answer to this question will be no. However, the issue may have to be addressed in order to avoid misunderstandings in the future.

What happens if your entity needs additional capital? When things first get started, you may believe that the initial capital contribution from each member plus the regular operating fees paid by members will suffice. However, this may not always be the case. The unexpected may occur and there may be a need to seek additional capital contributions from the various owners.

If you want to make provisions for such a contingency, there are a few things you might consider including in your agreements. First, the entity should provide advance notice to owners that includes: (1) the total amount of the additional contribution required, (2) the reason for the additional contribution, (3) each owner's share of the total additional contribution, and (4) the date each owner's share is due and payable. It is also advisable to set a reasonable cap on the amount of additional contributions that may be required. There are two good reasons to set such a cap if you are going to leave open the possibility of additional capital contributions. First, you want potential and current owners to have some comfort with the maximum additional amount of capital contributions they may be required to make. Second, you do not want to leave this type of provision open-ended ("The owners will contribute as much capital as necessary") because that sort of provision might permit creditors to break through and pursue the personal assets of individual owners in the cases of a corporation or LLC.

Profits and losses

For better or worse, most flying clubs do not generate a profit. This is largely because the owners will pay fees that match the costs of operating the aircraft. The object in most cases is to spread costs, not to make a profit. Nonetheless, it may be important for tax and/or other purposes to specify how you wish to distribute or allocate profits and losses among the various aircraft owners.

If a C corporation (see Chap. 2) owns your aircraft, profits and/or losses will not be distributed or allocated directly to shareholders for tax purposes. Any profits or losses will be treated as the profits or losses of the corporation, not the individual shareholders. On the other hand, if an S corporation, LLC, or partnership owns your aircraft, profits and losses will be allocated on the basis of the percentage of stock owned by each shareholder.

For LLCs, corporations, or partnerships, the usual presumption in the law is that profits and losses are allocated equally among the owners. However, you can alter that allocation by agreement.

Management: rights, powers, and duties

You and your fellow owners are going to have to spend some time considering how you want to divvy up responsibility and authority for running your new aircraft ownership arrangement. In some cases, the distribution of duties can be informal and accomplished on an as-needed basis. In other cases, responsibilities and authority will have to be carefully assigned and documented. To some extent, the amount of formality necessary may depend on the number of owners involved. As a general rule, the greater the number of owners, the more formal the delegation of duties and responsibilities must be. It stands to reason that larger numbers of owners may create the need for rotation of authority and responsibility.

One of the tricky things about all of this is the issue of liability, especially in the cases of an LLC or a corporation. Ostensibly, one of the reasons you may have decided to form an LLC or corporation was to reduce the personal liability exposure of the individual members or shareholders. If you start naming owners to positions of authority such as "maintenance officer" and "safety officer," do you expose them to personal liability in case of a mishap? It is difficult to answer that question. There is no developed area of case law on this issue as it results to aircraft ownership, LLCs, and corporations. However, you should carefully consider the pros/cons of formally naming people to responsibility for particular duties. By naming someone specifically to certain duties, you may be increasing his/her exposure to liability.

You should also keep in mind that for a corporation most states have designated required officer positions. In many instances, state law will require at least a president, secretary, and treasurer. Legally, there is usually no compelling reason to add to these basic offices with special officer or manager slots for aircraft-related issues.

For LLCs, the situation is a bit different. Most state LLC statutes do not specifically prescribe required officers for an LLC. Instead, they generally tend to refer to the generic term "manager." You can choose whether you want your LLC to be manager-managed or member-managed.

If your LLC is manager-managed, you will have to designate a manager or managers to deal with the day-to-day administration of your LLC. These managers will be the only parties authorized to act as an agent for your LLC. That means that the managers will be the only parties who can make contracts, set up bank accounts, and do any other chores necessary to running the LLC. However, you could limit the authority of selected manager(s). One way to limit their authority is to draft a provision limiting your LLC managers from purchasing, selling, or making any other commitments over a specified dollar amount without the express consent of the other members.

If your LLC is member-managed, any of the members may act as an agent for the LLC. In either case, you will have to inform the FAA how your LLC is managed (see Chap. 4) when you submit an application for aircraft registration.

In the end, the question of whether to directly name certain shareholders or members to specific aircraft-related duties is going to be up to you and your fellow aircraft owners. If the organization is small and nimble enough, you may be able to avoid the need to formally name particular owners to certain duties. If not, you may have to set aside certain responsibilities and formally name certain positions such as maintenance officer, safety officer, and/or accounting officer.

Transfers of interests and withdrawals

Planning for transfers, withdrawals, and death is critical to the smooth operation and survival of your multiple ownership enterprise. This is where you have to be extra careful in your planning. As usual, you can't plan for all contingencies, but it is not at all unlikely that at some point in the future, an owner is going to want to transfer or withdraw from the arrangement. Because we are all mortal it is also a reasonable possibility that an owner may die during the existence of your aircraft ownership arrangement. How do you want to deal with these contingencies? In most cases, you and the entity owning your aircraft will be best served by having a plan of action ready and in place in writing.

One of the first questions you will have to ask and answer is whether you want to give owners the right to transfer their interests to third parties. Unless stated otherwise, there is a general presumption that shares of stock, partnership interests, and LLC memberships are transferable. This could cause problems. In many cases you entered into an agreement to own an aircraft with a small group of similar-minded aviators whom you trust. Do you want any of your fellow owners to be able to sell off his/her interest in your aircraft to someone you have never met? This has to be considered and directly dealt with up front.

You may decide that you want your agreement to prohibit the transfer of any ownership interests to third persons. That solves the problem of the remaining owners. But what if you are the owner who for legitimate reasons wants to or has to terminate your interest in the aircraft. You may have been transferred to a new job or you may be interested in purchasing your own airplane. You probably don't want to be bogged down in your flying club arrangement forever.

One way to solve this problem is to require that the organization purchase the ownership interest of the withdrawing owner. That solution also raises the possibility of another problem. Will the organization have the funds to pay the withdrawing owner?

In some cases, you may want to try a provision that requires the organization to issue a promissory note for the withdrawing owner's interest. Many successful agreements have used a 5- to 7-year promissory note with interest payable at some benchmark rate such as *The Wall Street Journal* Prime Rate on the date the first payment is to be made. This approach allows the organization time, if needed, to put the cash together for the pending transaction.

Now to the next question—how much does your flying club pay for an owner's interest in the club? There are several approaches to dealing with this issue. None of them offer the perfect solution. However, you may find some of the approaches to be more practical and/or fair than others.

One approach is to use the value of the departing owner's capital account as the purchase price for the owner's interest. This is a relatively straightforward approach. However, it does require that the flying club maintain good books and records. Typically, provisions employing this method add any capital contributions made by the owner. This means that capital contributions must be clearly defined. Usually, capital contributions will be defined as initial contributions and subsequent assessments for the capital needs of the club. Capital contributions should not include amounts paid by the owner for the day-to-day use of the flying club's aircraft.

The simplicity of this approach is attractive. However, it might produce a result that does not closely match the fair market value of the owner's share of the club's assets. In some cases the owner's initial capital contribution was used to pay for his or her share of the aircraft. By the time that same owner is ready to withdraw—maybe years later, the aircraft may be worth substantially more (or less) than when it was first purchased.

Another approach is the "agreed value" approach. Here are the basics on how this approach works. On a periodic basis, the owners of the flying club gather to come to an agreement on the fair market value of the club's assets. This "agreed value" is then the value that is used until the next "agreed value" is agreed upon. In theory, this approach should allow owners to come to determine a fair value. When they are agreeing on the value, each of them has incentive to be fair because he or she may be the seller or the buyer depending on the circumstances.

This approach also has its drawbacks. Perhaps the biggest problem is the practical reality that many clubs will not follow through regularly on this valuation. This could mean that the most recent valuation is stale. One way around this problem is to devise an alternate valuation means if the latest agreed-upon value is older than a designated benchmark. Another potential drawback is the possibility that the owners reach an impasse in determining the agreed value. To some extent, this problem can also be dealt with using an alternate valuation method such as appraisal. In general, this method tends to work best when the club members are diligent about getting that agreed-upon value determined on a regular basis. When that happens, this is a solid approach to valuing each owner's interest in your club's aircraft.

Another approach is the appraisal method. The appraisal method requires that the club retain someone to value the aircraft once a withdrawal notice has been received. The appraisal method may provide the most accurate fair market value of any other method. However, there is always the possibility of disagreement when it comes to who will be the appraiser. One way to deal with this is to designate an appraiser ahead of time. Another approach is to average the appraisals made by the club's appraiser and the withdrawing owner's appraiser. In any event, the key here is to find qualified aircraft appraisers who are unbiased.

One other approach is the use of a reference manual for valuations. This can be a very simple and cheap approach. However, it is also likely to be inaccurate. Most aircraft valuation reference manuals are very useful tools. However, they are most useful as general guides to the valuation of your aircraft. They cannot factor in the subjective nuances of aircraft cosmetics and condition.

In the end, you will have to determine the fairest and most practical method for valuing a member's interest in the club's aircraft. This is an important decision for you and your club.

The discussion above has assumed that the club will not permit you to sell your ownership share(s) to anyone else. Sometimes the club may be amenable to the transfer of an owner's interest to a third party. If your club is willing to permit someone to transfer his or her ownership interest, it should be prepared to carefully regulate the process. The first step is often requiring a right of first refusal. Requiring the departing owner to first present his or her ownership interest for transfer to the club allows the club to call the first shots. Sometimes, the agreement will be set up to allow individual club owners a chance to purchase the departing shareholder's interest if the flying club declines the offer.

If both the flying club and the individual members of the flying club have declined the sale offer(s), the departing owner has the opportunity to offer his or her ownership interest to a third person or persons. The club can still maintain some control over this situation by insisting on the need for the departing owner to get the other owners' consent before any new owners come on board. Typically, this consent cannot be unreasonably withheld, but it can be totally left to the discretion of the owners.

Another difficult issue that your flying club may have to deal with is the death of a member. The first thing to remember is that the member's estate will most likely be the new owner of his or her interest in your flying club. The person who will be responsible for taking care of the affairs of the deceased member is called a personal representative or executor.

In most cases, the estate will want to convert the deceased member's interest in the aircraft to cash. The question is then whether the club has the funds available to fund a buyout of the deceased member's share. For all practical purposes, the death of a member can be treated as a transfer of a member's interest. That means the estate will first have to offer to transfer its interest in the flying club back to the flying club. This will set into motion all of the mechanisms for transferring an interest discussed above.

However, in the case of death, there is the possibility of funding the transfer immediately if careful planning is employed. Where the stakes are high enough, some clubs have taken out term life insurance policies on members. The club becomes the policyholder and beneficiary. The amount of insurance taken out should roughly track the amount of money needed to buy out the owner's share at any given time in the foreseeable future. This is a perfectly legitimate and efficient way to ensure that the club can fund a buyout of a deceased member's ownership share. To a large degree, this approach also protects the families and beneficiaries of each club member. It ensures that each club member's beneficiaries are able to immediately liquidate the deceased member's interest in the aircraft. In most cases, the beneficiaries would rather have cash than an ownership interest in an aircraft that they probably can't use. If the insurance proceeds are not adequate for a full buyout, the agreement can call for the remainder to be paid via a promissory note.

Termination

It is prudent to make provisions for the termination of your flying club. Perhaps the most important thing to consider is the event or events that would trigger the dissolution of the club. One of the most common triggers might be the vote of a majority of members/shareholders/partners indicating that they want to terminate the club.

The next thing to consider is who will manage the shutdown or winding up of the club. Usually that job falls to the remaining members, or the last person to be a member.

How will the assets of the club be distributed? Typically, the law requires the following order of distribution:

1. First, distribute any amounts owed to outside creditors.

2. Second, distribute amounts loaned by club members.

3. Third, distribute any remaining funds to members.

The last thing that may have to be accomplished is the formal filing of dissolution with state authorities. As a general rule, the filing of formal paperwork will only be required where formal written documents had to be filed to form the club, usually in the case of a corporation or LLC.

Accounting and tax issues

The club's documents should authorize the opening of bank accounts as necessary for the operation of the club. It may be best if the provisions for bank accounts are not too specific about the types of accounts and specific banking institutions. Flexibility will be required to open accounts on an as-needed basis.

There should be provisions for any special tax elections. For instance, if your flying club is a corporation, it may wish to authorize a subchapter S election. If your flying club is an LLC, it may wish to be treated as a corporation instead of a partnership. All of these decisions should be documented. A decision regarding an accounting year should also be documented. In most cases a calendar year with a closing date of December 31 is selected. However, for various reasons, some clubs may prefer a fiscal year (however, tax rules sometimes do not always permit a fiscal year without authorization). A number of these questions will have to be addressed when your club applies for a taxpayer identification number using an IRS Form SS-4.

General provisions

As with any written agreements, you will want to include some or all of the following general provisions:

- Rules for providing notice to ensure that there is a practical and efficient method available to notify members of important club matters.

- An acknowledgment that the agreement and rules are the complete and exclusive statement of the agreement and understanding between the club members.

- Amendment provisions to allow for the modification of the club's agreement or other documents under specified circumstances (e.g., majority vote of club members).

- A clear indication that any questions regarding the construction, validity, and interpretation of the club's agreement shall be governed by selected state law.

- An agreement that any disputes related in any way to the agreement will be subject to the jurisdiction of specified courts or arbitration associations.

- An agreement that the agreement is binding upon, and inures to the benefit of all the parties and their respective heirs, assigns, and personal administrators.

- If necessary, a provision indicating that the agreement can be executed in two or more counterparts to allow for the simultaneous execution of the documents without all the members needing to be present.

Aircraft Operating and Maintenance Rules

All of the issues addressed above involve matters that any multiple ownership entity should address. However, you are involved in a flying club. Therefore, there are special provisions you will have to consider that relate specifically to the needs of a flying club and aircraft operations. These provisions can be agreed to in any form. For corporations they might be included in bylaws or a shareholder's agreement. In an LLC, they might be included in the operating agreement. Regardless of the form these provisions take, they should be carefully thought through. These rules will provide day-to-day guidance on how you expect your aircraft to be operated and maintained. A review of some of the more common provisions you should consider is outlined below.

General operating rules

It might be advisable to remind every club member that compliance with the FARs is an absolute minimum standard for the operation of club aircraft. Violations of FARs will also be considered violations of club rules. There may be times when club rules may be more stringent than the FARs—the club rules should be abided by to the extent practicable. You can also provide that if club members violate any federal, state, or local law and/or regulations in the operation of your club's aircraft, they will be liable for any civil penalties or fines incurred.

Restrictions on aircraft use

In many cases, your aircraft's insurance policy will limit coverage to business and pleasure purposes only. You will want to make it clear to all members that aircraft use must be in conformance with the club's insurance coverage. All members should be familiarized with your club's insurance policy and what it does and does not cover.

Aircraft home base

The issue of where to base your club's aircraft can be contentious from the start. Depending on where you are located, you may have a number of different airports to choose from. Does the airport have available hangars? Is there an adequate maintenance facility? What about fuel prices? These are just some of the questions you and your fellow club members will have to sort out. However, once a choice is made, it should be documented in your club's rules. You may also want to place a provision indicating that any changes to the current aircraft home base must be approved by the club members through a majority or supermajority vote.

International operations

To play it safe with your insurance coverage, you may want to include an express provision prohibiting any international operations. If you want to give members some leeway in this regard, it may be prudent to require advance notification and approval by a majority of members before club aircraft can be operated internationally.

Authorized airports

You and your fellow owners may wish to restrict club members to certain runway and airport conditions. You should consider whether your aircraft could be safely operated out of grass or gravel fields. Are the pilots in the club experienced with operations on nonpaved surfaces? Although you may not adopt any restrictions, this is an area worthy of discussion when you are getting your flying club up and running.

Flight plans

Do you wish to require flight plans for all flights away from your home base? Should the requirement apply only to longer-distance flights? Again, certain club members might not appreciate the intrusion into their mode of operating an aircraft. You may at least want to consider this type of requirement.

Authorized pilots

At the very least, you will want to ensure that only pilots qualifying under your club's insurance coverage are authorized to operate club aircraft. However, most clubs will want to go further. It may also be prudent to insist that only club members be authorized to operate aircraft. Exceptions may be made for instructors or other special situations (if insurance requirements are met), but these exceptions can be subject to club members' approval.

Insurance

Aircraft insurance will be a big issue for your club. The members should periodically review the needs of the club and set insurance requirements that are appropriate. Giving notice of the coverage and amount of coverage in the club's rules is usually a good idea. Updating the requirements to suit your club's needs is critical.

Scheduling

Scheduling can be the downfall of a flying club if it is not properly handled. There are many different ways of handling scheduling. With the wide use of electronic calendars, the need for a person to act as a scheduling officer or liaison may no longer be necessary. In any case, you and your fellow club members will have to find a method that works best for the group. It may take a number of tries using different approaches, but this is one area where you will have to reduce the rules to writing and stick with them.

Maintenance deficiencies

With different people operating the same aircraft, it is important to carefully document any detected maintenance problems. One tried and true method is the squawk sheet with entries listing the pilot's name, date, and malfunction detected or suspected. A system should be put into place allowing the discrepancies to be acted upon by someone or some group within the club.

Securing aircraft

Rules regarding securing your club's aircraft may not be necessary. However, this has been a constant source of irritation and friction between club members. Sometimes tempers flare over the way the aircraft was secured by previous owners. This can lead to discord and that may weaken the fabric of your club. It may be best to remind everyone that they have an important obligation to the next user. That obligation includes cleaning up the aircraft and properly securing it after each use.

Charges

Sharing costs is one of the primary reasons you are forming or joining a flying club. The club has to come to grips with some methodology that will ensure that members will be able to fly at the lowest cost while allowing the club to meet its financial obligations. One method that has worked for many clubs is a combination of an hourly rate plus a fixed monthly or quarterly fee. The hourly rate will apply to only those hours (measure by Hobbs or Tach time) that a particular member flies the aircraft. Presumably, the club will put a record-keeping system in place that will permit each member to enter his name, the date, and the hours (or fraction of hours) flown on a particular flight. The member's hours will then be multiplied by a predetermined hourly rate and charged or invoiced periodically. This charge will ensure that the members who use club aircraft fairly compensate the club for the use. A fixed fee is designed in part to cover the known fixed costs of the aircraft, including storage, recurring maintenance items, taxes, insurance, and other costs that do not change or vary on the basis of the level of flight activity.

Of course, these charges and rates will need to be periodically reviewed. Both the fixed and variable costs of operations will change. Therefore, the club must be nimble enough to make changes and corrections as needed.

Finally, you may wish to provide for the unpleasant circumstance when a fellow club member is seriously delinquent on his or her financial obligations to the club. This issue can be dealt with in the body of a shareholder's agreement or operating agreement or directly in the aircraft operating and maintenance rules. There are many possible approaches to dealing with this type of problem. Most of them will likely result in forced termination of membership and practical attempts to collect amounts in arrears. In any case, the time to deal with this type of contingency is during the formation of your club. Waiting until a problem arises may be too late.

Fractional Ownership

In the last few years, there has been a lot of talk about fractional ownership. Fractional ownership has evolved and taken root in corporate aviation and generally involves business jets and turboprop aircraft. So what is fractional ownership? From a lawyer's standpoint, it is a form of aircraft co-ownership. It has become quite popular because many businesses would like to make use of corporate jets. However, they will not be able to use the aircraft enough to justify the high costs of sole ownership. With a fractional ownership arrangement, these businesses can own a fraction of a corporate jet, make use of a professional manager for the aircraft and crew, and even make use of other aircraft if the jet they own is not available. It sounds like a good deal and for many high-end aircraft users it is a good deal.

What smaller general aviation aircraft owners should be watching are current proposals to regulate fractional ownership arrangements. Whereas the safety record of corporate aviation and fractional ownership aircraft has been very solid, regulators have still been a bit uneasy because the arrangements often begin to resemble on-demand commercial operations.

The FAA is concerned because there is a difference in the standards applied to noncommercial operations in Part 91 versus the more stringent regulations applying to commercial operations in Parts 121 and 135. In essence, the FAA's proposed rules are an attempt to level the playing field by bringing fractional ownership arrangements to an equivalent level of safety standards applicable to commercial operations. FAA proposed new rules for fractional ownership arrangements in 2001. As of today, interested parties in the aviation industry have submitted comments and the FAA is studying these comments. It is difficult to say when the new rules will be finalized, but it should be soon.

How do these rules affect the typical light aircraft multiple ownership arrangement? The jury is still out on this. But on the basis of the proposed rules, and FAA's stated intention to focus the new rules on high-end, professionally managed programs, it seems likely that the new rules will have no negative impact on light general aviation aircraft pilot-owners. However, it is still advisable for light general aviation aircraft owners to keep an eye out for FAA's final rules.

Tax Issues for Aircraft Owners

14

Aircraft-Related Business Expenses

You may find that your aircraft can serve as a very valuable business tool. Aircraft owners who use their aircraft for business purposes can avoid the delays created by long drives to airports, long waits on security lines, and flight cancellations. It is also important to note that whereas only several hundred airports are serviced by scheduled commercial airlines, there are over 16,000 public-use airports throughout the United States. This vast network of airports allows private aircraft owners to arrive more directly at their destinations, saving even more valuable time.

This chapter is devoted to analyzing the tax implications of using your aircraft for business purposes. We'll review aircraft operating expenses, pilot training expenses, and depreciation. Some of the concepts will be best explained using the past Tax Court cases and rulings presented throughout this chapter.

Section 162

If you are an aircraft owner, you are sure to have wondered whether any of your aircraft-related expenses are tax deductible. To answer that question, you need to turn to the Internal Revenue Code, Section 162 (a). This law tells us (in pertinent part) that "[t]here shall be allowed as a deduction all the ordinary and necessary expenses paid or incurred during the taxable year in carrying on any trade or business." As you examine this legal provision, there are two questions that you need to address:

1. Is your activity a "trade or business"?

2. Are your expenses "ordinary and necessary"?

The "trade or business" test

Sometimes, it is not easy to determine if an expense is deductible as a "trade or business" expense. The term "trade or business" is not defined in the Internal Revenue Code or the IRS's regulations. A review of case law also indicates that the courts have not been able to provide clear guidance on this issue. Usually, you will have to consider one or more of the following questions to determine whether your aircraft-related expenses are incurred in the context of a trade or business:

1. Was your aircraft use related directly to a business activity?

2. Were your aircraft expenses incurred with the objective intent to make a profit or to generate income? (See the discussion of the "Hobby Loss" rule in the section "Hobby Loss" later in this chapter.)

3. Was your operation and management of the business-related activity extensive enough to indicate that you were really carrying on a trade or business?

You should be able to answer all three of these questions affirmatively in order to claim that your aircraft expenses were incurred while you were carrying on a trade or business. It is important to keep in mind that if you are an employee acting within the scope of your employment, you are engaged in the trade or business of being an employee. This may make it even easier to claim that your aircraft expenses were incurred in a trade or business.

The "ordinary and necessary" test

The next step is establishing that your aircraft expenses are "ordinary and necessary." For good reason, the law does not provide a cut and dried definition of this phrase. It would be very difficult to precisely pin down all of the possible conditions that would make an expense "ordinary and necessary." However, there is case law available to provide you and your tax counsel solid guidance on this issue.

First, you should note that the law says "ordinary *and* necessary," not "ordinary *or* necessary." This is important because it tells us that you will need to establish that your aircraft expenses are both ordinary and necessary, not just one or the other.

To be "ordinary," your aircraft expense must be an expense the courts would consider normal, usual, or customary in the type of business you conduct. However, the courts have also indicated that the expense does not have to be regular or recurring to qualify as an "ordinary" expense. Case law has been pretty clear on the point that the use of private aircraft for business and travel fits the bill when it comes to qualifying as "ordinary." Perhaps one of the most definitive statements on this issue is found in Case 14-1.

Case 14-1

David W. Marshall and Margaret Cole Marshall, Petitioners, v. Commissioner of Internal Revenue, Respondent
United States Tax Court
T.C. Memo 1992-65

OPINION BY: Pate
OPINION: MEMORANDUM FINDINGS OF FACTS AND OPINION
Respondent determined deficiencies in petitioners' Federal income taxes of $2,851, $6,194, and $2,070, for the years 1984, 1985, and 1986, respectively. After concessions

by both parties, the only issue for our decision is whether petitioner, David W. Marshall, is entitled to deduct, as a trade or business expense, the cost of using his private airplane in performing his duties as an officer in the United States Air Force.

FINDINGS OF FACT

Some of the facts have been stipulated and are found accordingly. The Stipulation of Facts and attached exhibits are incorporated herein by this reference.

David W. Marshall (hereinafter petitioner) and Margaret Cole Marshall (hereinafter Mrs. Marshall) are husband and wife and filed joint returns for 1984, 1985, and 1986. They resided in San Antonio, Texas, at the time they filed their petition.

Petitioner has been in the United States Air Force (hereinafter Air Force) for 25 years. During the years in issue he was a Lieutenant Colonel and Director of the AeroMedical/Casualty System Program Office (hereinafter ACS) at Brooks Air Force Base. As the program's Director, he was charged with developing equipment to be used to protect soldiers, airmen, and medical personnel, from the effects of chemical warfare. Specifically, petitioner was responsible for the research and development of, and later the production of, new products for decontamination, treatment, and evacuation of military personnel exposed to chemical warfare, and for the mass evacuation of the injured back to the United States. Because the Air Force had no inventory of equipment for this purpose at the time he was appointed, petitioner realized that he had to proceed with considerable urgency.

The program had an annual budget of approximately $10 million for research and development, and a $50 million budget for production of the equipment. Petitioner was responsible for finding qualified contractors to aid in the development of such equipment and technology. Because the concepts introduced by the program were so new, civilian contractors were not aware of the Air Force's requirements. As a result, petitioner had to seek out companies that were qualified and interest them in the program. In addition, petitioner was required to remain in contact with a number of Air Force commanders to keep them abreast of progress and problems in the program. To accomplish this, petitioner traveled to meet with both Air Force personnel and civilian contractors.

Petitioner started flying about 30 years ago. He holds an airline transport pilot license and instrument rating for all weather flying. During the years in issue, he owned a 1965 Piper Twin Comanche equipped for all weather flying. He used this airplane on many of his trips for the Air Force; during 1984 he made eleven business trips for the Air Force, using his airplane for eight of them; during 1985 he made a total of fourteen trips, using his airplane for eight of those; and during 1986 he made a total of eleven trips, using his private airplane for two of them. Sometimes, he carried other Air Force personnel. Petitioner did not conduct any personal business while on those trips, although on one trip to California he extended his stay for a short vacation.

Petitioner maintained a log and submitted expense vouchers to the Air Force for each trip. He was reimbursed by the Air Force at the rate of sixteen cents per mile, until July 1, 1986, when the reimbursement rate was raised to forty five cents per mile. Petitioner claimed a deduction for the amount that the business portion of his actual expenses and depreciation exceeded the amount of his reimbursements. These deductions totaled $9,295, $13,682, and $1,836, in 1984, 1985, and 1986, respectively. The business portion of depreciation included therein was $5,437, $6,999, and $766, respectively. The parties have stipulated that petitioner has substantiated all of the amounts he deducted.

As Director of ACS, petitioner was not required by the Air Force to own his own airplane or to fly other than by commercial means. However, on all of the trips, the use of his airplane was authorized by the Air Force and most of his travel vouchers acknowledged that, under the circumstances, "Travel by privately owned aircraft has been determined as more advantageous to the government."

Petitioner could have fulfilled his duties as Director of ACS without the use of a private airplane, but chose to fly his own airplane primarily for two reasons. First, it enabled him to land directly at the military airfields that were his ultimate destinations rather than to the airports which the commercial air carriers used. In addition, the use of his own airplane afforded petitioner flexibility; most of the conferences he attended were indefinite in length and such use allowed him to depart promptly upon their termination. This resulted in considerable time savings, especially when meetings ran late and the use of his airplane enabled petitioner to return that same night in order to report for duty at his office the next morning. In some instances, the use of commercial air carriers would have required petitioner to remain out-of-town overnight, precluding him from attending at least some of the meetings and appointments scheduled for the next day.

OPINION

Section 162 regulates the deductibility of all ordinary and necessary expenses paid or incurred by an employee during a taxable year in which the employee carries on a trade or business. To secure a deduction under this provision the employee must demonstrate that the expenditure was incurred in his trade or business, was an ordinary and necessary expenditure, and that he could not be fully reimbursed by his employer. *Flower v. Commissioner,* 61 T.C. 140, 154-155 (1973), *affd.* without published opinion 505 F.2d 1302 (5th Cir. 1974); *Fountain v. Commissioner,* 59 T.C. 696 (1973); *Westerman v. Commissioner,* 55 T.C. 478, 482 (1970); *Rink v. Commissioner,* 51 T.C. 746, 751 (1969); *Koree v. Commissioner,* 40 T.C. 961, 965-966 (1963); *Stolk v. Commissioner,* 40 T.C. 345, 356-359 (1963), *affd.* 326 F.2d 760 (2d Cir. 1964).

Recently, we considered the question of whether an employee can deduct the excess of the costs he incurred in using his private airplane for business over the amount of reimbursement received. *Noyce v. Commissioner,* 97 T.C. (December 16, 1991). In that case, we analyzed the question as follows:

we must decide if the expenses are ordinary and necessary expenses of petitioner's employment. An expense is necessary if it is appropriate and helpful in carrying on the trade or business. *Heineman v. Commissioner,* 82 T.C. 538, 543 (1984) (citing *Commissioner v. Tellier,* 383 U.S. 687, 689 (1966); *Welch v. Helvering,* 290 U.S. 111, 113 (1933)) * * *. An expense is ordinary when the paying thereof is the common and accepted practice in light of the time and place and circumstance. *Welch v. Helvering, supra* at 113-114. [Slip. opinion at page 25.]

Therefore, to allow petitioner a deduction for the cost of operating his private airplane, we must first find that petitioner's use of his airplane was "appropriate" and "helpful." Under the circumstances of this case, there is little doubt that the use of his airplane was appropriate and helpful to petitioner in the performance of his duties. It provided him with direct access to his destinations. In this respect, on many occasions, petitioner was able to arrive at his destination on time; whereas, had he used a commercial flight, he may not have been able to do so. Further, instead of having to wait for a commercial flight the next day, petitioner's direct access frequently enabled him to leave promptly after completing his assignment and report to work at his home base the next morning.

Secondly, we must find that such expenses were "ordinary," that is, a common or accepted practice. In this day and age, there is no doubt that the use of private airplanes by executives in charge of large projects is a common practice. The popularity of their use is reflected by the number of cases in which the expenses incurred in the use of a private airplane arises. See, e.g., *French v. Commissioner,* T.C. Memo. 1990-314; *Plante v. Commissioner,* T.C. Memo. 1987-355; *Sartor v. Commissioner,* T.C. Memo. 1984-274; *Sherman v. Commissioner,* T.C. Memo. 1982-582.

Finally, we must find that such use was reasonable in relation to its purpose. In this regard, in *Noyce v. Commissioner, supra,* we held that deductions for depreciation are not taken into account in meeting this criterion. Excluding the depreciation deduction, petitioner's expenses amount to $3,858, $6,683, and $1,070 for 1984, 1985, and 1986, respectively.

Petitioner was charged by the Air Force to supply new equipment urgently needed because it was not merely replacement equipment, but equipment needed to protect soldiers and medical personnel from the effects of chemical warfare, a serious threat in recent years. He was empowered to expend $10 million per year in research costs alone. To expend the money wisely and productively, petitioner decided that he needed the flexibility the use of his private airplane afforded him to maintain, on an expeditious basis, face-to-face contact with the persons he worked with to achieve the goals of the program. Although the Air Force regulations precluded reimbursing him for his total costs, when we compare the excess costs he incurred with the responsibilities he had to bear, we find them "reasonable."

Nevertheless, respondent points to the fact that the Air Force did not require the use of his airplane and argues that, therefore, the expenses he incurred in excess of his reimbursement were not necessary to petitioner's business of being an employee of the Air Force. Respondent cites *Noland v. Commissioner,* 269 F.2d 108 (4th Cir. 1959), *affg.* T.C. Memo. 1958-60, in support of this argument.

We find *Noland* distinguishable. In *Noland,* the taxpayer was an executive running six corporations. He personally paid for a Christmas party for employees, paid dues to various clubs, and made a donation to promote industrial development in his area. He deducted these expenditures as ordinary and necessary business expenses. In finding that these expenses were not "necessary" the Court reasoned that (*Noland v. Commissioner,* 269 F.2d at 113):

> When the corporation, reimbursing its officers and employees for direct expense incurred in furthering its business, does not reimburse an officer for a particular expense, that expense prima facie is personal, either because it was voluntarily assumed or because it did not arise directly out of the exigencies of the business of the corporation.

In this case, however, there is no question that petitioner's expenses arose from the "exigencies of the business." Each of petitioner's trips was approved and reimbursement authorized by the Air Force prior to his taking each trip. It is only because the rate of reimbursement was too low to cover petitioner's total costs that we have a dispute here.

In Revenue Ruling 70-558, 1970-2 C.B. 35 (cited in *Noyce v. Commissioner, supra*), the issue under consideration was whether the taxpayer, an employee of the Federal Government, could deduct the excess cost of using his privately owned airplane on his official business trips. In that ruling, respondent stated that:

> During 1969 the taxpayer was required to travel extensively in connection with his employment. Although he was not required to use his privately owned airplane for business travel, due to the urgency of his trips he was permitted to do so. He was issued overall travel authorizations which provided authority for travel by "private owned auto or aircraft" among other possible means of transportation.

The taxpayer was reimbursed for his travel at a standard rate based on the total miles traveled on official business. The taxpayer properly substantiated all the expenses incurred in operating the airplane on the business trips.

Held, the travel expenses incurred by the taxpayer in excess of his reimbursements are deductible as ordinary and necessary business expenses. * * *. Such expenses include depreciation on the airplane to the extent properly allocable to business use. [1970-2 C.B. 35.]

We can discern no meaningful distinction between petitioner's situation and the facts in Revenue Ruling 70-558.

Respondent next argues that petitioner's failure to seek reimbursement of actual expenses made his excess costs not "necessary" within the meaning of section 162. However, in making this argument, respondent assumes that the "actual expenses" authorized to be reimbursed by the Air Force would have exceeded the amount of the mileage reimbursement. That the "actual expenses" authorized to be reimbursed covered only out-of-pocket costs and did not cover a major portion of the expenses petitioner actually incurred, such as repairs, insurance, maintenance and hangar fees, casts doubt on the validity of respondent's assumption. Moreover, although not entirely clear, petitioner's Federal income tax returns for 1984, 1985, and 1986 indicate that respondent's assumption is incorrect. It appears that the business portion of petitioner's out-of-pocket costs were less than the amount of reimbursement he received from the Air Force.

Based on the facts in this record, we conclude that the amount petitioner expended to use his private airplane in the performance of his duties as Director of ACS was an ordinary and necessary business expense he incurred as an officer in the Air Force. Accordingly, he is entitled to deduct the excess of those costs over the amount of his reimbursement.

Because of concessions by both parties,

Decision will be entered under Rule 155.

[*Author's note:* Whenever a Tax Court case indicates that a decision will be entered under Rule 155, it means that the parties are required to recalculate any taxes due or refunds in accordance with the Tax Court's order. After the parties prepare a calculation (either in agreed or unagreed fashion), the calculation(s) must be submitted to the Tax Court for approval.]

Note the Tax Court's clear statement on the use of private aircraft. The court makes a point of saying, "[i]n this day and age, there is no doubt that the use of private airplanes by executives in charge of large projects is common practice." The truth is that in the current environment of heightened airport security and ensuing delays, the court's statement may be even more appropriate today. With this kind of case law, aircraft owners should, in most cases have the ability to establish the "ordinary" nature of their aircraft use in furtherance of business.

The next part of the test is proving that your aircraft expenses were "necessary." The courts have held that an expense will be considered "necessary" if a prudent businessperson would incur the same expense and the expense is anticipated to be helpful and appropriate in your business. Establishing the "necessary" part of the test is a bit more challenging than the "ordinary" test. However, there is case law supporting the contention that use of private aircraft is "necessary." Perhaps one of the most widely cited cases in this regard is Case 14-2, the *Sartor* case.

Case 14-2

Steven F. Sartor, and Gwenn Sartor, Petitioners v. Commissioner of Internal Revenue, Respondent
United States Tax Court
T.C. Memo 1984-274

OPINION: MEMORANDUM FINDINGS OF FACT AND OPINION

FAY, Judge: Respondent determined a deficiency of $8,283 in petitioners' 1980 Federal income tax. After concessions, the only issue is whether the expenses petitioners incurred in connection with their airplane are deductible.

FINDINGS OF FACT

Some of the facts are stipulated and found accordingly.

Petitioners, Steven F. Sartor and Gwenn Sartor, resided in Kaysville, Utah, when their petition was filed herein.

Except for a one-year period spanning 1977 and 1978 when he was employed by a competitor, since 1970 petitioner Steven F. Sartor (petitioner) has been employed by Dixico, Inc. (herein Dixico) as an outside salesman of "packaging materials." When petitioner returned to Dixico in 1978, he was assigned the sales territory of Utah, Idaho, and Colorado. In March 1980 his sales territory was expanded to include Washington, Oregon, Montana, and Vancouver, Canada.

Petitioner obtained a commercial pilot's license in 1965 and subsequently purchased a Cessna 182 airplane in 1975. He used this plane exclusively for personal purposes until 1979 when he began to use it for business. In August 1980 petitioner purchased a 50 percent interest in a newer and larger Cessna T-206 airplane (hereinafter referred to as the airplane) because it was capable of safe flights in instrument flight conditions. At this time, petitioner began aircraft qualification and instrument flight training in the airplane which he completed in February 1981.

During 1980 petitioner used the airplane almost exclusively for business purposes, flying to see customers in cities throughout Idaho, Utah, Oregon, and Washington. Petitioner determined that because there were less frequent commercial flights due to the deregulation of the airlines for him to cover his large sales territory, it was necessary to use the airplane so that he could maximize his sales. By using the airplane, petitioner was able to arrange a more flexible schedule which enabled him to reach customers before his competitors. Although Dixico only reimbursed petitioner for what it would have cost him to fly commercially, it approved of his using the airplane because Dixico expected it would allow petitioner to increase his sales.

Petitioner's bonus each year was based on the amount of his sales. After purchasing the airplane in 1980, his bonus increased from $4,263 in 1979 to $7,902 in 1980 and, by 1982, it was $12,591.

Petitioner incurred in 1980 the following three types of expenses in connection with the airplane: (1) travel expenses to visit clients (herein the travel expenses); (2) travel expenses to entertain his clients (herein the entertainment expenses); and (3) expenses for aircraft qualification, instrument flight training, and equipment and safety checks (herein the qualification expenses). The entertainment expenses included scenic airplane flights and fishing trips with clients and members of their families. Dixico did not reimburse petitioner for any part of the entertainment expenses.

In 1980 petitioner maintained a flight log wherein he recorded the origin, destination, date and travel time for each of his flights.

On their 1980 return petitioners deducted $14,913 of the airplane expenses as business expenses. In his notice of deficiency, respondent determined that petitioners were not entitled to this deduction because the airplane expenses were not ordinary and necessary business expenses and, with respect to the entertainment expenses, because petitioners failed to satisfy the substantiation requirements under section 274.

OPINION

The only issue is whether the expenses petitioner incurred in connection with the airplane are deductible. Conceding that all of the airplane expenses are ordinary within the meaning of section 162, respondent argues that the travel expenses are not deductible because they are not "necessary" nor "reasonable" as required by section 162, and that the entertainment expenses are not deductible because they have not been substantiated under section 274. Petitioner argues that the travel expenses were both necessary and reasonable because they allowed him to be a more successful salesman.

Respondent concedes that if we find that the travel expenses were necessary and reasonable business expenses then the qualification expenses are likewise necessary and reasonable business expenses.

Under section 162, the term "necessary" has been interpreted to mean appropriate and helpful in the development of the taxpayer's business. *Deputy v. duPont,* 308 U.S. 488 (1940). We find that the travel expenses were very helpful to petitioner in his business of being a salesman. The record shows that petitioner was given a large sales territory with customers located several hours from major airports. Due to the deregulation of the airlines, there were less frequent flights to petitioner's sales areas and alternate methods of transportation were inadequate and time consuming. By using the airplane, petitioner was able to arrange a more flexible schedule which enabled him to maximize his sales opportunities. Under these circumstances, we think it is clear that petitioner's additional travel expenses incurred by using the airplane were appropriate and helpful in his business as a salesman. Thus, we conclude that the travel expenses were necessary within the meaning of section 162.

Respondent's next contention is that petitioner's travel expenses were not reasonable. Generally, in order to be deductible a business expense must be reasonable in amount relative to its purpose. *United States v. Haskel Engineering & Supply Company,* 380 F.2d 786, 788 (9th Cir. 1967); see also sec. 1.162-2(a), Income Tax Regs. Respondent thus argues that the travel expenses were not reasonable because they exceeded petitioner's bonus in 1980. Although as it turned out petitioner's expenses exceeded his bonus in 1980, we find that the circumstances show that the expenses were reasonable. It was reasonable for petitioner to expect his bonuses to increase each year because the airplane enabled him to have a more flexible schedule to maximize his sales. Petitioner's bonuses did in fact increase substantially from $7,902 in 1980 to $12,591 in 1982. In addition, by using the airplane to increase his sales for Dixico petitioner assured himself of continued employment with the company. Thus, we find nothing to suggest that the amount of travel expenses in 1980 were unreasonable in relation to their purpose. Accordingly, we conclude that petitioner's travel expenses are necessary and reasonable business expenses deductible under section 162.

* * *

Decision will be entered under Rule 155.

The *Sartor* case is very instructive to aircraft owners who wish to be able to deduct their aircraft expenses. One of the most valuable lessons from the *Sartor* case is the need to demonstrate to the IRS that your use of an aircraft was "appropriate and helpful" to meet your company's scheduling needs and your scheduling needs. As indicated earlier, there are over 16,000 airports in the United States. However, it is estimated that only 437 are served by regularly scheduled air carrier flights. Sartor was successful because he was able to make it very clear to the Tax Court that his airplane was a valuable tool in getting him to and from his customers. Even if your clients or customers are located close to bigger cities, you can still make a case that your time is valuable and that your airplane permits you to operate more efficiently on your schedule, not the airlines' schedule. In order to make these arguments work, you must be prepared to carefully document the nature and purpose of your business flights. The purpose of the trip should be recorded and significant positive outcomes should be highlighted. The Tax Court has been persuaded by the argument that the aircraft have been an integral part of maintaining a business or increasing a business or individual's revenue.

If you are using your aircraft as an employee, you'll also need to work with your employer to satisfy the "necessary" part of the test. Before you use your aircraft on company business, you should get formal written approval from your employer. In many cases it might be even more persuasive if your employer has a written policy on the use of private aircraft for business. Without company acknowledgment of your private aircraft use, you risk having the IRS take the position that your use of a private aircraft was solely for your own personal benefit and not intended for your employer's benefit. That will make the "necessary" part of the test very difficult to pass.

You also need to carefully work out a cost reimbursement plan with your employer. Obviously, it would be best for you if your employer picked up the full cost of your business aircraft use. However, this kind of a deal is the exception to the rule. It is much more common to see arrangements where an employer is willing to partially reimburse you for the use of your business-related aircraft expenses. This partial reimbursement is very important. It is an affirmation of the employer's position that the use of your aircraft is appropriate and helpful in the conduct of your business. This goes a long way to establishing the necessity of your private aircraft use.

One trap to avoid is not seeking reimbursement where a reimbursement policy exists. In some cases, your employer may have a reimbursement policy for the use of private aircraft. However, the policy or budget of your employer may allow for reimbursement only up to an annual budget cap. Even if you should exceed the budget cap, you must seek reimbursement. The rejection of your claim for reimbursement will be very important in establishing the deductibility of your aircraft expenses—especially if your employer approved of your aircraft use beyond the budgetary allowances. (See *Nye v. Commissioner,* T.C. Summary Opinion 1999-119; *Lucas v. Commissioner,* 79 T.C. 1, 7 (1982); and *Stolk v. Commissioner* 40 T.C. 345.) Essentially, these cases advance the IRS position that an employee is not entitled to a deduction for an expense if the employee has a right to reimbursement from his or her employer and has failed to seek such reimbursement.

Reasonableness

Beyond the "ordinary and necessary" test, you still have one more hurdle to clear for your aircraft expenses to be deductible. The IRS also needs to be convinced that your aircraft expenses are "reasonable." The Internal Revenue Code refers to reasonableness only in the context of salaries and other compensation for services [see Internal Revenue Code Section 162(a)(1)]. However, the courts have required that for any business expense to be deductible, it must be reasonable.

As you might guess, there is never going to be a definitive definition of what "reasonable" means in any context. However, the key consideration may be a comparison of benefits and expenses. If the aircraft expenses seem "reasonable" when compared to the benefits (or potential benefits) obtained from the use of the aircraft, your chances of successfully meeting the test are increased.

You should be aware that in at least two Tax Court cases, the court has indicated that when determining whether expenses are "reasonable," depreciation will not be considered. This approach was evidenced in the *Marshall* case (see Case 14-1) and Case 14-3 involving a business executive for Intel. Notice that in the *Noyce* case, the executive's aircraft expenses, including a large amount of depreciation, were approximately the same as his annual corporate salary.

Case 14-3

Robert N. Noyce and Ann S. Bowers Noyce, Petitioners v. Commissioner of Internal Revenue, Respondent
United States Tax Court
97 T.C. No. 46

OPINION BY: Ruwe
OPINION: The primary issue for decision is whether petitioners may deduct operating expenses and depreciation with respect to use of the airplane for petitioner's travel on behalf of Intel. Section 162(a) allows a taxpayer to deduct ordinary and necessary expenses paid or incurred in carrying on a trade or business. Section 168 allows a deduction for depreciation of tangible property used in a taxpayer's trade or business. Petitioners have the burden of proving their entitlement to these deductions. Rule 142(a); *Welch v. Helvering,* 290 U.S. 111, 78 L. Ed. 212, 54 S. Ct. 8 (1933).

A taxpayer is considered to be in the trade or business of being an employee separate and apart from the trade or business of his corporate employer. *Leamy v. Commissioner,* 85 T.C. 798, 809 (1985); *Lucas v. Commissioner,* 79 T.C. 1, 6-7 (1982); *Primuth v. Commissioner,* 54 T.C. 374, 377 (1970). See Rev. Rul. 62-180, 1962-2 C.B. 52. However, the voluntary payment or guarantee of corporate obligations by corporate officers, employees, or shareholders may not be deducted on the taxpayer's personal return. *Noland v. Commissioner,* 269 F.2d 108, 111 (4th Cir. 1959), *affg.* a Memorandum Opinion of the Court; *Gantner v. Commissioner,* 91 T.C. 713, 726 (1988), *affd.* 905 F.2d 241 (8th Cir. 1990). In *Noland,* the Court stated that:

> When the corporation, reimbursing its officers and employees for direct expense incurred in furthering its business, does not reimburse an officer for particular expense, that expense prima facie is personal, either because it was voluntarily assumed or because it did not arise directly out of the exigencies of the business of the corporation. [*Noland v. Commissioner, supra* at 113. Citations omitted.]

Respondent contends that petitioner voluntarily used his airplane and assumed the airplane expenses and, therefore, is not entitled to the claimed deductions. Petitioners argue that there was no voluntary assumption of corporate obligations because the airplane was used and the expenses were incurred pursuant to written corporate policy. We agree with petitioners. Respondent's Revenue Ruling 57-502, 1957-2 C.B. 118, acknowledges that a corporate resolution or policy requiring a corporate officer to assume such expenses indicates that they are his expenses as opposed to those of the corporation. In *Gantner,* we disallowed the taxpayer's deduction for expenses and depreciation related to a computer system which he purchased and used in the operation of the Northstar Driving School, Inc. The taxpayer was an officer and a 50-percent shareholder of Northstar. Even though the taxpayer used the equipment in his employment with the corporation, we found that the deductions were not attributable to the taxpayer's role as an employee noting that "There was no corporate resolution or requirement that petitioner, as an employee, incur those expenses." *Gantner v. Commissioner,* 91 T.C. at 726. Similarly, in *Stolk v. Commissioner,* 40 T.C. 345 (1963), *affd.* 326 F.2d 760 (2d Cir. 1964), we disallowed a corporate officer's deduction of expenditures for entertainment and gifts, in part, for the taxpayer's failure "to prove that as part of his duties the corporation expected or required him to assume and pay from his own funds any of the disputed expenses, without payment." *Stolk v. Commissioner,* 40 T.C. at 357.

In *Lockwood v. Commissioner,* T.C. Memo. 1970-141, the taxpayer was an officer of Momex, Inc., but not a shareholder. The taxpayer's wife was a 25-percent shareholder in Momex but not an employee. The shareholders and officers of Momex agreed that the officers would pay out of their own pocket, without corporate reimbursement, certain travel and entertainment expenses incurred by them on behalf of the corporation. We held that the taxpayer's travel expenses incurred on behalf of the corporation were deductible.

In the instant case, Intel had a written travel reimbursement policy explicitly stating that it expected its officers to incur certain expenses for Intel's benefit, despite the fact that such expenses would not be reimbursed. Reimbursement of such expenses was considered inappropriate in light of the corporate culture and the officers' overall compensation. An example of such expenses is the excess cost of first-class airfare over coach airfare when first-class travel was necessary for business purposes. We find that Intel expected petitioner, as a corporate official, to incur and pay travel expenses in excess of the amount which was reimbursable under its policy. Therefore, we hold that petitioner's use of the airplane and payment of the attendant expenses do not constitute the voluntary assumption of corporate expenses.

Respondent argues that by virtue of petitioner's position as founder and chief executive officer of Intel and his involvement in the development of its corporate culture, petitioner, in effect, required himself to assume the travel expenses and, therefore, such assumption was voluntary.

The Intel culture and travel policies that precluded Intel's payment of petitioner's total travel costs were in place prior to petitioner's purchase and use of the airplane and were clearly policies that were established for business purposes. In order to find that petitioner required himself to assume the travel expenses, we would have to ignore the corporate entity of Intel. Courts have consistently interpreted *Moline Properties, Inc. v. Commissioner,* 319 U.S. 436, 87 L. Ed. 1499, 63 S. Ct. 1132 (1943), to preclude ignoring the corporate form when adoption of that form served a business purpose. *Moncrief v. United States,* 730 F.2d 276, 280 (5th Cir. 1984). Respondent has not argued, nor could it be seriously contended, that Intel served no business purpose.

Therefore, despite petitioner's participation in establishing the Intel policy, we find he did not "voluntarily assume" the travel expenses.

Having found that the use and expenses of the airplane were not corporate obligations which were voluntarily assumed, we must decide if the expenses are ordinary and necessary expenses of petitioner's employment. An expense is necessary if it is appropriate and helpful in carrying on the trade or business. *Heineman v. Commissioner*, 82 T.C. 538, 543 (1984) (citing *Commissioner v. Tellier*, 383 U.S. 687, 689, 16 L. Ed. 2d 185, 86 S. Ct. 1118 (1966); *Welch v. Helvering*, 290 U.S. at 113). An expense is ordinary when the paying thereof is the common and accepted practice in light of the time and place and circumstance. *Welch v. Helvering, supra* at 113-114.

In Revenue Ruling 70-558, 1970-2 C.B. 35, respondent held that a Federal Government employee who used his privately owned airplane on official business trips was entitled to deduct expenses and depreciation allocable to such use to the extent those amounts exceeded the standard rate travel reimbursements received from the Government. The taxpayer in Revenue Ruling 70-558 was not required by his employer to use his private airplane but was permitted to do so because of the urgency of his trips. For purposes of determining whether petitioner's use of his airplane was ordinary and necessary, we can discern no meaningful distinction between his situation and the facts in Revenue Ruling 70-558.

It is undisputed that petitioner's duties as vice chairman of Intel required frequent and extensive travel, some of which was not regularly or easily scheduled. Furthermore, the parties agree that petitioner's access to the airplane enabled him to reduce significantly his traveling time, thereby allowing him to attend an increased number of meetings and make an increased number of appearances. Consequently, there can be little dispute that use of the airplane was "appropriate and helpful" to the execution of petitioner's duties. Furthermore, respondent's Revenue Ruling 57-502, 1957-2 C.B. 118, states that "a resolution requiring the assumption of such expenses by * * * [a corporate officer] would tend to indicate that they are a necessary expense of his office." Based on all of the foregoing, we find the airplane expenses for the Intel flights were a necessary expense of petitioner's business as a corporate official.

The expenses of using the airplane must also be ordinary in order to be deductible. *Welch v. Helvering, supra* at 113. The principal function of the word "ordinary" in section 162(a) is to clarify the distinction between expenses which are currently deductible and expenses which are capital in nature. *Commissioner v. Tellier*, 383 U.S. 687, 689, 16 L. Ed. 2d 185, 86 S. Ct. 1118 (1966). The expenses at issue here were not incurred in the acquisition of a capital asset but in the conduct of petitioner's duties as an employee of Intel.

Petitioner traveled by private aircraft only when there was business advantage in doing so. The cost of replicating petitioner's travel schedule and time savings via commercial charter carrier would have exceeded the costs of operating petitioner's airplane. In light of Intel's policies and petitioner's travel requirements, we hold that payment of the excess travel expenses arising from petitioner's use of the airplane was ordinary under the circumstances. See *Lockwood v. Commissioner, supra*.

It has been held that for an expense to be considered ordinary and necessary, it must also be reasonable in amount. *Commissioner v. Lincoln Electric Co.*, 176 F.2d 815, 817 (6th Cir. 1949); see sec. 1.162-2(a), Income Tax Regs. Respondent argues that the expenses are not ordinary and necessary because petitioners' claimed deduction of $139,369 in connection with the airplane is unreasonable in light of the $105,076 salary petitioner was paid as vice chairman of Intel in 1983. The issue here is whether the Intel-related airplane expenses are reasonable in amount. Petitioners argue that in determining reasonableness of the amount of business expenses for purposes of section 162, only out-of-pocket expenses should be considered and that the statutory allowance for depreciation should be excluded from consideration. For purposes of this issue, the $139,369 figure, which respondent relies on, can be broken down as follows:

	Expenses	Depreciation
Intel Flights	$23,140	$96,722
ACM Charter Flights	$3,766	$15,741
	$26,906	$112,463

Whether depreciation should be included in the amount of expenses in making a reasonableness determination depends on whether it is a "business expense" under section 162. The regulations under section 162 provide in relevant part that "Business expenses deductible from gross income include the ordinary and necessary expenditures directly connected with or pertaining to the taxpayer's trade or business, except items which are used as the basis for a deduction or a credit under provisions of law *other* than section 162." Sec. 1.162-1(a), Income Tax Regs. (Emphasis added.) Depreciation is not really an "expenditure" but an allowance based on a presumed wasting of a previous capital investment. See 5 Mertens, Law of Federal Income Taxation, sec. 23A.03, at 14 (1974 rev.). The authority for deducting an allowance for depreciation in this case is section 168. Therefore, depreciation does not fall under the regulatory rubric of trade or business expense. As such, the section 168 depreciation deduction amount should not be included in the amount of business expense, the reasonableness of which is to be determined.

To hold otherwise would raise serious problems since allowable deductions for depreciation will often not be reflective of economic depreciation. If we were to look simply at the combination of allowable depreciation deductions and other expenditures for a particular year, the combination of these amounts might often seem exorbitant in amount, especially in the early years of operation. Respondent has not argued that we should consider actual economic depreciation. In the instant case, however, the $96,722 depreciation allowance did not reflect actual economic depreciation. The airplane did not decrease in value but increased in value by approximately $340,000 from 1983 to 1989. Therefore, petitioner did not suffer an actual economic loss in addition to his out-of-pocket expenditures of $23,140.

The issue then is whether the $23,140 expended for Intel flights is a reasonable amount of expense under the circumstances. Whether an expenditure is reasonable or not for purposes of section 162 is a question of fact. *Boser v. Commissioner,* 77 T.C. 1124, 1133 (1981), as amended 79 T.C. 11(1982), *affd.* by unreported order (9th Cir., Dec. 22, 1983). We do not find the amount of business expense to be unreasonable. The parties have stipulated to the fact that replication of petitioner's private airplane flights through a commercial service would have been more costly. Therefore, we hold petitioners are entitled to deduct petitioner's out-of-pocket expenses of operating the airplane with respect to the Intel-related flights to the extent they exceed the reimbursable expenses.

With respect to depreciation, respondent states on brief that he "agrees that if the Court finds that the petitioner flying his own plane is ordinary and necessary that the amount of depreciation is reasonable." Nevertheless, respondent appears to argue that whether depreciation is deductible at all is dependent on whether the asset's use is "ordinary and necessary," and that the combined amount of deductions for depreciation and business expense must be "reasonable" in order to be "ordinary and necessary." Such a position is without support in the law.

Availability of deductions for depreciation on tangible property in this case is dependent solely upon compliance with section 168, which has only two requirements for deduction of depreciation. First, the asset (tangible) must be of a type which is subject to wear and tear, decay, decline, or exhaustion. Sec. 168(c); sec. 1.167(a)-2, Income Tax Regs. Second, the property must be used in the taxpayer's trade or business or held for the production of income. Sec. 168(c)(1). The language of the section is unequivocal.

IRC Sec. 168. ACCELERATED COST RECOVERY SYSTEM.

(a) Allowance of Deduction.—There shall be allowed as a deduction for any taxable year the amount determined under this section with respect to recovery property.

(b) Amount of Deduction.—

(1) In general.—Except as otherwise provided in this section, the amount of the deduction allowable by subsection (a) for any taxable year shall be the aggregate amount determined by applying to the unadjusted basis of recovery property the applicable percentage determined in accordance with the following table:
    * * *

(c) Recovery Property.—For purposes of this title—

(1) Recovery Property Defined.—Except as provided in subsection (e), the term "recovery property" means tangible property of a character subject to the allowance for depreciation—

(A) used in a trade or business, or

(B) held for the production of income.

Nowhere in the language of section 168 is there a suggestion that availability of the depreciation deduction is dependent on satisfaction of the requirements of section 162. There simply is no requirement that the use of the depreciable property be "ordinary" or "necessary." The only requirement is that it be used in the taxpayer's trade or business.

Subsequent legislative enactments and the accompanying legislative history support our finding that there are no requirements for deducting depreciation allowances other than those imposed by section 168. In 1984, as part of the Deficit Reduction Act of 1984, Pub. L. 98-369, 98 Stat 494, Congress added section 280F to the Internal Revenue Code. This section places limitations on the investment tax credit and depreciation deductions that may be taken for luxury automobiles and certain listed property used by taxpayers in their trade or business. Sec. 280F.

With respect to depreciation of luxury automobiles, the House Report stated that under the then current law:

A taxpayer who acquires an automobile for use in a trade or business and uses it for business purposes is entitled to an investment tax credit and cost recovery deductions under the Accelerated Cost Recovery System (ACRS), in addition to deductions for operating and maintenance expenses.
    * * *

The report then goes on to state that:

the extra expense of a luxury automobile operates as a tax-free personal emolument which the committee believes should not qualify for tax credits and deductions. [H. Rept. 98-432 (Part II), 1387 (1984).]

Subsequently, section 280F was enacted containing a schedule designating maximum automobile depreciation that a taxpayer may deduct. See sec. 280F(a)(1). In addition, section 280F(d)(3) provides that property used as a means of transportation by an employee shall not be considered used in a trade or business for purposes of determining deductible depreciation unless such use "is required for the convenience of the employer and as a condition of employment." H. Rept. (Conf.) 98-861 (1984), 1984-3 C.B. 1, 281 (Vol. 2). The apparent intent behind section 280F is to impose a type of necessary and reasonableness requirement on the deductibility of depreciation allowances for property used in a trade or business.

Finally, we note that respondent has unsuccessfully made this argument before this Court on a prior occasion. In *Hoye v. Commissioner,* T.C. Memo. 1990-57, the language of respondent's statutory notice conditioned allowance of the deduction for depreciation and investment tax credit upon a motor home being "ordinary or necessary to [taxpayer's] trade or business, in accordance with section 162." We characterized that position as untenable. At trial and on brief, respondent argued instead that the taxpayers were not entitled to the depreciation deduction or investment tax credit because the amounts claimed were unreasonable. We observed that the cases on which respondent relied involved section 162, and that the taxpayers were entitled to depreciation if the property was used in their trade or business. Since respondent did not challenge the fact that the motor home was used in the taxpayer's trade or business, we held that the taxpayers were entitled to deduct depreciation to the extent the motor home was used in his trade or business. *Hoye v. Commissioner, supra.*

Based on the foregoing, we find that petitioners are entitled to deduct depreciation on the airplane to the extent of its use in petitioner's employment.

* * *

Decision will be entered under Rule 155.

The United States Court of Appeals recently affirmed the analysis in *Noyce* in *Kurzet v. Commissioner,* 222 F.3d. 83 (10th Cir. 2000). With the *Kurzet* case in place, the bottom line results in these types of cases will largely depend on an aircraft owner's ability to demonstrate that the use of a private aircraft is appropriate and helpful in his or her business.

Investment expenses

Section 212 of the Internal Revenue Code permits deductions for "ordinary and necessary" expenses paid for:

- The production or collection of income
- The management, maintenance, or conservation of property you hold for the production of income
- Any expenses you pay in connection with the determination, collection, or refund of any tax

It is not unusual for aircraft owners to use their aircraft in the furtherance of real estate rental activities and/or other activities involving the production of income. In applying Section 212, the "ordinary and necessary" tests applicable are the same as the tests described above under Section 162. Additionally, the courts will apply a test of reasonableness to any aircraft expenses claimed as related to income-producing activities.

While you may be able to take a deduction under Section 212, you must be careful in supporting your claim. The IRS tends to carefully scrutinize this type of deduction. In the case of *J. M. French v. Commissioner of the Internal Revenue*, T.C. Memo 1993-314, the Tax Court sided with the IRS in holding that expenses claimed by an aircraft owner for trips to his rental resort condominium were really disguised family vacations and, therefore, not deductible.

Educational and proficiency expenses

As an aircraft owner and pilot, you may be looking to upgrade your flying skills. One of the benefits of owning your own aircraft is having a measure of control over how the aircraft is scheduled for your instructional time. Many aircraft owners take advantage of this situation to pick up advanced ratings. When it comes to tax time, many of these owners want to know whether they can take a deduction for their flight training expenses.

Beyond upgraded certificates and ratings, you want to be proficient and you must also keep your certificates and ratings current. This costs money. Are any of these expenses deductible?

This section will review the IRS rules regarding the deductibility of flight training expenses and proficiency flying. We'll review both the rules and some of the more significant cases interpreting the rules.

Internal Revenue Service Regulation Section 1.162-5(a) provides that educational expenses can be deducted if the education (1) maintains or improves the skills required in your present job or (2) meets the express requirements of your employer or the requirements imposed by law to retain your employment status. However, Internal Revenue Service Regulation Section 1.162-5(b)(2) and (3) provide that educational expenses are not deductible if the education you received either:

1. Allows you to meet the minimum educational requirements or standards for qualification in your existing position.

2. Qualifies you for a new trade or business.

These rules and their application are illustrated in Fig. 14-1.

Following the flowchart, here are the questions you will need to answer and a discussion of each question's meaning and importance.

Is your flight training required to meet the minimum educational requirements for your current job?   For most light aircraft owners, the answer to this question will be "no." If the answer is "yes," you can stop here; your flight training expenses are not deductible. In the typical situation, the aircraft owner is a professional or businessperson seeking to increase revenue through the use of an aircraft. Working toward a private or instrument rating will not be necessary to meet the minimum educational requirements for his or her job. This provision was intended to make expenses for things like bar review prep courses and CPA review prep courses (for future lawyers and CPAs) and similar "minimum qual-

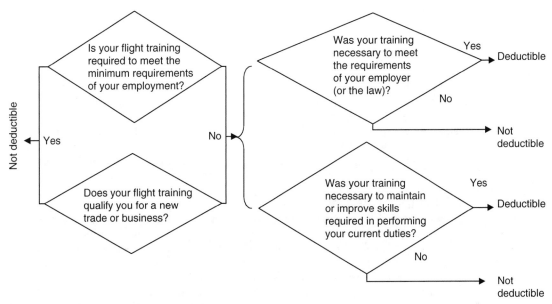

Figure 14-1 Tax deductibility of flight training expenses.

ification" tests and degrees nondeductible. If your flight education incidentally results in a raise, promotion, or extra revenue, you may still take a deduction as long as your flight training did not qualify you for a new trade or business as discussed below.

Did your flight training qualify you for a new trade or business? The seemingly obvious answer for most aircraft owners is "no." You haven't worked on new certificates or ratings to become an airline or corporate pilot. However, before you jump to the conclusion that this question can be answered in the negative, you need to examine the words a bit more carefully. The question does not ask whether you engaged in the flight training in order to qualify for a new trade or business. It merely asks whether the flight training qualified you for a new trade or business. This has been a stumbling block for aircraft owners and pilots who improved their skills through flight training for commercial certificates, ATPs, and certified flight instructor certificates. You may have engaged in the advanced training to simply improve your skills. You probably had no intention of changing professions or jobs. However, once you earn a commercial, ATP, or CFI certificate, you have qualified for a new trade or business. Therefore expenses you incur in obtaining these certifications are generally not deductible. However, once you've gotten your commercial rating, ATP, or CFI and you tack on additional type ratings or instructor ratings, you may have a strong argument that you are merely maintaining or improving skills necessary in your current profession. That would make the expenses deductible. The *Galbreath* case (Case 14-4) is an example of how these rules muddied the waters for a well-meaning air traffic controller taking flying lessons.

Case 14-4

Michael F. Galbreath, Petitioner v. Commissioner of Internal Revenue, Respondent
United States Tax Court
T.C. Memo 1982-540

STERRETT, Judge: By notice of deficiency dated October 16, 1980, respondent determined deficiencies in petitioner's Federal income taxes for the taxable years 1976 and 1977 in the amounts of $285 and $2,494, respectively. The issues for decision are: (1) whether expenses incurred by Mr. Galbreath for flight training during the years in question are deductible as educational expenses; and (2) if such expenses are so deductible, whether the deduction must be reduced to the extent Mr. Galbreath was reimbursed for such expenses by nontaxable benefits from the Veterans' Administration.

FINDINGS OF FACT

Some of the facts have been stipulated and are so found. The stipulation of facts and exhibits attached thereto are incorporated herein by this reference.

Petitioner resided in Oberlin, Ohio at the time of filing the petition herein. He filed individual Federal income tax returns for the taxable years 1976 and 1977 with the Internal Revenue Service Center, Cincinnati, Ohio.

During the years 1976 and 1977, petitioner was employed by the Federal Aviation Administration as an air traffic control specialist. In this capacity, he was responsible for the safe, orderly and expeditious flow of air traffic over an assigned area. The Federal Aviation Administration provided petitioner with on-the-job training as an air traffic control specialist.

During 1976 and 1977, petitioner was enrolled in a private pilot's course at Midwest Aviation, Inc. This course led to petitioner's receiving his private pilot certificate on May 19, 1977. Petitioner then was enrolled from May 19, 1977 to October 6, 1977 in a course entitled "Commercial." Upon completion of that course, petitioner received his instrument rating certification. The costs of these courses were $770 in 1976 and $7,039 in 1977. Of these expenses, $1,192.10 was attributable to the private pilot's course and the remaining $6,616.90 was attributable to the course for which petitioner received his instrument rating certification. Petitioner received reimbursement from the Veterans' Administration in the amount of $5,955.21 during 1977 for his expenses related to the instrument rating course.

It is stipulated that pilot training and instrument rating certification are not required in order for one to be an air traffic control specialist. Petitioner stated at trial that his primary reason for taking these courses was to upgrade his ability as an air traffic controller.

On his Federal income tax returns for 1976 and 1977, petitioner deducted as educational expenses $780 and $7,039, respectively. In his notice of deficiency, respondent disallowed these deductions in their entirety on the ground that it had not been established that the expenses were incurred primarily to maintain or improve skills required in petitioner's employment in 1976 and 1977. Alternatively, respondent asserted that, if the amounts claimed are determined to qualify as educational expenses, such expenses still should be disallowed to the extent petitioner was reimbursed by the Veterans' Administration for such courses.

OPINION

Section 162(a), I.R.C. 1954, allows a deduction for all ordinary and necessary expenses paid or incurred by a taxpayer during the taxable year in carrying on his trade or business. Since the statute does not address directly educational expenses, the regulations take on an added significance. *Bradley v. Commissioner,* 54 T.C. 216, 218 (1970). Section 1.162-5(a), Income Tax Regs., sets forth objective criteria for determining whether amounts expended for education are ordinary and necessary expenses incident to a taxpayer's trade or business. As a general rule, educational expenses are deductible if the education maintains or improves the skills required by the taxpayer in his employment or other trade or business, or if the education meets the express requirements of the taxpayer's employer imposed as a condition to the retention of his employment relationship, status, or rate of compensation. Sec. 1.162-5(a), Income Tax Regs. However, a deduction is not allowable, even though the education maintains or improves skills, or is expressly required by an employer, if such education is part of a program which will lead to qualifying the taxpayer in a new trade or business. Sec. 1.162-5(b), Income Tax Regs.

Since it was stipulated that pilot training and instrument rating certification are not requirements of an air traffic controller, petitioner must establish, in the first instance, that the flight training courses maintained or improved the skills required in his occupation to be entitled to a deduction.

Petitioner contends that he took the courses primarily to provide himself with an understanding of what the pilots he was hired to service were doing so that he might tailor his communications accordingly. He testified that flight training was encouraged by the Federal Aviation Administration and that the knowledge he acquired improved his performance as an air traffic controller. In support of this assertion, petitioner related to this Court several examples of how his knowledge of an experience in handling airplanes obtained from these courses enabled him, as an air traffic controller, to help people in trouble and better perform his duties.

Respondent, however, contends that the Federal Aviation Administration provided petitioner with the training needed to perform his occupation as an air traffic controller, and that the courses in question did not maintain or improve skills required in his trade or business.

Whether education maintains or improves skills required by the taxpayer in his employment must be determined from all the facts and circumstances involved. *Boser v. Commissioner,* 77 T.C. 1124, 1131 (1981); *Baker v. Commissioner,* 51 T.C. 243, 247 (1968). Petitioner has the burden of proof to establish that he is entitled to such deductions. *Wassenaar v. Commissioner,* 72 T.C. 1195, 1199 (1979). In order to satisfy this burden, petitioner must show that there was a direct and proximate relationship between the flight training courses taken by him and the skills required in his employment as an air traffic controller. *Boser v. Commissioner, supra; Carroll v. Commissioner,* 51 T.C. 213, 218 (1968), *affd.* 418 F.2d 91 (7th Cir. 1969). A precise correlation is not necessary, and the educational expenditure need not be for training which is identical to petitioner's prior training so long as it enhances existing employment skills. *Boser v. Commissioner, supra; Lund v. Commissioner,* 46 T.C. 321, 331-332 (1966).

Because the question of whether the flight training courses maintained or improved petitioner's skills as an air traffic controller is essentially factual, previously decided cases are of little precedential value. We have examined the cases relied upon by respondent and find each one sufficiently distinguishable on its facts so that none controls the instant case. Petitioner described how in several instances his flight training helped him to better perform his duties as an air traffic controller. Based on his credible testimony in this respect, we believe that there is a sufficient nexus between flight training and the skills required of an air traffic controller. Since an air traffic controller offers direct assistance to pilots in the air, a knowledge of flight capabilities of aircraft would appear to be invaluable in emergency situations. Thus, we hold that the flight training courses taken by petitioner maintained and improved his skills as an air traffic controller.

However, even though there was a direct and proximate relationship between the flight training courses and the skills required of an air traffic controller, a deduction for educational expenses is not allowable if such education will lead to qualifying petitioner for a new trade or business. Sec. 1.162-5(b)(3), Income Tax Regs. Qualification, in this instance, contemplates not only the actual entering into a new trade or business but also the mere capacity to do so. *Weiszmann v. Commissioner,* 52 T.C. 1106, 1111 (1969), *affd. per curiam* 443 F.2d 29 (9th Cir. 1971). An individual might continue his activities in his existing trade or business and never enter into the new one for which he has qualified. Nevertheless, his educational expenses in merely becoming qualified for a new trade or business are nondeductible. Thus, if the flight training courses taken by petitioner qualify him for a new trade or business, his educational expenses are nondeductible even though he continues in his profession as an air traffic controller.

The standard for determining whether education qualifies a taxpayer for a new trade or business is objective rather than subjective. *Robinson v. Commissioner,* 78 T.C. 550, 557 (1982); *Bodley v. Commissioner,* 56 T.C. 1357, 1361 (1971); section 1.162-5(b)(1), Income Tax Regs. Such determination is made by comparing the types of tasks and activities the taxpayer was qualified to perform before the education in question and those which he is qualified to perform afterwards. *Glenn v. Commissioner,* 62 T.C. 270, 275 (1974). If substantial differences exist in the tasks and activities of various occupations or employments, then each such occupation or employment constitutes a separate trade or business. *Davis v. Commissioner,* 65 T.C. 1014, 1019 (1976).

Petitioner points out that the instrument rating which he received did not by itself qualify him for a new trade or business. However, the regulations also [deny] a deduction for expenditures "for education which is part of a program of study * * * which will lead to qualifying him in a new trade or business." Section 1.162-5(b)(3), Income Tax Regs. The course which petitioner took entitled "Commercial" included specific maneuvers which are needed for a commercial pilot certification. Although petitioner did not actually receive his commercial pilot certification in the years in question, he did fulfill many of the requirements for obtaining such certification. It is not necessary that every requirement for qualification in a new trade or business be satisfied in order for a deduction to be disallowed. In the present case, petitioner's education certainly formed part of the program leading to qualification as a commercial pilot. We held in a companion case that the occupation of a commercial pilot is a separate trade or business from that of an air traffic controller, and here again we sustain respondent's determination that the $6,616.90 attributable to the "Commercial" course is nondeductible.

With respect to the private pilot's course, we believe that petitioner is entitled to a deduction for the $1,192.10 expended. This conclusion is based on the fact that a private pilot's license does not, by itself, qualify or lead to qualifying an individual for a new trade or business. Many people each year take flight training to become private pilots simply for their personal enjoyment. A private pilot's license is thus an end in and of itself rather than an integral part of a program leading to commercial pilot certification. Since petitioner did not receive any reimbursement from the Veterans' Administration for the amounts spent obtaining his private pilot certification, the entire $1,192.10 is deductible.

Accordingly,

Decision will be entered under Rule 155.

Notice how the court disallowed Galbreath's instrument training expenses because they were incurred under the umbrella of a training program titled "Commercial." This case should drive home how important it is to understand how literally the courts have looked at the question of whether your flight training might qualify you for a new trade or business.

Again, if you respond to this question with a "yes," your flight training expenses are not deductible. If you can answer "no," your expenses may still qualify for deductibility if you can pass one of the following two tests.

Was your flight training necessary to meet the requirements of your employer (or the law or regulations) for keeping your salary, status, or employment? If you've gotten this far in the regulatory test and you can answer "yes" to this question, your flight training expenses are most likely deductible. This test is designed for professionals who need recurrent training to retain their certificates, licenses, or professional credentials (or just to keep their employer happy). In the context of aviation training, this test is most likely to affect professional pilots who are required to undergo periodic proficiency training and/or flight instructors who generally must be recertified every 2 years. If you are not a professional pilot or flight instructor, the most relevant test may be the next one.

Will your flight training be helpful in maintaining or improving the skills required in your current job? The answer to this question may be your best hope if you use your aircraft for business. If you want to use an aircraft for business, you will most likely have to get at least a private pilot certificate. It is also a good bet that you'll eventually need an instrument rating. As indicated in the *Sartor* case (see Case 14-2), the courts recognize that if your aircraft expenses are "ordinary and necessary," your training expenses should also be deductible. In the *Knudtson* case (Case 14-5), the Tax Court clearly indicates that Knudtson's instrument training improved skills helpful to his current job.

Case 14-5

Kenneth L. Knudtson and Wadena F. Knudtson, Petitioners v. Commissioner of Internal Revenue, Respondent
United States Tax Court
T.C. Memo 1980-455

FEATHERSTON, Judge: Respondent determined deficiencies in the amounts of $13,877.90 and $2,581.38 in petitioners' Federal income taxes for 1975 and 1976, respectively. After concessions by petitioners, the "[issue remaining for decision is] . . . whether any portion of the hours that the airplane was flown during the instrument flight training of petitioner Kenneth L. Knudtson in 1975 and 1976 may be included in determining the percentage of business use of the airplane during those years.
FINDINGS OF FACT
Some of the facts in this case have been stipulated.
Petitioners Kenneth L. Knudtson (hereinafter petitioner) and Wadena F. Knudtson, husband and wife, filed joint Federal income tax returns for 1975 and 1976 with the Internal Revenue Service Center, Ogden, Utah. At the time the petition herein was filed, they resided in Redmond, Washington.
During 1975 and 1976, and for a number of years prior and subsequent thereto, petitioner owned and operated Knudtson Auto Electric, a sole proprietorship engaged

in the business of rebuilding automobile windshield wiper motors. In this business, maintaining a source of supply of used wiper motors to be rebuilt is more of a problem than selling the rebuilt motors. Therefore, it has been important that petitioner maintain a close association with his suppliers. In this connection, it has been necessary for petitioner to travel to the places of business of his suppliers and to the biannual conventions held by them. It has also been necessary for him to travel in connection with sales of rebuilt wiper motors.

Petitioner's suppliers, for the most part consisting of automobile wrecking yards, are located throughout the United States. Most of petitioner's sales of rebuilt wiper motors are made to warehouse distributors on the West Coast of the United States.

On November 18, 1975, petitioner purchased a model A-36 Beechcraft Bonanza airplane (the plane or the A-36). The plane had an estimated useful life of 7 years and cost $89,700. The purchase price included some $20,000 for special instruments to be used for flying in inclement weather.

At the time that he purchased the A-36, petitioner had a private pilot's license, but he did not have an instrument rating. Between November 18, 1975, and April 13, 1976, he flew a total of 53.6 hours in the A-36 in connection with instrument flight training. On April 13, 1976, petitioner received an instrument rating on his pilot's license. Without this rating, it would have been difficult, if not impossible, for petitioner to maintain any kind of a schedule in using the plane for business purposes.

During 1975, petitioner paid an excise tax on the airplane in the amount of $4,485. In 1976, he paid $10,882.64 in operating expenses for the airplane. The parties agree that the number of hours flown by petitioner in the A-36 during the years in question can be characterized as follows:

Year	Business	Personal	Maintenance	Training	Total hours
1975	1.0	3.7	0	4.7	9.4
1976	100.2	25.1	15.8	48.9	190.0

On his 1975 and 1976 Federal income tax returns, petitioner claimed deductions for the depreciation and operating expenses of the A-36. In addition, on the 1975 return, he claimed an investment credit and additional first-year depreciation for the airplane. Although it is not entirely clear from the record, it appears that the investment credit and the deductions for depreciation and operating expenses were all computed without any adjustment for personal use of the airplane.

In the notice of deficiency, respondent determined that only 11 percent and 58 percent of the total hours flown in the A-36 during 1975 and 1976, respectively, were for business purposes. The remainder of the hours flown, including all hours for instrument flight training, were treated as having been flown for personal purposes. To the extent that he determined that the airplane was used for personal purposes, respondent disallowed the investment credit and the deductions for additional first-year depreciation, annual depreciation, and operating expenses claimed by the petitioner for 1975 and 1976.

[*Author's note:* Respondent's computation of the percentage of business use of the airplane is based on the figures shown in the table, set forth in the above Findings, which gives a breakdown of the hours and purposes for which the plane was flown in 1975 and 1976. With respect to 1976, he computed that 100.2 hours flown for business purposes is equivalent to 53 percent of the total hours flown for all purposes during that year. He arrived at a final figure of 58 percent business usage by apportioning the hours designated as "maintenance" between the business and personal use of the airplane.]

OPINION

Section 162(a) permits a taxpayer to deduct "all the ordinary and necessary expenses paid or incurred during the taxable year in carrying on any trade or business." Section 167(a)(1) provides that "a reasonable allowance for the exhaustion, wear and tear * * * of property used in the trade or business" shall be allowed as a deduction for depreciation. In the case of certain tangible personal property, "the term 'reasonable allowance' as used in section 167(a) may, at the election of the taxpayer, include an allowance" for additional first-year depreciation of 20 percent of the cost of the property. Sec. 179(a).

It is undisputed that petitioner is entitled to deduct annual depreciation and operating expenses of the airplane for each of the years in question, under sections 167 and 162, and that he is also entitled to...a deduction for additional first-year depreciation for 1975, under sections 38 and 179. However, respondent contends that the various deductions are allowable only to the extent of the actual business use of the airplane in each of the years 1975 and 1976. In this connection, he argues that use of the plane for instrument flight training constituted personal, rather than business, use.

Petitioner concedes that he is not entitled to the deductions and investment credit to the extent that he would have been had the plane been used 100 percent of the time for business purposes. He contends, however, that the ratio of business to personal use of the plane during 1975 does not represent the use that he intended to make of the plane when it was purchased and that, therefore, the percentage of business use for 1976 should be used in determining the amount of deductions and investment credit for 1975.

As we understand the law, petitioner's original intention with respect to the plane's use does not control the amount of the deductions or investment credit to which petitioner is entitled. Original intentions often change.

Section 162 allows a deduction for business expenses "paid or incurred during the taxable year." It is clear that the percentage of business use of the plane during 1976 has no relation to the amount of the expenses paid or incurred in operating the plane for business purposes during the taxable year 1975. Further, the allowance for depreciation is to be determined in accordance with the facts as they exist in the period for which a tax return is made. See *Pasadena City Lines, Inc. v. Commissioner,* 23 T.C. 34, 39 (1954); cf. *Roy H. Park Broadcasting, Inc. v. Commissioner,* 56 T.C. 784, 805, fn. 10 (1971); *Philadelphia Quartz Co. v. Commissioner,* 13 B.T.A. 1146 (1928).

We think it clear that a similar rule must be applied with respect to the investment credit and the allowance for additional first-year depreciation, since computation of both of these items depends upon the extent to which a deduction for depreciation is allowable under section 167. See secs. 1.48-1(b)(2) and 1.179-1(a), Income Tax Regs. Accordingly, we hold that the investment credit and the deductions for annual depreciation, additional first-year depreciation, and operating expenses for the year 1975 must be computed with reference to the actual business use of the airplane during that year.

Petitioner further contends that respondent erred by failing to include the hours that the plane was used for instrument flight training in computing the percentage of business use for 1975 and 1976. It is his position that operation of the plane for instrument flight training falls squarely within section 1.162-5(a)(1), Income Tax Regs., which permits a deduction for the expense of education that maintains or improves skills required by an individual in his trade or business. Accordingly, petitioner asserts that flight training hours should at least be apportioned between the agreed business and personal use of the plane, as respondent did with respect to maintenance hours.

Respondent argues that:

Training and proficiency flights are personal in nature, and the operating costs, investment credit, and depreciation attributable to them are not deductible.

In support of this proposition, he cites *Gibson Products Co., Inc. v. Commissioner*, 8 T.C. 654 (1947), and several Memorandum Opinions of this Court. The holding in *Gibson*, however, was simply that a corporate taxpayer failed to show that the expense of training its president to fly, largely for his personal benefit, was a business expense of the corporation. In the other cases relied on by respondent, training and proficiency flights were found to be personal in nature, but the general use of airplanes by the taxpayers in those cases was almost entirely personal and it does not appear that knowledge of flying was among the skills "required" in their respective businesses.

The record in the instant case makes it clear that the expense of owning and operating the A-36 constitutes an "ordinary and necessary" expense of the petitioner's business, within the meaning of section 162. Petitioner's suppliers, and his customers to a somewhat lesser extent, are located throughout a very large geographic area. It is particularly important to petitioner's business that he maintain a close working relationship with his suppliers. He is, therefore, required to travel extensively in connection with his business. Petitioner aptly describes the plane as a business tool. Obviously, the expense of owning and operating the plane is "necessary" in the sense that it is "appropriate and helpful" to the development of petitioner's business. It is also "ordinary" in the sense that it is a "normal and natural response" to the specific conditions under which petitioner conducts his business. *Commissioner v. Tellier*, 383 U.S. 687, 689 (1966); *Welch v. Helvering*, 290 U.S. 111, 113 (1933); *Graham v. Commissioner*, 35 T.C. 273, 278-279 (1960). Respondent does not argue otherwise.

Since the expense of owning and operating the plane is an ordinary and necessary expense of petitioner's business under section 162, it follows that knowledge of flying is one of the skills "required" by petitioner in his business. Clearly, the instrument flight training maintained or improved that skill. During a substantial part of the year, weather conditions make instrument flying essential for efficient use of the plane for petitioner's purposes. Of course, petitioner could have hired a pilot with an instrument rating, but there is nothing in the tax law that would require him to do so. On the record before us, we hold that the expense of operating the plane for instrument flight training is deductible as the expense of education that maintained or improved skills required by petitioner in his business and that all hours flown for this purpose must be included in computing the percentage of business use of the plane during each of the years 1975 and 1976.

To reflect the foregoing,

Decision will be entered under Rule 155.

Another question that you might run into is whether your expenses for proficiency flying are deductible. The answer to this question often corresponds to the answer to the question of whether your flying is helpful to your current job. If your flying is helpful to accomplishing your current job duties, it is most likely that your proficiency flying should also be deductible. But what exactly counts as your proficiency flying?

On the basis of case law, it seems that if your flying is otherwise deductible, you should be able to deduct the cost of proficiency flying that allows you to meet FAA minimum requirements for currency (for VFR or IFR flying). However, you will probably have to allocate your proficiency hours to account for personal versus business use. The current IRS approach to allocation follows the overall approach taken in the *Knudtson* case. For instance, let's say you had the following breakdown of flight hours in your logbook for the previous tax year:

Total time: 110 hours

Business flights: 60 hours

Personal flights: 40 hours

Proficiency flights: 10 hours

Let's further assume that your proficiency flights cost you $1,500.00. Because some of your aircraft use was personal, you will not be able to deduct the full amount. You will have to allocate the proficiency flying between your business and personal aircraft usage.

The Tax Court approach in *Knudtson* suggests that you first calculate a ratio of total business flight hours to all flight hours, excluding proficiency flights. In our case that would equate to a business flight hours percentage of 60 percent (60 business flight hours divided by 100 total flight hours, excluding proficiency flights). Next, you multiply the ratio calculated by the proficiency flight expenses. This will result in the classification of $900 (60 percent of $1,500) of your proficiency flying as deductible.

Can you go further than that by including costs for proficiency flying beyond the legal FAA minimums for currency? Maybe—but you'd better be careful. You might be able to make a strong argument if your insurance policy (life or liability) requires proficiency beyond FAA minimums. Some aviation counsel and accountants have attempted to go further by citing *Colangelo v. Commissioner of Internal Revenue,* T.C. Memo 1980-543. In the *Colangelo* case, the Tax Court allowed a Navy flight surgeon to deduct the cost of up to 300 hours a year for proficiency flying. However, this case may be the exception to the rule rather than the rule. The Tax Court makes it clear that *Colangelo* was in a very unique position as a Navy flight surgeon. On the basis of the caveats in *Colangelo* and other IRS interpretations, the best advice may be to stick with FAA and/or insurance proficiency requirements when considering proficiency flying expenses.

Depreciation Expense

If you use your aircraft for business purposes, one of the more significant deductions available is for depreciation expenses. Depreciation is available for your aircraft only if it is used at least partially for business purposes. Depreciation rules allow you to annually expense a portion of your aircraft's cost, therefore reducing your net income by the amount of depreciation and, in turn, reducing your tax liability.

Depreciation basics

The tax law and accounting rules won't allow you to take a deduction for the full amount of your aircraft purchase price in the year of purchase, even if you use it 100 percent for business. The underlying theory behind the law and rules is that your aircraft is expected to provide you with an economically useful life of several years (with most aircraft it can be many years). Therefore, your expenses for wear and tear, deterioration, and normal obsolescence should be spread over the estimated years of the aircraft's economic usefulness.

In the 1980s the principles guiding depreciation for tax purposes changed dramatically. Although the tax code still required depreciation for assets with expected useful lives of over 1 year, the rate of the depreciation was no longer tied to any reasonable measure of the assets' true useful life. Instead, in many cases, new government rules allowed for an accelerated depreciation for certain assets under certain circumstances. For aircraft, this means that more of your aircraft can be expensed as depreciation in the earlier years of its life. This is generally advantageous because it reduces your tax burden more significantly in those early years and therefore leaves more money in your pocket. With inflation and the time value of money, money you save on taxes in the earlier years of your aircraft's life is worth more than money saved in its later years.

The IRS depreciation system developed for assets placed into business use (tax people call this "placing into service") after 1986 is called MACRS (modified accelerated cost recovery system). Under this system, your aircraft's costs (excluding any anticipated salvage value) are expensed ("written off") over a period of time specified by IRS guidelines for aircraft depreciation. IRS rules also indicate that the system of depreciation in effect when you place your aircraft into service is the system that will govern your aircraft's depreciation through the time you own your aircraft. Because virtually every aircraft placed in service before 1987 should be fully depreciated by 2003, this discussion reviews only the MACRS rules.

Bonus depreciation

Internal Revenue Code Section 179 allows you to take a deduction for "bonus depreciation." For tax years starting after 2002, the maximum amount of bonus depreciation available is $25,000 per year. The purpose of Section 179 is to stimulate capital spending by smaller businesses. This policy objective is achieved by allowing greater depreciation in the early years of an asset's life. This reduces taxable income. In turn, the amount of money that must be spent to pay taxes is reduced. The theory is that the less money you spend on taxes, the more money you'll have available to spend on other equipment and business needs.

You should keep in mind that bonus depreciation is only available if you purchased an aircraft you are using in a trade or business more than 50 percent of the time. It will not apply to an aircraft you acquired other than by purchase (i.e., by gift). Similarly, bonus depreciation is not available if your aircraft is held as investment property.

The amount of bonus depreciation you can expense for your aircraft in its first year is reduced dollar for dollar if your aircraft (or your aircraft plus other business property purchased) placed in service during any year costs over $200,000. For example, if you purchase an aircraft for $215,000 during 2002 you would be allowed to expense up to $10,000 [$25,000 − (215,000 − 200,000)] as bonus depreciation for year 2002.

You may only expense bonus depreciation in the year you first put your aircraft into service. Once you've elected to take bonus depreciation on your aircraft, you will depreciate the aircraft using either accelerated MACRS depreciation or straight-line depreciation (all discussed below under "Mechanics of Calculating Depreciation for Aircraft").

If you purchase a new aircraft between September 10, 2001, and January 1, 2005, you may be eligible for additional bonus depreciation. The bonus depreciation for an aircraft should be calculated at 30 percent of its purchase price. This special 30 percent bonus depreciation can be taken in addition to the Section 179 bonus depreciation and normal depreciation in the first year of your aircraft purchase.

For example, if you purchased a new aircraft for $100,000 on January 7, 2003, that you intend to use entirely for business, you could claim depreciation expense as follows:

1. Section 179 depreciation of $25,000

2. Plus 30 percent bonus depreciation of $22,500 [($100,000 − $25,000) × 30 percent]

3. Plus normal first-year depreciation calculated on the basis of the remaining cost of your aircraft ($52,500) or ($100,000 − $25,000 − $22,500)

Mechanics of calculating depreciation for aircraft

In calculating depreciation on your aircraft, it is important to note that aircraft are classified as "listed property" by the IRS. Listed property generally includes property that can be used for personal as well as business purposes.

If you do not use your aircraft more than 50 percent of the time for business, you will not be able to qualify for regular MACRS depreciation or bonus depreciation under Section 179. In determining whether the 50 percent test is met, you should use hours for the measurement of aircraft usage. If you use your aircraft in conjunction with income-producing property or investments, you cannot count this time toward the 50 percent test. However, you can still depreciate your aircraft if it is used in investment activities. Once you determine the appropriate depreciation method, that method will apply to both the business and investment use of your aircraft.

If your aircraft is used less than 50 percent of the time for business use, you will have to use ADS (alternative depreciation system). For listed property such as aircraft, that means you will have to use the straight-line method of depreciation.

If the percentage of business use for your aircraft drops below 50 percent in any year, you must switch to ADS. The switch has to be permanent. On top of that, you will have to recalculate what accumulated depreciation would have been for the previous years if you used straight-line depreciation. Any difference between accumulated depreciation already taken on your aircraft and accumulated depreciation you would have taken on the aircraft has to be reported as income for the tax year when your aircraft business usage falls below 50 percent.

Before getting into some examples of how to calculate depreciation on your aircraft, you should also be aware that the IRS requires the use of averaging methods or "conventions" when performing your calculations. The two conventions available are the half-year convention and the midquarter convention. As a general rule, the half-year convention will apply when you place 60 percent or more of your business personal property (including any aircraft) in service during the first 9 months of the year. The quarter-year convention will apply when you place 40 percent or more of your business personal property (including any aircraft) in service during the last 3 months of the year.

The IRS provides tables indicating the depreciation percentages you should be using depending on what type of property you are depreciating and the convention you will be using. The IRS classifies most tangible personal property (this category includes aircraft) as 3-, 5-, 7-, and 10-year property. An aircraft you use for personal and business purposes classifies as 5-year property. If you use your aircraft to carry persons or property for compensation, then your aircraft will be classified as 7-year property. Figure 14-2 identifies the percentages the IRS requires for MACRS depreciation calculations using both the half-year and midquarter conventions.

The half-year convention assumes that you placed your aircraft in service halfway through the tax year. This convention also assumes that you sell your aircraft halfway through a tax year. The midquarter convention assumes that you placed your aircraft into service in the middle of the quarter when it was placed in service. This convention also assumes that you disposed of your aircraft in the middle of a quarter, regardless of when it was actually sold.

Let's try to put some of this information into practice with a few examples.

Example 1. On March 1, 2003, you purchase and place in service an aircraft that costs $40,000. This is the only property you place in service in 2003. Therefore, the half-year convention will apply. You use the aircraft 60 percent of the time for business, 15 percent of the time for investment purposes, and 25 percent of the time for personal use. Since your business use of the aircraft exceeds 50 percent, you can use MACRS on the 75 percent combined business and investment uses.

First, you could elect Section 179 bonus depreciation equaling $15,000 ($25,000 maximum × 60 percent business use). Your depreciable basis in the aircraft is now $25,000 ($40,000 purchase price less $15,000 Section 179 bonus depreciation).

Half-Year Convention

	Year 1	Year 2	Year 3	Year 4	Year 5	Year 6	Year 7	Year 8
5-Year	20.00%	32.00%	19.20%	11.52%	11.52%	5.76%	none	none
7-Year	14.29%	24.49%	17.49%	12.49%	8.93%	8.92%	8.93%	4.46%

Midquarter Convention

For Aircraft Placed in Service in First Quarter

	Year 1	Year 2	Year 3	Year 4	Year 5	Year 6	Year 7	Year 8
5-Year	35.00%	26.00%	15.60%	11.01%	11.01%	1.38%	none	none
7-Year	25.00%	21.43%	15.31%	10.93%	8.75%	8.74%	8.75%	1.09%

For Aircraft Placed in Service in Second Quarter

	Year 1	Year 2	Year 3	Year 4	Year 5	Year 6	Year 7	Year 8
5-Year	25.00%	30.00%	18.00%	11.37%	11.37%	4.26%	none	none
7-Year	17.85%	23.47%	16.76%	11.97%	8.87%	8.87%	8.87%	3.33%

For Aircraft Placed in Service in Third Quarter

	Year 1	Year 2	Year 3	Year 4	Year 5	Year 6	Year 7	Year 8
5-Year	15.00%	34.00%	20.40%	12.24%	11.30%	7.06%	none	none
7-Year	10.71%	25.51%	18.22%	13.02%	9.30%	8.85%	8.86%	5.53%

For Aircraft Placed in Service in Fourth Quarter

	Year 1	Year 2	Year 3	Year 4	Year 5	Year 6	Year 7	Year 8
5-Year	5.00%	38.00%	22.80%	13.68%	10.94%	9.58%	none	none
7-Year	3.57%	27.55%	19.68%	14.06%	10.04%	8.73%	8.73%	7.64%

Figure 14-2   MACRS depreciation rates for 5- and 7-year property.

Next, you can depreciate the remaining depreciable basis in the aircraft using the MACRS tables. For the first year of aircraft use, the MACRS table indicates a 20 percent rate of depreciation. Therefore, MACRS depreciation for year 1 will be $3750 ($25,000 depreciable base after Section 179 deduction × 75 percent business and investment use × 20 percent rate of depreciation).

Your total of Section 179 bonus depreciation plus MACRS depreciation will be $18,750 in the first year you place your aircraft into service. In future years, you will continue to apply the appropriate MACRS percentages to the depreciable base of $25,000.

Example 2.   Assume the same facts as in the example above, except that you use your aircraft 40 percent for business, 15 percent for investment purposes, and 45 percent for personal use. Because business use does not exceed 50 percent, you cannot take Section 179 bonus depreciation and you must use the straight-line depreciation method under ADS.

Under ADS guidelines, your aircraft must be depreciated over a 6-year period (if your aircraft is used for commercial purposes the ADS recovery period is 12 years). This means your total depreciation for 2003 will be $2291.66 ($50,000 cost × 55 percent combined business and investment use × ⅙ for the first of 6 years × ½ to factor in the half-year convention for the first year of use).

Example 3. You pay $30,000 for a used aircraft on February 23, 2000. For years 2000 and 2001, you use your aircraft 60 percent for business and 40 percent for personal use. You decided to forgo the Section 179 bonus depreciation. For 2000 and 2001 you used regular MACRS depreciation. For 2000 you had a depreciation deduction of $3600 ($30,000 × 60 percent × 20 percent). For 2001 you had a depreciation deduction of $5760 ($30,000 × 60 percent × 32 percent).

In 2002, your business use drops below 50 percent to 45 percent. At this point, you must permanently switch to the straight-line method. Your depreciation deduction for 2002 will be $2250 ($30,000 × 45 percent × ⅙).

You must also compute what your depreciation expense would have been in 2000 and 2001 using the straight-line method. Your depreciation in 2000 would have been $1500 ($30,000 × 60 percent × ⅙ × ½). Your depreciation in 2001 would have been $3000 ($30,000 × 60 percent × ⅙). In 2002 you'll have to include an extra $4860 in income. This amount represents the difference between depreciation that you previously took using MACRS ($9360) and the depreciation you would have taken had you been using straight-line depreciation ($4500).

Effect of depreciation on aircraft sales

Before we leave the topic of depreciation, we'll need to see it through to the end. Depreciation may create a substantial tax impact when you sell your aircraft. Here's why.

The IRS requires you to track the "adjusted basis" of your aircraft throughout its life. The adjusted basis is the amount you paid for the aircraft plus any major additions or improvements (see "Capitalize versus Expense" later in this chapter) minus the amount you have depreciated your aircraft over its life (also known as "accumulated depreciation"). When you sell your aircraft, your gain or loss is calculated by comparing the sale price of the aircraft and its adjusted basis. If you sell the aircraft for more than its adjusted basis, you have a gain. If you sell it for less, you have a loss.

Therefore, it is possible that you could sell your aircraft for less than what it cost you, and you would still have to report a gain and pay tax on that gain. This tax phenomenon is known as "depreciation recapture." It could be thought of as the IRS's way of "recapturing" the depreciation deductions you've been taking if you sell your aircraft for a gain.

Here's an example of how depreciation recapture works: Let's assume that you purchased an aircraft in February 1999 and it cost you $60,000. You have used the aircraft 100 percent for business purposes and have used regular MACRS

5-year depreciation up until the date the aircraft was sold in December 2002 (generating a total of $49,632 in depreciation expenses). Your selling price for the aircraft is $40,000. Here's how you would compute the gain or loss on the sale of your aircraft:

Sale price		$40,000
Purchase price	$60,000	
Less: accumulated depreciation	$49,632	
Adjusted basis		$10,368
Gain on sale		$29,632

The $29,632 gain on this sale consists entirely of recaptured depreciation. This type of gain on an aircraft will be classified as ordinary income and taxed at your applicable tax rate on ordinary income.

If the aircraft was sold for $70,000 (and that happens from time to time during periods when aircraft are appreciating in value), the gain on the sale would now be $59,632. Of that gain, $49,632 would be from the recapture of depreciation and taxed as ordinary income. The remaining $10,000 would be treated as capital gain income and taxed at applicable capital gain rates. The capital gain income is also known as "Section 1231 gain."

After giving all of this complexity some thought, many aircraft owners may decide that it is best to avoid this discussion altogether by not depreciating their aircraft. That could be a big mistake. Here's why. Depreciation expense reduces the adjusted basis in your aircraft whether you claim it on your tax return as a claimed expense or not. The law requires that the adjusted basis of your aircraft be reduced by "allowed or allowable" depreciation that you would be entitled to deduct. If you fail to take depreciation deductions, and wait too long to amend past tax returns (over 3 years), you could lose your depreciation deductions and, to make matters worse, you may have to deal with the payment of tax on the recapture of depreciation you never even claimed.

Can you be required to pay taxes on gains when you sell a nonbusiness aircraft for a profit? Yes; the gain on this type of sale would be taxed entirely as a capital gain. Therefore, if you purchased an aircraft for $50,000 in 1997 that was never used in business and sold it for $65,000 in 2002, you would have to report a $15,000 capital gain for 2002. There would be no recapture of depreciation— you never used the aircraft for business and therefore you were not entitled to claim depreciation. If you sold the same aircraft for $35,000 and incurred a $15,000 loss, you would not be able to claim a capital loss. The law recognizes only capital gains on your personal use property; it ignores capital losses.

Capitalize versus expense

Another question that you and your tax adviser will have to wrestle with from time to time is whether to depreciate the cost of aircraft repairs and/or improvements or expense these types of costs when incurred. As stated earlier, Section 162 of the Internal Revenue Code provides that ordinary and necessary expenses of your trade or business are deductible. However, Internal Revenue Code Section 263 states that costs of repairs or improvements must be "capitalized" (added to the adjusted basis of your aircraft) if they:

- Increase your aircraft's fair market value.
- Extend the life of your aircraft.
- Permit your aircraft to be adapted for a different use.

So what's the difference between expensing repairs and improvements when incurred and capitalizing the expenditures so they can be subsequently depreciated? There could be a big difference between the two approaches because of the timing of the deductions. If you can expense the costs for aircraft repairs and improvements when made, you will get an immediate reduction of your tax liability. If you must capitalize the costs, you will have to spread the costs over the applicable depreciation period. As a general rule, you would prefer to reduce your tax liability now rather than later because of the time value of money (a dollar in hand today is worth more than a dollar next year). However, in certain instances, you will be forced by the tax rules to capitalize your costs.

This question frequently comes up in the context of engine overhauls. If you had your aircraft engine(s) overhauled, you may want to expense the full cost of the overhaul in the year you paid for the overhaul. However, if the overhaul meets any of the Section 263 tests, you will have to capitalize the costs and depreciate them over time. Although there is no direct guidance on this question when it comes to light general aviation aircraft, the IRS has provided some pretty clear guidance through Tax Advice Memorandum (TAM) 9618004 for turbine aircraft owners.

In the TAM, the IRS ruled that FAA-mandated overhauls for turbine engine aircraft must be capitalized instead of being expensed in the year the cost of the overhaul was incurred. It may be fair to surmise that if the IRS believes that these FAA-mandated overhauls should be capitalized for tax purposes, the overhaul expenses that you may incur for your lighter general aviation aircraft should also be capitalized. After all, one of the reasons for an overhaul is to extend the life of your aircraft engine, and therefore, your aircraft. Additionally, the IRS might have a credible argument that the overhaul of your aircraft engine(s) also has the effect of increasing the value of your aircraft. In any case, it seems prudent to advise that when it comes to engine overhauls, the best approach is to capitalize the costs. It also seems that the cost of long-lived improvements to avionics or your aircraft's interior should also be capitalized rather than expensed.

Other repair, maintenance, and improvement costs may not be so easy to classify. For instance—getting a new paint job for your aircraft. Does it increase the aircraft's value? Does it extend the useful life of your aircraft? Again, if the answer to either of these questions is "yes," the costs of the paint job must be capitalized.

On the other hand, repairs that simply return your aircraft to its ordinary operating condition are expenses that should be currently deducted. This would include most routine maintenance such as tire repairs and replacement, repairs to avionics, and repairs to dings or scratches in your aircraft's airframe.

Hobby Loss

This chapter has been devoted to a review of tax treatment of aircraft-related business expenses. Sometimes, the big issue raised by the IRS is whether or not your aircraft activities are really a disguised hobby. If your aircraft activities can be viewed as having elements of personal pleasure, your activity may be subject to IRS scrutiny and a determination that it is really a hobby, not a business. This IRS rule has become better known as the "Hobby Loss" rule and is found in Internal Revenue Code Section 183.

As indicated above, if your activity is viewed as a business, all of your expenses can be deducted when calculating taxable income. If your activity is viewed as a hobby, your expenses may be deducted only to the extent that you have income. In other words, if the IRS views your losses as a hobby loss, you will not be able to deduct any net losses from your hobby. Further, most hobby-related expenses must be reported as miscellaneous itemized deductions, subjecting the expenses to an offset equal to 2 percent of your adjusted gross income.

Typically, the hobby loss rule may be applied to aircraft owners engaged in an aviation-related business. Usually businesses will involve activities such as flight instruction, banner towing, aerobatic instruction, aerial photography, and any other activity involving the use of aircraft. The rule applies to activities engaged in through individual owners, partnerships, LLCs, and S corporations.

The primary test for determining whether your aviation activity will be characterized as a hobby is whether or not you have entered into your aviation-related business with a profit motive. Although the IRS won't second-guess the reasonableness of your profit-making expectation, you'll still have to bear the burden of proving that you had an "actual and honest" intent to make a profit. However, the burden of proof that your activity is a hobby will shift to the IRS if you have been able to show a profit in any 3 of 5 consecutive years in your activity. You do have the ability to file an election that will force the IRS to wait until the completion of 5 years before the IRS can challenge your profit-making motivation. However, as a matter of strategy, filing such an election may draw undue attention to your aviation enterprise.

Determining whether your aircraft activities are a hobby or a business requires a judgment call. The IRS will make a determination on your activity on the basis of the nine following factors outlined in the IRS's regulations:

1. Is your activity conducted in a businesslike manner? You will advance your claim that you are engaged in a bona fide business if you conduct your opera-

tions in a businesslike manner and keep solid, traceable books and records. If there are other similar operations to yours that are profitable, and you have carried on your activity in a "substantially similar" manner, the case that your business is not a hobby is also advanced. The IRS will also look to see that you have changed your operating practices or methods from time to time to react to the market and improve your chances of profitability.

2. What is your expertise and/or the expertise of your advisers? Your case that your activity is a business will be a lot stronger if you can provide evidence that you consulted with others who are business or legal professionals and/or experts in your type of activity. However, you'll also have to establish that you conducted yourself and your activity consistent with your preparation and the counsel you have been provided.

3. How much time and effort do you put into your activity? When evaluating this factor, more is better. The more personal time you devote to your aviation activity, the stronger your case that it is not merely a hobby. If you've withdrawn wholly or partially from another activity that was profitable, your case is even stronger. However, the IRS does point out that the fact that you do not devote yourself full time to your activity does not necessarily mean that it is merely a hobby, especially if you have competent persons assisting you.

4. Do you expect that the assets used in your activity will appreciate in value? More specifically, this regulation would be asking you if you anticipate that your aircraft will increase in value. Depending on the market, this could be a positive factor for you in establishing that your activity is driven by the motivation to make a profit.

5. Have you had any success in carrying on other activities? If you've engaged in similar aviation activities in the past with some measure of success, it will be very helpful in establishing your profit motive for your current activity.

6. What is your activity's history with respect to profits or losses? The IRS indicates that losses in the early or start-up phases of your business will not necessarily indicate that your activity is a hobby. However, if your losses continue with no realistic hope of turnaround in sight, and you continue in the same manner, your activity will start to look more and more like a hobby. The IRS also advises that it will consider unforeseen circumstances such as weather, theft, or depressed market conditions when reviewing your profit and loss history.

7. How much profit has been earned, even if occasional? This test indicates that the IRS will not just look at how many years you have profit versus how many years you show losses. A string of years where you show small losses that is punctuated by a year with substantial profit will go a long way to establishing profit motive. One the other hand, a large investment in assets such as aircraft, with a relatively small showing of occasional profit may not be very persuasive in establishing your profit motive.

8. What is your financial status? If you have a substantial income from other sources, it will make it more difficult to establish that your unprofitable aviation activities are not a hobby. The fact that the losses from your aviation

activity generate substantial tax benefits will tend to indicate to the IRS that you are not engaged in a profit-motivated activity.

9. How much personal pleasure is derived from the activity? The fact that you enjoy your aviation activities may indicate to the IRS that your activity is a hobby. However, the mere fact that you gain personal satisfaction or enjoyment from your aviation activities is not enough for the IRS to conclude that you are engaged in a hobby.

In the end, these factors are used by the IRS as a guide to determining whether your activity should be classified as a hobby. The IRS regulations indicate that it will not simply weigh the number of factors indicating a hobby versus the number of factors indicting a profit-motivated business. However, before you embark on any aviation-related business, you should carefully review these factors with your tax counsel to determine the best way to establish and operate your new venture.

Compliance Issues

Now that we've had an opportunity to review the substance of the rules, it's time to review the basics of how to implement the rules. Most people will need help from a qualified tax return preparer to make sure they get through the maze of IRS forms necessary to properly comply with the law. However, in the end, it is your tax return that will be filed with the IRS. You should have a fundamental understanding of how your aircraft-related activities are being reported.

How you report your aircraft-related business activities will be largely dependent on whether the activities flow through to you as an individual or through a separate business entity. Therefore, the discussion on compliance has been broken down to zero in on the reporting basics for:

- Employees
- Sole proprietorships
- Partnerships
- Corporations
- LLCs
- Hobbies

Employees

If you are an employee who uses your aircraft for your employer's business, you will report your aircraft and aviation-related expenses as employee business expenses. The reporting form for this type of expense is IRS Form 2106 (Employee Business Expenses).

Your entries on Form 2106 for aviation-related expenses will most likely (and most appropriately) include your actual aviation expenses incurred on behalf of your employer. The big question you will have to wrestle with is how to capture all of your business-related costs on this reporting form.

The biggest issue may be the recording of interest expenses on your aircraft. If you have no interest payments, and you personally own the aircraft (it is not owned by a separate entity such as an LLC or a corporation), you should be able to record all of your aviation expenses directly on Form 2106. The form has a place for calculating actual expenses, the percentage of your business use of an aircraft, and depreciation on your aircraft.

However, if you are like many aircraft owners, one of your larger carrying costs is interest. If you are still making interest payments, you will not be able to deduct them as a part of your aircraft's costs when you are completing your Form 2106. The tax law will not permit you to deduct interest paid on an aircraft loan as an employee business expense.

That could pose a problem. After all, interest may be one of the larger costs you incur as an aircraft owner. To ignore it means that you will be incurring costs that you cannot deduct in any manner even though you are making use of your aircraft in your business as an employee.

One solution to this problem may be to lease your aircraft. The way you accomplish this is to transfer title to your aircraft (initially or once you are aware it will be used for business) to another entity. A C corporation may be the entity of choice in situations like this. Once the aircraft is titled to the C corporation, you will then lease the aircraft from the C corporation whenever you use the aircraft for business or personal purposes. The rent you pay should cover the operating costs of the aircraft. Therefore, your rent costs are your actual costs for operating the aircraft. Now all you will have to do is calculate the percentage of aircraft business use and apply that percentage to the total of your aircraft expenses to get to your actual expenses for business.

Here's an example of how this might work. You set up a C corporation when you purchase your aircraft. Title to the aircraft is held by the C corporation. You use the aircraft about 60 percent of the time for business purposes. During the year, the aircraft generates a total of $20,000 in cash expenses excluding depreciation. In turn, you pay the corporation $20,000 in rental fees to cover all cash expenses. You can report $12,000 of your rental fees paid as employee business expense ($20,000 × 60 percent) less any reimbursements from your employer for the use of your aircraft (remember that if you are eligible for reimbursement, you must request reimbursement). The C corporation will likely show a loss equal to the amount of depreciation taken for the aircraft (remember that you must take depreciation allowed or allowable). The losses will accumulate and can be carried forward into the future as net operating losses (NOLs). The NOLs may come in handy someday to offset capital gains if your C corporation sells the aircraft at a later date.

This sort of arrangement must be established and maintained with care. Your books and records and the books and records of the corporation or LLC must be kept separately. There must be separate bank accounts set up for the LLC or corporation. You must also create a clear and true flow of funds from yourself to the corporation or LLC and in turn from the corporation or LLC to the various vendors related to your aircraft (e.g., banks, insurance companies, FBOs, etc.). You'll have to coordinate very carefully with your tax adviser to make this arrangement work. There is no "one size fits all" solution.

Sole proprietorships

If you use your aircraft in a business where you are the sole proprietor, your aircraft and aviation-related expenses will be reported on an IRS Schedule C. Although there is no specific box or line on the Schedule C for aircraft or aviation expenses, there are lines for travel expenses, insurance expenses, interest expenses, depreciation expenses, and other expenses.

All of your depreciation-related expenses, including Section 179 bonus depreciation and regular depreciation, should first be calculated and reported on an IRS Form 4562 (Depreciation and Amortization). The result of that calculation is what you report on your Schedule C.

In the end, the net profit or loss from your sole proprietorship will include all of your aviation-related expenses. Your sole proprietorship's net profit or loss will then be carried to the first page of your IRS Form 1040 and included in your income as a positive or negative depending on whether you had a gain or loss.

Partnerships

As stated earlier, a partnership is an association or combination of two or more persons for the purpose of making a profit and sharing profits or losses. If your aircraft is owned by a partnership and used for partnership purposes, you will report aircraft or aviation-related expenses on IRS Form 1065 (U.S. Partnership Return of Income). This tax return reports your partnership's revenues and expenses, including anything related to the aircraft owned by the partnership. It also provides an accounting for each partner's share of income or losses.

Just as with a sole proprietorship, a Form 1065 includes most of the necessary lines for reporting expenses related to aircraft. Again, depreciation must be separately computed and reported on an IRS Form 4562.

Each partner's separate share of income or loss is reported to the partners on an IRS Schedule K-1. Your K-1 will indicate the amount of income or loss that you must report on your individual IRS Form 1040 Schedule E. This income or loss will then be carried to your IRS Form 1040.

Corporations

If a corporation owns your aircraft, you have two choices in how to go with respect to the taxation of the corporation. One approach is to make the corporation a regular C corporation. The second approach is to file an election with the IRS to make the corporation an S corporation. Each approach has advantages and disadvantages that we'll review.

By default, a corporation will be considered a regular C corporation for tax purposes. That means the corporation's earnings will be taxed at corporate tax rates unaffected by its shareholders' tax status. Corporations report revenue and expenses on an IRS Form 1120. Any net income of the corporation is taxed at corporate tax rates. On the other hand, if your corporation is an S corporation, taxable income will generally flow through to its owners in the proportion of their share ownership and be taxed at their individual tax rates.

Currently (and for the last 14 to 15 years), individual tax rates have generally been lower than corporate tax rates. Therefore, it has been an advantage to be an S corporation if profits have been generated. For instance, a C corporation that has taxable income of $20,000 in 2002 will pay taxes of $3000. If the corporation was an S corporation and had one married 100 percent shareholder, that shareholder would pay $2400 in taxes (related solely to the corporation's earnings). However, this example does not take into account a myriad of other issues you and your personal tax adviser will need to consider before making a decision on the tax structure of your corporation. Because of your personal circumstances, there may be more to consider than incremental tax rates.

However, if you and your tax adviser decide that an S corporation is more suitable for your situation, you will still have to qualify and file an election. The election is fairly simple to file. It must be filed on an IRS Form 2553 (Election by a Small Business Corporation). In order to qualify as an S corporation, your corporation must meet the following conditions:

- It must be a domestic corporation created under the laws of the United States or any of its states.

- It must not be: (1) a financial institution using a reserve method of accounting for bad debts; (2) taxable as an insurance company; (3) allowed a tax credit for income from Puerto Rico or any U.S. possession; (4) a DISC (domestic international sales corporation); or (5) a TMP (taxable mortgage pool).

- It can't have more than 75 shareholders.

- Every shareholder must be an individual, decedent's estate, bankruptcy estate, trust, or tax-exempt charitable organization.

- No shareholder may be a nonresident alien or be married to a nonresident alien who has an ownership interest by local property law unless the nonresident alien spouse elects to be taxed as a U.S. resident.

- It must have only one class of stock.

The law requires that you make your election for S corporation status during the year before you want it to be effective, or by the sixteenth day of the third month in the tax year you want the election to become effective. If the first tax year is less than 2 months and 15 days, your election must be made no later than 2 months and 15 days after the first day of that year. However, you can get your election accepted as timely filed even if you are past the statutory due dates outlined above, but within 12 months of the due date of filing the first tax return for the corporation (usually March 15 of each calendar year). In order to make this late election work, you should file a Form 2553 and on the top of page 1 write "FILED PURSUANT TO REV. PROC. 98-55" and attach a statement explaining your reason for the late filing.

If your corporation is recognized as an S corporation, revenue and expenses for the corporation will be reported on an IRS Form 1120S. Any net income or loss will then be passed through to you as an individual shareholder (much as it is in a corporation). Your share of the net income or loss will be your share ownership percentage in the corporation multiplied by the corporation's net income or loss. Similar to a partnership, your share of the income or loss will be transmitted to you by the corporation on an IRS Form K-1 (1120S).

LLCs

LLCs are very flexible entities when it comes to tax treatment. An LLC can elect either corporation or partnership status on IRS Form 8832 (Entity Classification Election). If you are the sole member of an LLC, the IRS will disregard the LLC as an entity and treat your activity as a sole proprietorship.

Hobbies

If your aviation activity is a hobby, you have to treat your expenses differently than if your activity had a profit motive. Expenses from a hobby can be deducted. However, they can generally be deducted only to the extent of revenue from the activity. Additionally, expenses that are related to your aviation hobby (except any real estate taxes, casualty losses, or mortgage interest) must be deducted as a miscellaneous itemized deduction. That means that your expenses will be subject to a reduction to the extent of 2 percent of your adjusted gross income for the year in question. Your income from the hobby activity will be reported as "other income" on your form 1040.

Summary

A summary of the basic tax compliance issues is found in Fig. 14-3.

Charitable Use of Your Aircraft

Whether you use your aircraft for business or solely personal purposes, you may be able to take some deductions for expenses incurred in using your aircraft for charitable purposes. You must first remember that for any expenses to be deductible, they must be rendered for the benefit of an entity recognized by the IRS as a charitable organization. Unfortunately, the current tax code will allow only that direct expenses be deducted. This means that you may be limited to deducting fuel and oil expenses directly related to the charitable flights your aircraft has undertaken. This will obviously fall well short of the

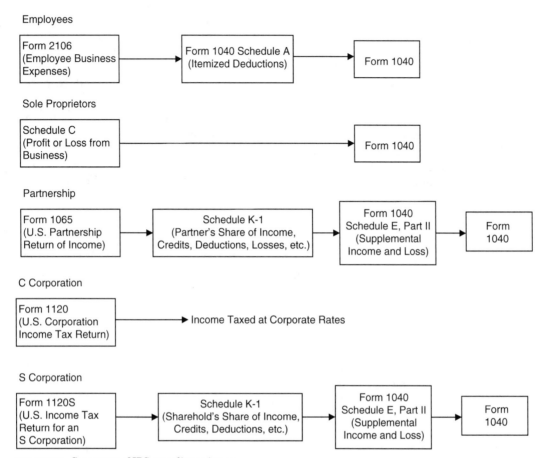

Figure 14-3 Summary of IRS compliance issues.

actual expenses incurred when you calculate indirect costs such as insurance, hangar or tie-down fees, maintenance, and other costs. If you use your aircraft substantially for charitable purposes, you might consider placing it a separate entity such as a corporation or LLC. When you pilot the aircraft for charitable purposes (or for any other purpose), you will pay the corporation or LLC a rental fee. The rental fee can approximate your true costs (usually on a per hour basis) to run the aircraft. This will allow you a charitable deduction that more closely approximates the true value or cost of the flight.

Nontaxable Exchanges

If you own an aircraft for business purposes, there may be a time when you want to (or must) acquire a different aircraft to suit your needs. You may be looking for a faster, newer aircraft. Or you may want an aircraft with a larger load-carrying capacity. There may also be the unfortunate occasion where you must replace your old airplane because it was seriously damaged or destroyed in an accident.

This chapter explores two of the basic tax saving opportunities that might be available to you if you're looking to exchange your old aircraft with a replacement. The first part of the chapter will focus on "like-kind" exchanges (when you want to replace your old aircraft) and the second part will look at "involuntary conversions" (when you need to replace your aircraft because it was damaged or destroyed). For purposes of this chapter we will often refer to your current aircraft as the "old aircraft" and the aircraft you are seeking to acquire as the "replacement aircraft."

Like-Kind Exchanges

If you recall from the last chapter, an aircraft used for business purposes will be depreciated for tax purposes. For most business aircraft users, this will occur over a 5- or 6-year period of time. While you own your aircraft, the depreciation helps your tax situation by generating an expense that reduces your tax liability. However, when you sell your aircraft, depreciation that you've taken on your old aircraft can come back to bite you. As discussed in Chap. 14, depreciation you've taken reduces the tax basis on your aircraft. In turn this reduced basis increases the amount of tax you will have to pay if you sell your old aircraft.

Is there any way to avoid paying a heavy tax on gain from the sale of your old aircraft? Although you might not be able to avoid the tax forever, there may be a way to defer or delay the tax if you are looking to exchange your old aircraft with a replacement aircraft. Section 1031 of the Internal Revenue Code outlines the three basic requirements you'll need to meet in order to make your exchange a tax-free exchange:

- First, you must be replacing your old aircraft with a replacement aircraft through an exchange.
- Second, both your old aircraft and replacement aircraft must be held for productive use in a trade or business or for investment purposes.
- Third, you must be replacing your aircraft with another aircraft.

The underlying tax philosophy behind this provision is that you are really replacing one piece of business property for another (your old aircraft is being replaced by your replacement aircraft). Therefore, your investment in a replacement aircraft can be looked upon as a continuation of your investment in your old aircraft. As long as your old aircraft is exchanged for a replacement aircraft, you will not be taxed on the gain from the transaction. However, if you fail to execute a legal tax-free exchange, you may have a current tax liability for any gains on the disposal of your old aircraft. In order to properly structure a tax-free exchange, you will undoubtedly need professional help from competent tax counsel.

Nonetheless, you should be aware of the basics regarding tax-free exchanges so that you and your tax counsel can work together to make the exchange work to your advantage. Each of the three requirements for Section 1031 is examined below to give you an overview of what you will have to do (and what you should not do) if you want to put together a tax-free exchange of your old aircraft for a replacement aircraft. This section is concluded with an example of how many tax-free exchanges, are accomplished in the world of general aviation; in the accompanying CD is a sample agreement that tracks this illustration.

Exchange requirement

In order for your transaction to qualify for tax-free treatment, it must involve the actual and direct exchange of your old aircraft for a replacement aircraft. If you simply sell your old aircraft and use the proceeds to purchase a new aircraft, it will not meet the requirements for a tax-free exchange.

Does your exchange have to be simultaneous? Most likely it does not. In the past, the IRS required a simultaneous transaction. However, changes to the tax law in the 1984 Tax Reform Act permitted nonsimultaneous transactions within the parameters of two important time limitations:

- First, counting from the day you transfer your old aircraft, your replacement aircraft must be positively identified within 45 days.
- Second, counting from the day you transfer your old aircraft, your replacement aircraft must be transferred to you within 180 days.

Overall, the IRS's regulations indicate that your exchange transaction must be integrated and mutually dependent. Essentially, the IRS will not bless the tax-free nature of your transaction unless you can show that you really engaged in a legitimate exchange, not just a sale of your aircraft and a subsequent purchase.

Trade or business held for investment

The second requirement for a tax-free exchange is that your old aircraft and your replacement aircraft must be held for productive use in a trade or business or held for investment. Whereas some aircraft may be held for investment, it is more common to see aircraft held for productive use in a trade or business. How involved must your aircraft be in your trade or business? There is no direct answer to this question in the tax code or regulations. However, in a private letter ruling, the IRS has stated that property used entirely for business purposes, with the exception of some minimal amount of time, is likely to qualify as being used in a trade or business for purposes of meeting the requirements for tax-free exchange treatment (P.L.R. 8103117, October 27, 1980).

This seems to indicate that your aircraft must be used substantially for trade or business purposes in order to qualify for a tax-free exchange. Can an exact percentage of business use be identified? No. However, the IRS seems to be making the statement that your business use should be substantial if you are thinking about taking advantage of the tax-free exchange provisions.

How long must you have held onto your old aircraft? How long will you have to hold onto your replacement aircraft? Again, there are no specific responses in the code or regulations. However, the IRS again provides some guidance in the form of private letter rulings. In one private letter ruling, the IRS indicates that 2 years of business use is adequate to qualify for tax-free treatment (P.L.R. 8429039, April 17, 1984). In another case, the IRS acquiesced to tax-free treatment of an exchange when circumstances required a taxpayer to dispose of his replacement property when less than 6 months had passed because of an emergency (P.L.R. 8126070, March 31, 1981).

Like-kind requirement

Inasmuch as we're talking about aircraft, it is generally a pretty straightforward process to make sure that your exchange is an exchange of like-kind assets. You cannot pursue tax-free exchange treatment if you are exchanging an aircraft for a computer or a computer for an aircraft.

The IRS has established certain classes of assets that can be used to determine if exchanges of personal property are like-kind. These classes can be found listed in Revenue Procedure (Rev. Proc.) 87-56, 1987-2 C.B. The relevant business class for our purposes includes "Airplanes (airframes and engines), except those used in commercial or contract carrying of passengers or freight, and all helicopters (airframes and engines)." Therefore, as a general matter, you can exchange an airplane for an airplane, or an airplane for a helicopter (or vice versa).

If you've met all of these requirements, you will have a legal, tax-free exchange. Next, you'll have to deal with the details of the accounting for your tax-free exchange and how the numbers will work.

Accounting issues

Beyond meeting the requirements for a tax-free exchange, you should also be aware of the accounting effects. The two most practical issues you and your tax adviser will need to address are the giving or receiving of cash or other property to "even out" an exchange and your basis in the replacement aircraft transferred to you. Both of these issues are addressed below.

Giving or receiving cash or boot. In the simplest of situations, you will be exchanging your old aircraft for a replacement aircraft as an even swap with no cash changing hands. For better or worse, this is rarely the case in aircraft exchanges. Usually cash or other property (collectively referred to by tax practitioners as "boot") also changes hands in exchange transactions because of disparity in the fair market values of your old aircraft and its replacement.

If you receive boot and you have a gain on the transfer of your aircraft, you will be taxed on the amount of gain that is the lesser of the amount of boot you received or the gain on the transfer. Here's an example of how this rule will work:

You and Sam exchange aircraft. The exchange qualifies as a tax-free exchange. You purchased your aircraft for $260,000 and depreciated it $60,000 to date so it now has an adjusted tax basis of $200,000. As far as current market value goes, your aircraft is worth $240,000 and Sam's aircraft is worth $190,000. Because of this difference, Sam has to transfer his aircraft plus $50,000 in cash to you. On this transaction, you must recognize and pay tax on $40,000 of gain. The $40,000 represents your gain on the transaction ($240,000 of value you received less the adjusted tax basis of $200,000 on your old aircraft). Because this is less than the $50,000 you received in boot, you only need to pay tax on the $40,000.

If Sam's aircraft was worth $210,000 (instead of $190,000), and he had to give you $30,000 in cash, your taxable gain would be $30,000. Under this circumstance the amount of the boot received ($30,000) is less than your actual gain ($240,000 less your $200,000 adjusted basis in the old aircraft).

If you have a loss situation, you do not recognize a gain or loss even though you might receive boot. Assume the same facts as the situation just described above, except that now your adjusted basis in your old aircraft is $300,000 rather than $200,000. This means your loss on the transfer is $60,000 ($240,000 of value you received less the adjusted tax basis of $300,000 on your old aircraft). Because this is a loss situation, the receipt of boot of $50,000 does not trigger the need for you to pay tax on the boot you received. The recognized loss on this deal is zero because it is structured as a tax-free exchange (this is a good point to be reminded that the nonrecognition rules of Section 1031 are mandatory and there may be times when you would prefer to reduce your taxes by recognizing a loss).

If you give boot, you generally will not recognize a gain or loss on your tax return. For example, you receive a replacement aircraft with a current fair market value of $250,000. In exchange, you transfer your old aircraft worth $210,000 (with an adjusted tax basis of $150,000) plus cash of $40,000. Based on these facts, you have a gain of $60,000 ($250,000 of value you received less $150,000 of adjusted tax basis on your old aircraft, less $40,000 in cash you paid for the transaction). However, because of the nonrecognition provisions in Section 1031, you will not have to recognize any gain.

Your basis in the replacement aircraft. If your exchange is not qualified as a tax-free exchange under Section 1031, gain or loss will be recognized and the basis of your replacement aircraft would be its acquisition cost. If your transaction qualifies as a tax-free exchange, you'll have to adjust the tax basis of your replacement aircraft to reflect any postponed or deferred gains or losses. The basis of your replacement aircraft will be its fair market value less postponed gain or plus postponed loss.

For example, suppose your old aircraft has an adjusted basis of $300,000 and a current market value of $380,000. You exchange your old aircraft for a replacement aircraft with a current market value of $380,000. As long as the exchange qualifies as a tax-free exchange, you do not have to pay tax on the $80,000 tax gain you enjoy as a result of this transaction. However, the basis of your replacement aircraft will be $300,000 ($380,000 current market value of replacement aircraft less $80,000 postponed gain).

Assume the same facts, with the exception that the basis of your old aircraft is now $480,000 with a current market value of $380,000. Now the basis of your replacement aircraft will be $480,000 ($380,000 current market value of replacement aircraft plus $100,000 in postponed loss). If you turned around and sold your replacement aircraft for its current fair value of $380,000, you would recognize a $100,000 loss for tax purposes.

Illustration

It is not practical to illustrate every possible example of how a tax-free exchange might be arranged in a general aviation aircraft exchange. However, the following illustration is fairly typical of many arrangements for tax-free exchanges involving light to corporate class aircraft.

You own a Beech Bonanza that you have used almost exclusively for business purposes. Your Bonanza has a current fair market value of $250,000 and you have depreciated the Bonanza to the point where its adjusted tax basis is now $50,000. You are now looking to get a bit more space and speed with a Piper Malibu. You have attempted to work an exchange directly with several Malibu owners, but none are able or willing to work the deal with you. So you turn to an aircraft dealer and broker named Busy Broker Limited (Busy Broker) to assist you with this transaction. Busy Broker does not have any suitable Malibu aircraft in their inventory. However, they are able to locate a seller (Sam Seller) who owns just the Malibu you've been looking for and is willing to sell it for $500,000. Busy Broker also has a waiting buyer (Buddy Buyer) who is very interested in purchasing your Bonanza for its current fair market value ($250,000).

Busy Broker could simply act as a broker in this transaction. If the commission percentage were 5 percent, Busy Broker would earn a commission of $12,500 on the sale of your Bonanza to Buddy Buyer and another $25,000 commission on the sale of Sam Seller's Malibu to you. However, you are concerned because of the rather steep taxes you would have to pay on the gain from the sale of your Bonanza. You'd rather try to legally avoid that tax with a tax-free swap of aircraft.

In order to make this deal work, Busy Broker agrees to accommodate the transaction by assisting in the following chain of transactions designed to qualify your transaction for tax-free treatment:

1. You transfer title to your Bonanza to Busy Broker.

2. You set an agreed-upon value to your Bonanza of $237,500. You are willing to establish this agreed-upon value because you no longer have to pay a sales commission as part of this deal.

3. Busy Broker sells the Bonanza and transfers title to the Bonanza to Buddy Buyer in exchange for $250,000.

4. Busy Broker purchases Sam Seller's Malibu for $475,000. Sam Seller is willing to sell his Malibu for this price because he no longer has to pay a sales commission to Busy Broker.

5. After receiving title to the Malibu, Busy Broker transfers title to the Malibu to you in exchange for $262,500 (the difference between the $500,000 value of the Malibu less the $237,500 agreed-upon value of your Bonanza).

6. You pay any applicable sales or use taxes (see Chap. 17).

If this transaction is properly coordinated in accordance with the law and the time limits set by the law (i.e., the 45- and 180-day time limits for identifying and transferring title to the replacement aircraft), you will not be required to pay any tax on your gain for this transaction. However, you will now have an adjusted basis of $312,500 ($500,000 fair market value of Malibu less postponed gain of $187,500) in your new Malibu. As you might note above, Busy Broker still comes out whole on his commissions. If the transaction were a series of traditional sales, Busy Broker would have earned $37,500 in commissions. Busy Broker still nets $37,500 in this series of transactions ($250,000 sale of Bonanza less the purchase price of $475,000 for Malibu plus the $262,500 you pay in cash for the Malibu).

Despite the fact that this type of transaction is commonplace and appears to be meeting IRS approval, you still need to be very cautious. The primary concern is that Busy Broker may be looked upon as your agent in the transaction, especially in view of the fact that Busy Broker received the cash proceeds from the sale of your aircraft to Buddy Buyer. A carefully worded agreement with "antiagency" language (see the sample agreement included in the CD in the back of the book) may help lessen this potential problem. It is also helpful if you can structure your deal so that Busy Broker actually receives title to your aircraft. However, none of this guarantees a "clean" bill of health from the IRS. That is why a number of cautious aircraft owners enlist the assistance of "qualified intermediaries" to help ensure the success of their tax-free exchanges. An overview of how qualified intermediaries can work for your transaction is outlined below.

Use of qualified intermediaries

As illustrated above, it can get a bit complicated to make a tax-free exchange work. The IRS recognized this issue in the early 1990s and in 1991 it issued regulations allowing for certain "safe harbors" to permit greater ease in tax-free exchange transactions. One of the more relevant safe harbors for aircraft transactions is the IRS's grant of permission for you to use a qualified intermediary to facilitate a tax-free exchange.

Under the IRS regulations, a qualified intermediary is a person who:

1. Is not the taxpayer or a "disqualified person" (discussed below).

2. Enters into an exchange agreement with you in which the qualified intermediary will acquire your old aircraft, transfer the old aircraft to a third person who will take title to the aircraft, and then acquires a replacement aircraft and transfers it to you.

A "disqualified person" is generally defined by the regulations as any one of the following:

1. Your agent at the time of the transaction (an "agent" is defined as anyone who may have acted as your employee, attorney, accountant, investment banker, or broker during the 2-year period immediately preceding the transfer of title to your old aircraft). However, if someone provided professional assistance or services with respect to your tax-free exchange or other routine title or escrow services, he or she will not be considered "agents."

2. Anyone who bears a husband-wife or brother-sister relationship.

3. An entity or trust in which you or an entity you own has a greater than 10 percent interest.

If you want to play it safe (and this is probably a good occasion to exercise caution), you may want to contact an industry trade group such as the Federation of Exchange Accomodators (FEA) (current Web address: www.1031.org; current telephone number: 916-388-1031) to get a referral to a qualified intermediary. There are no professional credentials at this time for qualified intermediaries. In view of this it may be a good idea to ensure that your qualified intermediary has insurance for fraud and errors and omissions that will protect you in case things go wrong with your transaction.

As alluded to above, a tax-free exchange using a qualified intermediary is a two-part transaction. In the first part of the transaction you will transfer your old aircraft to the qualified intermediary. The qualified intermediary will then transfer your old aircraft to a buyer in exchange for cash. In the second part of the transaction, a seller will transfer a replacement aircraft to the qualified intermediary in exchange for cash. Title to the replacement aircraft will then be transferred from the qualified intermediary to you. An illustration of how this process works is found in Fig. 15-1.

As always, it is prudent to check with qualified tax counsel before you embark on a tax-free exchange. The first question you will have to resolve is whether it will be beneficial to you (there may be times when you prefer to recognize a gain or loss on your aircraft sale). If you still want to move forward with a tax-free exchange, you will need professional assistance to ensure your transaction is properly executed.

Involuntary Conversions

Are there tax consequences when you receive insurance proceeds on an aircraft that has been destroyed or stolen? There may be. Especially if the

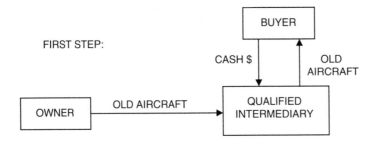

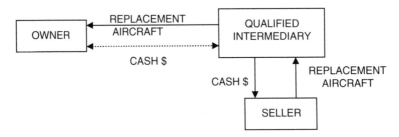

Figure 15-1 Tax-free exchanges of aircraft.

amount of insurance proceeds you receive exceeds the adjusted tax basis of your aircraft.

For instance, let's say that you've owned an aircraft that you use entirely for personal reasons. Ten years ago you purchased the aircraft for $80,000 and over the years you spent $10,000 in long-lasting improvements. Your aircraft is totally destroyed in a hangar fire on October 19, 2002. Thankfully, you carried replacement value insurance. The value of your aircraft has increased over the years and on January 23, 2003, your insurance company pays you $120,000 on your loss. Are you required to pay tax on a $30,000 gain ($120,000 insurance proceeds minus $90,000 adjusted tax basis) in this situation?

Under the general tax rules, you will have to pay tax on the gain of $30,000. However, you may be able to postpone the gain if you invest all the proceeds from your insurance recovery in "qualified replacement property" within a specified period of time.

In the context of aircraft, the only property that will qualify for replacement property will be another aircraft. For aircraft, the specified period of time for replacement will generally be 2 years after the close of the tax year (usually December 31) in which you receive your insurance proceeds.

Applying this to the example above, you can avoid paying any tax on your $30,000 gain if you reinvest at least $120,000 in a replacement aircraft. If you purchase a replacement aircraft for $110,000, you will have to pay tax on $10,000 ($120,000 in insurance proceeds less $110,000 for the cost of a replacement aircraft) while electing to postpone the remaining $20,000 of gain. This will reduce your basis in the replacement aircraft by $20,000 to $90,000. As far as timing is concerned, you will be allowed to postpone the gain as long as you purchase another aircraft between October 19, 2002, and December 31, 2005 (2 years after the year that you received the insurance proceeds).

16

Tax Implications of Aircraft Leasing

Like many other aircraft owners, you might have considered the possibility of leasing your aircraft. Typically, you would pursue this option in order to reduce the costs of operating your airplane. The concept is simple: If you can make some income from the rental of the aircraft, you reduce your costs of aircraft ownership.

Some aircraft owners are also persuaded that there are tax benefits associated with leasing. Several airplane manufacturers tout the benefits of purchasing one of their airplanes and then leasing it through a flight school or fixed-base operator. The aircraft sellers suggest that you can cash flow your aircraft operating expenses while you get the benefits of certain tax deductions.

Before you enter the business of aircraft leasing, you should familiarize yourself with the tax implications of your proposed business. There are certain tax rules you and your tax adviser will have to address. The purpose of this chapter is to explain those rules so you'll better understand the tax benefits and drawbacks to aircraft leasing.

As you might guess—if your aircraft leasing revenue exceeds your expenses, your profits will be taxable. Typically, the income will be taxed the same as your ordinary income. Essentially, this means that you will be taxed on your leasing income at the same rate as your other income is taxed. If you make a profit on your aircraft leasing business, it makes the tax picture a lot less complicated. Generally, the IRS will simply collect the taxes on the income generated by your leasing activity. You'll also be satisfied that your airplane is actually generating a cash flow and a profit.

However, if your aircraft leasing activity is operating at a loss, the situation becomes a bit more complex. The losses from the leasing activity may or may not provide you with an immediate tax benefit. The flowchart in Fig. 16-1 provides an overview of the tax analysis necessary for losses related to aircraft leasing activities.

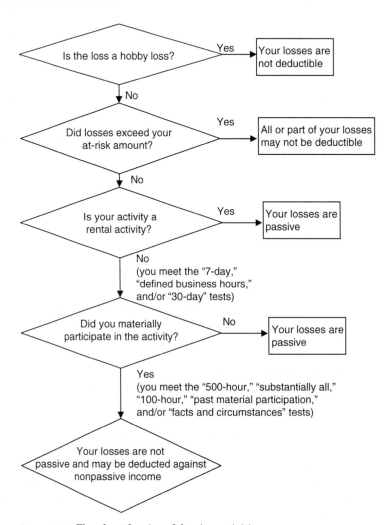

Figure 16-1 Flowchart for aircraft leasing activities.

The first thing you have to keep in mind is that the aircraft leasing activity must be entered into and supported by your honest objective to make a profit. As discussed in Chap. 14 in the section on hobby losses, you have to honestly intend that your leasing activity will make a profit. If you incur losses, you may not be able to take advantage of them for tax purposes if the IRS sees your leasing activities as a mere excuse to enjoy the use of your aircraft while writing off aircraft-related expenses. The IRS will classify the activity as a "hobby" and any losses will be disallowed. Therefore, it would be a mistake to view your aircraft leasing activity merely as a way of creating a shelter from taxation. To do so will only invite trouble.

Even if you treat your aircraft leasing activity as a business and you run the activity in a businesslike manner, it may still lose money—that's okay from a tax perspective. However, you will still have to determine how and when the losses you incurred can be taken as tax deductions.

The second set of rules you will have to deal with is the IRS "at-risk" rules. Essentially, the at-risk rules limit losses from your aircraft leasing activity to the amount that you are financially at risk in the activity. Generally speaking, you are considered at risk to the extent of cash and the adjusted basis (tax value) of the property you contributed to your aircraft leasing activity. You will also be considered at risk for amounts you borrowed for the aircraft leasing activity if (1) you are personally liable for the repayment of the borrowed amounts and/or (2) the amounts borrowed are secured by property other than your aircraft (i.e., a mortgage on your residence). On the basis of these rules, if you purchased an aircraft for either cash, credit, or a combination of both, you would generally be considered at risk for the amount of your aircraft's purchase price.

These at-risk rules apply to individuals (including partners and S corporation shareholders) and to certain closely held corporations other than S corporations. For purposes of the at-risk rules, a C corporation is a closely held corporation if, at any time during the last half of the tax year, more than 50 percent in value of its outstanding stock is owned, directly or indirectly, by five or fewer shareholders.

If you have allowable losses under the at-risk rules, the losses reduce the amount you are at risk for later years. When losses you incur bring your at-risk amount to zero, you may not deduct any further losses. If your losses take your at-risk amount to below zero, the rules permit the IRS to tax some or all of the amount you previously claimed as losses.

If your losses do not exceed your at-risk amounts, you must next determine whether the IRS's passive activity loss rules will apply to your aircraft leasing activity. The IRS put these rules into effect in 1987 in an effort to curb what it perceived as being abusive tax shelters.

The passive activity rules apply to individuals, estates, trusts, personal service corporations, and closely held corporations. Even though the rules do not apply to partnerships and S corporations directly, they do apply to you if you are a partner or shareholder of any such entity.

Here's how the rules can affect you. If you have losses from your aircraft leasing activities, and the losses are considered passive, you can only deduct the losses against your passive activity income for the same year. This may sound fair enough at first glance. However, most taxpayers do not have passive income that can be used to offset passive losses. Therefore, the losses generated by your aircraft leasing activity will not be available for offset against your usual income sources such as salary, wages, interest, dividend, or capital gain income. This typically takes the wind out of the sails of those who tout the immediate tax benefits of leasing your aircraft.

So how does the IRS determine whether your aircraft leasing activity is a passive activity? The IRS defines a passive activity as a trade or business that is a (1) rental activity or (2) an activity in which you do not "materially participate."

The first part of the IRS definition causes immediate concern if you are entering the aircraft leasing business. If the IRS considers rental activities to be passive, how can you possibly avoid having your aircraft leasing activity labeled as a passive activity? After all, the whole point of the business is to rent your airplane.

Nonetheless, there may be some hope on this front. The IRS rules indicate that your aircraft leasing activity may avoid being labeled as a rental activity if you meet the requirements of any of the "tests" described below:

1. *Seven-Day Test.*   The average period of customer use is 7 days or less. You figure the average period of customer use by dividing the total number of days in all rental periods by the number of rentals in the tax year. Under this test, if you lease your aircraft for 180 days in 60 distinct and continuous rental periods, your average period of customer use would be 3 days.
2. *Defined Business Hours Test.*   You customarily make the aircraft available during defined business hours for nonexclusive use by various customers.
3. *Thirty-Day Test.*   The average period of your customers' use is 30 days or less (as calculated in exception 1 above), and you render significant personal services in connection with the leasing activity. The IRS instructs us that significant personal services include only services performed by an individual and do not include:
   a. Services required to permit the lawful operation of your aircraft (i.e., annual or 100-hour inspections).
   b. Services to repair or improve your aircraft that would extend its useful life (i.e., an engine overhaul).
   c. Services that are similar to those you would commonly provide for long-term aircraft rentals (i.e., cleaning or maintenance and/or routine aircraft repairs).

What these rules clearly imply is that the only way you can avoid having your aircraft leasing activity labeled as a rental activity (and therefore passive), is to ensure that your aircraft is rented to a large customer base on a frequent basis. This in turn implies that you probably must seek a rental arrangement through a flight school or fixed-base operator (FBO) for all practical purposes.

This sounds easy enough. If you rent your airplane through a flight school or FBO, it will be rented by lots of different users for short periods of time. It will also be made available for use during defined business hours. On the face of it, it looks like these exceptions will work quite neatly in the context of a leasing arrangement through a flight school or FBO.

However, this is not necessarily the case. In two fairly recent Tax Court cases, the IRS has taken the position that aircraft rental arrangements through FBOs or flight schools should still be considered passive activities. The Tax Court agreed with the IRS in both cases. You can review both the *Frank* case (Case 16-1) and the *Kelly* case (Case 16-2) here.

Case 16-1

Andrew Wesley Frank and Joy Mary Frank, Petitioners v. Commissioner of Internal Revenue, Respondent

United States Tax Court
Tax Court Memo 1996-177

OPINION: Memorandum of Findings of Fact and Opinion

COLVIN, Judge: Respondent determined deficiencies in petitioners' Federal income tax of $51,944 for 1991 and $12,672 for 1992 and accuracy-related penalties of $10,389 for 1991 and $2,534 for 1992.

* * *

After concessions, the issues for decision are:

* * *

5. Whether petitioner's losses from their airplane leasing activity are passive losses under section 469. We hold they are.

FINDINGS OF FACT

* * *

Airplane Leasing

Petitioners bought an airplane for $7,700 on January 22, 1991. Mr. Frank did not know how to fly when they bought the airplane. He learned to fly 2 or 3 years later.

Petitioners bought the airplane with the intent of making a profit. Petitioners leased it to General Aviation Pilots Association (GAPA) from January 22, 1991, to July 15, 1991, and to Slant Alpha Inc., Flight Training (SAFT) from July 1991 to the end of 1991. GAPA and SAFT rented the airplane to people learning to fly. GAPA and SAFT paid for fuel, oil, maintenance, inspection fees, insurance, and parts. GAPA and SAFT sent statements to petitioners showing the airplane's income and expenses. GAPA or SAFT paid petitioner the profit if the airplane's income exceeded its expenses. If the airplane's expenses exceeded its income, petitioners paid the difference to GAPA or SAFT.

Mr. Frank, with a mechanic, changed the airplane's oil, removed screws from the inspection plates, and tied down and washed the airplane. The record does not indicate how often Mr. Frank did those things. Petitioner spoke with flight schools (not otherwise identified in the record) to find the best rate for leasing, maintaining, and insuring the airplane.

Petitioner reported a $2,773.52 loss for the airplane in 1991 and a $3,672.39 loss in 1992.

OPINION

* * *

D. Whether Petitioner's Airplane Leasing Activity is a Passive Activity

Individuals generally may not deduct losses from a passive activity. Sec.469(a). A passive activity is any activity involving a trade or business in which the taxpayer does not materially participate. Sec.469(c)(1). Rental activities are per se passive. Sec.469(c)(2),(4);sec.1.469-1T(e)(1), Temporary Income Tax Regs., 53 Fed.Reg.5702 (Feb 25, 1988). Rental activity is any activity in which payments are principally for the use of tangible property. Sec.469(j)(8). An activity is a rental activity for a taxable year if:

(A) During such taxable year, tangible property held in connection with the activity is used by customers or held for use by customers; and

(B) The gross income attributable to the conduct of the activity during such taxable year represents (or, in the case of an activity in which property is held for use by customers, the expected gross income from the conduct of the activity will represent) amounts paid or to be paid principally for the use of such tangible property (without regard to whether the use of the property by customers is pursuant to a lease or pursuant to a service contract or other arrangement that is not denominated a lease).

Sec. 1.469-1T(e)(3)(i)(A),(B), Temporary Income Tax Regs., 53 Fed.Reg. 5702 (Feb. 25, 1988).

Petitioners lease the plane to GAPA and SAFT. GAPA and SAFT rented the plane to customers who were learning to fly. The airplane was held by GAPA and SAFT for use by customers. The amounts to be paid to petitioners and SAFT were for the use of the airplane. Services were not the dominant element of the relationship. This is a rental activity for purposes of section 469. Rental activity is passive whether or not the taxpayer materially participates in it. Sec.469(c)(4). Petitioners may not use the $25,000 offset under section 469(i) because it only applies to rental real estate activities. We hold that losses from petitioner's airplane leasing activity are passive losses.

Case 16-2

William Warren Kelly, Petitioner v. Commissioner of Internal Revenue, Respondent
United States Tax Court
T.C. Memo 2000-32

OPINION: MEMORANDUM OPINION

POWELL, SPECIAL TRIAL JUDGE: Respondent determined deficiencies in petitioner's 1993 and 1994 Federal income taxes in the amounts of $1,631 and $2,014, respectively. The issue is whether petitioner's aircraft leasing activity is a passive activity under section 469. Petitioner resided in Springfield, Oregon, at the time he filed his petition.

BACKGROUND

The facts maybe summarized as follows. During 1993 and 1994, petitioner was employed full time by a logging company as an equipment operator and mechanic. In 1993, petitioner owned six fixed-wing light aircraft, two of which he purchased in December 1993. In 1994, petitioner purchased an additional aircraft. During 1993 and 1994, petitioner entered into aircraft leasing agreements with Friendly Air Service, Inc. or other fixed-base flight schools (collectively Friendly Air) in the Eugene, Oregon, area.

Under the lease agreements, Friendly Air leased the aircraft from petitioner. Friendly Air would in turn use the aircraft for flight instructions or rent them to other pilots at hourly rates. Petitioner does not have a commercial pilot's license and cannot give flight instructions or transport paying passengers. The leases were for 1 year but could be canceled with a 30-day written notice. Friendly Air scheduled all flights and was responsible for routine cleaning, maintenance, and fueling of the aircraft. Petitioner received $34 per hour of flying time. Petitioner was responsible for the payment of all fuel, maintenance, repair costs, and premiums for commercial insurance. Friendly Air maintained financial records for the leasing of the aircraft. Petitioner did not keep any contemporaneous logs or records of the aircraft activities. The parties, however, agree that petitioner did spend at least 500 hours each year in conjunction with the activity.

Petitioner claimed losses from his aircraft leasing activities in the amounts of $11,274 and $27,014 for 1993 and 1994, respectively. Respondent disallowed those losses as passive activity losses.

DISCUSSION

Section 162 allows deductions for ordinary and necessary expenses incurred in carrying on a trade or business. Section 212 allows deductions for ordinary and necessary expenses incurred for the production of income or the management or maintenance of property held for the production of income. Section 469, however, limits the deductions for losses from any "passive activity."

A passive activity is any activity involving the conduct of a trade or business in which the taxpayer does not materially participate. See sec. 469(c)(1)(A) and (B). All rental activities are generally passive. See sec. 469(c)(2). Furthermore, a rental activity is passive whether or not the taxpayer materially participates. See sec. 469(c)(4); *Frank v. Commissioner,* T.C. Memo 1996-177. An activity is a rental activity if (1) during the taxable year the tangible property held in connection with the activity is used or held for use by customers and (2) the gross income attributable to the conduct of the activity represents amounts paid for the use of the tangible property. See sec. 469(j)(8); sec. 1.469-iT (e)(3)(i), Temporary Income Tax Regs., 53 Fed. Reg. 5702 (Feb. 25, 1988). Under the literal language of the statute, petitioner is engaged in a rental activity and section 469(a) applies.

The regulations provide several exceptions where activities involving tangible property will not be considered rental activities. See sec. 1.469-1T(e)(3)(ii), Temporary Income Tax Regs., 53 Fed. Reg. 5702 (Feb. 25, 1988). Petitioner, however, has not directed us to any specific provision of the regulations. Morever, we have examined these provisions and do not find any relief for petitioner. For example, section 1.469-1T(e)(3)(ii)(A) and (B), Temporary Income Tax Regs., 53 Fed. Reg. 5702 (Feb. 25, 1988), provides that, if the period of customer use is 7 days or less (or 30 days or less and there are significant personal services provided by the taxpayer), the activity involving the use of tangible personal property is not a rental activity. But, under the facts here, the lessee is Friendly Air, and the leases were on a yearly basis. Even if petitioner satisfied the other requirements, the exceptions in the regulations would not apply.

Petitioner also may contend that the exception contained in section 469(i) is applicable because he "actively participated" in the activity. Sec. 469(i)(l). But that section applies only to "rental real estate activities." *Id.*; see also *Frank v. Commissioner, supra.*

In sum, petitioner's leasing of the aircraft is a rental activity and, as such, is a passive activity under the statute and the regulations. While petitioner may have materially participated in the activity, material participation does not exempt the activity from the passive loss rules contained in section 469.

Decision will be entered for respondent.

Although it might appear that renting an aircraft to a flight school or FBO sets the stage for meeting one or more of the exceptions noted above, the Tax Court ruled that the airplanes were really only being rented to one customer—the flight school or FBO. On the basis of that finding of fact, the Tax Court ruled that the aircraft owners could only have passive rental income because they had only one customer (the FBO or flight school) at a time.

Obviously, these rulings create a difficult situation for you if you desire to offset your usual income sources with losses from your aircraft leasing activity. There is no easy or guaranteed way to resolve this problem. One possible approach is to utilize a carefully worded agreement with the FBO or flight school your aircraft will be leased through. An example of an agreement that considers tax implications can be found in the CD that accompanies this book.

Any tax-sensitive agreement should attempt to capture an arrangement where the FBO or flight school is charged with mere safeguarding, scheduling, and collection functions for your aircraft. A well-drafted agreement should also attempt to emphasize that you will be responsible for a lion's share of the work related to the aircraft. Most important, the agreement must make it clear that the renters of the aircraft are your customers, not the FBO's or the flight school's customers.

The agreement provided in the CD and similar agreements have not been tested yet in the courts. It is extremely difficult to determine with any certainty whether these tax-sensitive agreements could withstand IRS and/or Tax Court scrutiny. However, it appears clear from the *Frank* and *Kelly* cases that you must be able to establish that you are not simply renting your airplane to a flight school or FBO. If the IRS views the flight school or FBO as your customer, your aircraft leasing activity has a high probability of being characterized as a passive rental activity.

Even if you are able to get over the rental activity hurdle, you still have one more test to overcome. The IRS rules require that you must "materially participate" in your aircraft leasing activity in order to avoid having the activity classified as passive.

The IRS considers you to have materially participated if you satisfy any one of the following tests:

1. *Five Hundred Hour Test.* You participated in the aircraft leasing activity for more than 500 hours in the tax year.

2. *Substantially All Test.* Your participation is substantially all of the participation in the aircraft leasing activity of all individuals for the tax year, including the participation of individuals who did not own any interest in the activity.

3. *One Hundred Hour Test.* You participated in the aircraft leasing activity for more than 100 hours during the tax year, and you participated at least as much as any other individual (including individuals who did not own any interest in the activity) for the tax year.

4. *Past Material Participation Test.* You materially participated in the aircraft leasing activity for any 5 (whether or not consecutive) of the 10 immediately preceding tax years.

5. *Facts and Circumstances Test.* Based on all the facts and circumstances, you participated in the activity on a regular, continuous, and substantial basis (please note that participation under this rule must exceed 100 hours per year).

The IRS rules also take the position that your participation in managing your aircraft leasing activity does not count in determining whether you materially participated if (1) any person other than you received compensation for managing the activity or (2) any individual spent more hours during the tax year managing the activity than you did (regardless of whether the individual was compensated for the management services). A carefully worded agreement should assist you in proving your material participation by expressly detailing your duties with respect to the aircraft. The agreement may also require an acknowledgment by the FBO or flight school you are working with that none of its employees has invested more than 100 hours in the aircraft activity. This provision will help support your claim that you materially participated in the activity.

IRS publications indicate that you are not required to keep a contemporaneous daily time report or log to prove your material participation. This would allow you to show the services you performed by using a calendar appointment book, or narrative summary. However, it would be prudent to keep detailed and contemporaneous records of all your time related to an aircraft leasing activity.

You might sense by now that being able to write off losses from your aircraft leasing activity may not be so easy after all. If you fail to meet the requirements of the tests described above, your losses will be considered passive. Therefore, you can't use them to offset salaries, wages, capital gains, interest, or dividend income. However, all is not lost. You can store up or suspend these passive losses through the life of your airplane. These losses may be used to offset future passive income.

Obviously, this is of little help if you don't have any passive income. If that's the case, you may take some consolation in the fact that all of your suspended passive losses are generally usable to offset any gains when you ultimately dispose of your aircraft leasing activity. This could be a big tax help to aircraft sellers who are pleased to find out that their old airplane has actually appreciated in value over time. Although the appreciation on the aircraft is welcome, the capital gains tax that you may be subject to on the sale of your aircraft will not be so welcome. The ability to offset capital gains on the sale of your aircraft with stored up passive losses may take some of the sting out of a situation that could otherwise result in a big tax bill.

17

State and Local Taxes

So far, our discussions related to taxation have focused entirely on federal income tax issues. In this chapter, we'll switch gears and discuss some of the often-overlooked, but nonetheless significant state and local tax issues facing most aircraft owners. As an aircraft owner, you are probably aware that your aircraft makes you an inviting target for tax collectors. With tight state and local budgets, you can bet that there will be a sustained effort by taxing authorities to ensure that aircraft owners are complying with applicable state and local tax provisions. Typically, you can't avoid most state and local taxes related to your aircraft. However, you should be aware of the general rules so you can plan, and perhaps reduce or eliminate your liability. We'll first take a look at the basics of sales and use taxes. After that, we'll examine personal property taxes and registration fees.

Sales and Use Taxes

Sales and use taxes originate with state laws. Although the laws related to sales and use taxes may vary widely from state to state, there are some common threads that run through the laws.

Sales tax

The sales tax is customarily imposed on the sale or transfer of tangible personal property. Tangible personal property includes any property that is touchable and moveable. Your aircraft certainly qualifies as tangible personal property. With tax rates approaching 9 percent in some states, it is essential that you do your homework before you make your aircraft purchase in order to avoid unpleasant surprises. Sales taxes are a part of daily life in most states and these taxes apply across the spectrum of sales transactions. These taxes apply to your purchase of a newspaper as well as to your purchase of an aircraft.

The first thing you and your tax counsel should investigate is whether the state where you intend to purchase your aircraft has a sales tax. If you are purchasing an aircraft in the following states, no sales or use tax applies (as of the date of publication):

- Alaska
- Delaware
- Massachusetts
- Montana
- New Hampshire
- Oregon

Keep in mind that state sales and use tax laws change regularly, so you should double-check current law before assuming that taxes do not apply to your transaction.

You may be wondering how states track aircraft sales and use tax transactions. Probably the most prevalent means of tracking is through FAA records. Many states obtain updates from the FAA showing aircraft purchasers located in that state. If your name or your company's name shows up on these lists, you are likely to be contacted by your neighborhood tax collector. Another typical tracking method is for the state to require the registration of aircraft based and used within its borders. These laws may require FBOs or airport operators to ensure that based aircraft comply with registration requirements. When your name or your company's name shows up on a list of registered aircraft, it then becomes a relatively easy task to determine if you have paid a sales or use tax to that state. State and local officials have also been known to review air traffic control tower records and check aircraft on airport ramps to determine if aircraft are subject to sales and/or use taxes. The bottom line is that states are becoming more aggressive in seeking revenues and aircraft owners are relatively easy targets.

Use tax

Usually, the concept of sales tax is pretty easy to sort through. You make a purchase and you pay the tax based on a percentage of the sales price. In the context of an aircraft sale, you are simply paying a percentage of the sales price to the state where you purchased the airplane. It's no different from your purchase of a pack of chewing gum from the local convenience store—except for the dollars involved.

On the other hand, people often have a harder time understanding the concept of a use tax. A use tax is usually imposed on the storage of an aircraft within a state. Wherever a sales tax is imposed, you are going to see a complementary use tax. The use tax is designed, in part, to capture revenue that could be lost by a state when purchases are made outside the state in a no-sales tax or low-sales tax jurisdiction. In essence, the use tax was meant to equalize a state's position relative to other states that had no sales or use tax or lower sales or use taxes. Here's how it works:

Example 1. You own a corporation formed under Delaware law. When your company purchases an aircraft that will be used for your personal and business use, it takes delivery of the aircraft in Wilmington, Delaware. Because Delaware has no sales tax, you have no liability to pay for sales tax as a result of this transaction. If you base and use the aircraft at a Delaware airport, you will likewise avoid any liability for a use tax. However, if shortly after your purchase, you move the aircraft to a new base of operations in Frederick, Maryland, you will be liable for the payment of use tax in the state of Maryland.

Many aircraft owners have been led to believe that if they form a corporation or LLC in a state with no sales or use tax, they can legally avoid the payment of sales or use tax to the state where they keep their aircraft. This is simply not the case. As indicated in the example above, the use tax exists to capture taxes on goods brought into a state from another state with a lower sales tax rate or no sales or use tax.

What happens when you pay a sales tax in a state with a low sales tax and then move the aircraft to your home base in a state where sales and use taxes are charged at a higher rate? The following example provides an illustration of this situation:

Example 2. You purchase an aircraft in Virginia with the intention of moving it immediately to your home base in Maryland. You take delivery of the aircraft in Virginia and fly it to your home base in Maryland the same day. Because Virginia does not have a flyaway exemption, you will be legally required to pay a sales tax of 2 percent to Virginia. With the aircraft in Maryland, you are now subject to a use tax of 5 percent (the normal sales and use tax rate for the state). However, the state of Maryland must give you credit for the 2 percent tax you paid Virginia. Therefore, you will owe Maryland a use tax of only 3 percent (5 percent Maryland tax minus 2 percent Virginia tax rate).

As a general matter of constitutional law, states must give you credit for taxes paid to another state. It behooves you to keep careful records of sales or use taxes you have paid in order to avoid double taxation from another jurisdiction as you move your aircraft from state to state. There are a few states that have tried to avoid giving a credit for sales or use taxes paid to another state. One recent state tax court opinion involving an automobile takes the state of Indiana to task for trying to withhold a credit to an Indiana resident who had already paid a 6 percent sales tax to the state of Florida when he purchased his auto in Florida. When the taxpayer returned to Indiana with the car, the state tax collectors wanted him to pay a 5 percent Indiana use tax while refusing to provide any credit for the taxes already paid in Florida. Indiana relied on a state law that provided for credits to be allowed on everything except for use of cars, boats, and aircraft. The Indiana Tax Court wisely found that the state law relied on by Indiana was unconstitutional because it violated the

Commerce Clause of the United States Constitution. (See *Rhoade v. Indiana Department of State Revenue,* Cause No. 49T10-9902-TA-6.)

Another difficult issue raised by the use tax is the question of sufficient contacts. Just how much "use" or "storage" has to occur in order to trigger a use tax. In one extreme example, tax authorities in New York State attempted to levy the use tax on an aircraft that stopped in the state only momentarily because of weather or emergencies. Thankfully, the State Tax Commission of New York State put the brakes on this effort and ruled that such transitory stops in the state would not create sufficient contacts, or nexus to trigger the imposition of a use tax.

So just how much contact must your aircraft have in a state before you incur a use tax liability? For better or worse, there is no single answer to this question. The question must be answered on a case-by-case basis. Further, the answer may depend on the court and the state law being applied.

In the following case, the taxing authorities in the state of Missouri took the position that even though an aircraft was clearly based and used in Ohio, it was still subject to Missouri use tax because it had enough regular contact with the state to justify the tax. A review of Case 17-1 should give you some idea of the difficulty in making this type of determination.

Case 17-1

Director of Revenue, Appellant, v. Superior Aircraft Leasing Company, Inc., and The Administrative Hearing Commission of the State of Missouri, Respondents
Supreme Court of Missouri
734 S.W.2d 504

OPINION BY: Donnelly
OPINION: This is an appeal by the Director of Revenue from an order of the Administrative Hearing Commission invalidating the assessment of Missouri use tax, section 144.610, RSMo 1986, by the Department of Revenue against Superior Aircraft Leasing Company, Inc. This Court has exclusive jurisdiction of the cause pursuant to Mo. Const. art. V, § 3, because it involves the construction of a revenue law of the state.

There is no dispute as to the facts. Respondent-taxpayer, Superior Aircraft Leasing Co., is a Missouri corporation with its office in Lebanon, Missouri, and its principal place of business in Troy, Ohio. On April 2, 1980, the president of Superior Aircraft, Kenneth Low, purchased and took delivery of an airplane at Beech Field, Wichita, Kansas. The airplane was a 1980 Model 58 Beechcraft Baron (FAA Registration No. N427KL). Low intended to fly the plane directly to Dayton International Airport in Vandalia, Ohio, but because of darkness and his concern for safety in piloting a new aircraft during darkness, he decided to stay overnight in Lebanon, Missouri. He completed his trip to Ohio on the following day.

The aircraft was purchased in order to lease it to Ohio Aviation Company, a company that is in the business of selling Beech aircraft and providing an air charter service. Ohio Aviation leased the aircraft for use in the charter service and for promoting the sale of other aircraft. Upon its arrival in Ohio, the aircraft in question was immediately placed into service pursuant to the oral lease. When not in use by Ohio Aviation, respondent was allowed to use the aircraft for its own business.

Respondent admittedly took delivery in Kansas to avoid liability for Missouri sales and use taxes. No sales or use tax was ever paid on the purchase, use or storage of the aircraft in Kansas, Ohio or in any other state. No personal property tax has ever been paid on the aircraft.

The Department of Revenue assessed respondent for unpaid tax, penalties and interest for the period January 1, 1981, to September 30, 1981. On appeal, the Administrative Hearing Commission held that respondent was not liable for the use tax assessed by the Department of Revenue.

Section 144.610, RSMo 1986, in pertinent part, reads:

1. A tax is imposed for the privilege of storing, using or consuming within this state any article of tangible personal property....This tax does not apply with respect to the storage, use or consumption of any article of tangible personal property purchased, produced or manufactured outside this state until the transportation of the article has finally come to rest within this state or until the article has become commingled with the general mass of property of this state.

2. Every person storing, using or consuming in this state tangible personal property purchased from a vendor is liable for the tax imposed by this law, and the liability shall not be extinguished until the tax is paid to this state....

For purposes of the use tax, section 144.605(7), RSMo 1986, defines "storage" as "the keeping or retention in this state of tangible personal property purchased from a vendor for any purpose, except sale or subsequent use solely outside the state." The term "use" is defined as "the exercise of any right or power over tangible personal property incident to the ownership or control of that property, except that it does not include storage or the sale of the property in the regular course of business." Section 144.605(10), RSMo 1986.

The use tax is a levy on the privilege of using within the taxing state property purchased outside the state, if the property would have been subject to the sales tax had it been purchased at home. *Southwestern Bell Telephone Co. v. Morris,* 345 S.W.2d 62, 66 (Mo. banc 1961). Use taxes have been consistently upheld by the United States Supreme Court and are recognized as serving a dual purpose. The primary function is to "complement, supplement, and protect the sales tax." *Management Services v. Spradling,* 547 S.W.2d 466, 468 (Mo. banc 1977). It achieves this goal by imposing the tax on the "exercise of incidents of ownership of property that was not subject to the sales tax at the time of its acquisition" because of the commerce clause of the United States Constitution. White, *State Sales and Use Taxes-Variations, Exemptions, and The Aviation Industry,* 45 Journal of Air Law and Commerce 509, 515-516 (1979-80). In doing so, it eliminates "the incentive to purchase from out-of-state merchants in order to escape local sales taxes thereby keeping in-state merchants competitive with sellers in other states," and it also provides a means to augment state revenues. *Management Services v. Spradling, supra* at 468.

For determining the applicability of the "use" tax, this Court has adopted and consistently followed the "taxable moment" analysis first recognized in *Southern Pacific Company v. Gallagher,* 306 U.S. 167, 59 S. Ct. 389, 83 L. Ed. 586 (1939). *See Management Services v. Spradling, supra,* at 469, and *King v. L & L Marine Service,* 647 S.W.2d 524, 527 (Mo. banc 1983). The main premise of this theory is that there is no burden on interstate commerce when a period of time is found during which the property has reached the end of its interstate movement and has not yet begun to be consumed in interstate operations. That period of time is the "taxable moment." Under this theory, it is the period when the property ceases to be moved in interstate commerce that is taxable, rather than the transaction of interstate commerce itself.

In *Gallagher,* the United States Supreme Court upheld application of a California use tax where supplies and equipment were transported to California from out-of-state and then consumed in the operation and maintenance of the interstate railway system. The Court explained:

> We think there was a taxable moment when the [goods] had reached the end of their interstate transportation and had not begun to be consumed in interstate operations. At that moment, the tax on storage and use—retention and exercise of a right of ownership respectively—was effective. The interstate movement was complete, the interstate consumption had not begun. *Southern Pacific Co. v. Gallagher,* 306 U.S. at 177.

If we could assume that the "taxable moment" doctrine has continuing viability, we would have no difficulty in finding that a "taxable moment" did not occur in the present case. However, the Director of Revenue contends that the "taxable moment" analysis, adopted by this Court in *Management Services* and applied in *King,* is no longer the proper method for determining the validity of a state tax and we must agree.

The contemporary approach of the United States Supreme Court was announced in *Complete Auto Transit, Inc. v. Brady,* 430 U.S. 274, 97 S. Ct. 1076, 51 L. Ed. 2d 326 (1977). The Supreme Court therein "attempted to clarify the apparently conflicting precedents it has spawned," in determining the effect of the Commerce Clause on state taxation of interstate commerce. *Mobil Oil Corp. v. Commission of Taxes,* 445 U.S. 425, 443, 100 S. Ct. 1223, 1234, 63 L. Ed. 2d 510 (1980). The basic principle that interstate commerce is immune from state and local taxation was rejected. Today, "interstate commerce may constitutionally be made to pay its way." *Maryland v. Louisiana,* 451 U.S. 725, 754, 101 S. Ct. 2114, 2133, 68 L. Ed. 2d 576 (1981) (citing *Complete Auto Transit, supra,* and *Western Live Stock v. Bureau of Revenue,* 303 U.S. 250, 58 S. Ct. 546, 82 L. Ed. 823 (1938)).

The State's right to tax interstate commerce is limited, however, and no state tax may be sustained unless the tax: (1) has a substantial nexus with the State; (2) is fairly apportioned; (3) does not discriminate against interstate commerce; and (4) is fairly related to the services provided by the State.

Maryland v. Louisiana, 451 U.S. at 754. Vermont and Nevada, when confronted with a challenge to the imposition of use tax on an aircraft, have applied the *Complete Auto Transit* test. See, *Whitcomb v. Commissioner of Taxes,* 144 Vt. 466, 479 A.2d 164 (Vt. 1984); *Great American Airways v. Nevada State Tax Commission,* 101 Nev. 422, 705 P.2d 654 (Nev. 1985). See also, *Delta Air Lines, Inc. v. Dept. of Revenue,* 455 So. 2d 317 (Fla. 1984) (test used in imposing state sales tax on purchase of aviation fuel); and *Burlington Northern Railroad Co. v. Ragland,* 280 Ark. 182, 655 S.W.2d 437 (Ark. 1983) (test used in denying application of use tax to railway equipment).

Here, even though the plane was hangared and repairs, if needed, were made in Dayton, Ohio, there were contacts with Missouri sufficient to create a substantial nexus. During the period April 2, 1980, through September 1981, 17.7 percent of the total flight hours were logged on flights to Missouri solely for Superior Aircraft's business. All of these flights, with the exception of one for inspecting a construction site, were recorded as being for board meetings of Superior Aircraft. The time spent in Missouri for each of those trips ranged from several days to approximately a week.

Because Superior Aircraft is a corporation organized and licensed under the laws of Missouri and maintains a business office in Missouri, it is reasonable to infer that the board meetings were conducted in accordance with Missouri law. Additionally, if necessary, Superior Aircraft could have used Missouri state courts to enforce resolutions arising from such board meetings. Such evidence shows both that a "substantial nexus" exists with Missouri and that the tax is "fairly related" to the services provided by the state.

The use tax imposed herein is also fairly apportioned. Multi-state tax burdens can constitutionally be avoided by either allowing an offset or credit for sales or use taxes paid in another state, or by a system of apportionment. *International Harvester Co. v. Department of Treasury,* 322 U.S. 340, 60, 64 S. Ct. 1019, 1035, 88 L. Ed. 1313 (1944). Superior Aircraft has not paid sales or use tax in any other state and even if it had done so Missouri has a system of tax credits for taxes paid in other states. See Section 144.450, RSMo 1986.

The remaining element of the test, whether the tax is discriminatory, requires that the state place no greater burden upon interstate commerce than it places upon competing intrastate commerce of like character. *Complete Auto Transit,* 430 U.S. at 282, 97 S. Ct. at 1080. Missouri's use tax makes no distinction between interstate and intrastate businesses. Any business, intrastate or interstate, which makes an out-of-state purchase of an airplane to be stored, used or consumed in Missouri, is liable for use taxation. There is no discrimination since interstate and intrastate commerce are equally burdened.

Under the test prescribed in *Complete Auto Transit,* imposition of the Missouri use tax in this case is permissible under the Commerce Clause of the United States Constitution. *Management Services v. Spradling* and *King v. L & L Marine Service* should no longer be followed.

The decision of the Administrative Hearing Commission is reversed.

Notice that in this case, the aircraft was in Missouri approximately 17.7 percent of its logged time. This was enough for the Missouri court to tip the scales in favor of imposing the use tax. It should be noted that one judge dissented in this case and indicated that he did not believe that the time and contacts of the aircraft in the state of Missouri were adequate enough to justify the use tax.

On the basis of case law and experience, it seems that a mere stopover for weather or even for extended maintenance may not be enough to create a sufficient contact for use tax purposes. However, beyond that, it is often difficult to determine exactly what threshold the tax collectors and courts will use to establish contact sufficient for use tax liability.

Exemptions

If you are not planning to buy and use your aircraft in one of states where sales and use taxes don't apply, you'll have to gather some additional information. One of the first bits of information you should seek relates to exemptions. Does your aircraft purchase qualify for any exemptions in the state where you plan to make your purchase? A review of some common exemptions is found below.

If you intend to make your purchase in one state and then move the aircraft to another state for use, you may first want to determine if the state where you will make the purchase has a flyaway exemption. A flyaway exemption is an exemption that will allow you to purchase an aircraft in a state with sales tax without having to pay sales tax as long as you fly the aircraft out of the state within a specified period of time. This type of exemption is a commonsense provision recognizing the fact that many aircraft sales transactions take place over state borders with only a temporary connection between the purchaser and the state where the aircraft sales transaction takes place. However, a review of sales and use tax laws from the various states indicates that many states do not have a flyaway exemption. Before you take delivery of an aircraft in a state other than the state where the aircraft will be based, take a close look at that state's laws to determine if it has a flyaway exemption. If not, you may want to change your plans regarding the location where the sale should take place or prepare to deal with two taxing jurisdictions—the one where the purchase took place and the one where you are eventually taking the aircraft to be based and used.

Another possible exemption from sales or use tax is the casual, isolated, or occasional sale exemption (we'll call it the "casual sale exemption"). This exemption is no longer found in most states, but it is worth a look at the applicable state law to determine if it is available. In its purest form, the casual sale exemption fully exempts any aircraft sales transaction where the seller is not regularly engaged in the trade or business of selling aircraft. In modified forms, the casual sales exemption only exempts sales up to a certain dollar limit. For instance, only the first $1000 of your aircraft purchase may be exempt from sales tax in jurisdictions with a limited exemption for sales and use tax. In any case, you should be aware of this exemption and you should check to see if it is available where you will be buying or using your aircraft. Case 17-2 illustrates the typical considerations when evaluating whether a sale is an "isolated or casual" sale.

Case 17-2

Nevada Tax Commission, Appellant, v. Alexander K. Bernhard, Respondent
Supreme Court of Nevada
683 P.2d 21

OPINION BY: Per Curiam
OPINION: This is an appeal from a judgment of the lower court holding that respondent's purchase of an airplane constituted an "occasional sale" and was therefore exempt from a use tax under the provisions of the Sales and Use Tax Act, 1955, Nev. Stats. ch. 397. The Nevada Tax Commission has appealed and contends that the lower court erred in holding that the transaction constituted an occasional sale. We disagree.

Christen Industries, Inc. (hereafter Christen) is a California corporation engaged in the business of selling aerobatic airplane kits and is a registered retailer with the California State Board of Equalization. Christen used a twin-engine Cessna aircraft in the course of its operations for the purpose of corporate transportation.

In November of 1981, Christen sold the Cessna in California to Alexander K. Bernhard, the respondent. Bernhard based the Cessna in Nevada, but he did not remit a use tax to the State of Nevada.

The Nevada Department of Transportation subsequently determined that Bernhard owed a use tax on the airplane and sent him a deficiency notice to this effect. Bernhard petitioned for a redetermination of the assessment. A hearing officer for the Department of Taxation upheld the assessment, which decision was affirmed on appeal to the Nevada Tax Commission.

In December of 1982, Bernhard petitioned the district court to review the decision of the tax commission. The district court held that the sale of the Cessna constituted an occasional sale which, under NRS 372.320, exempted the transaction from a use tax.

On appeal, the tax commission contends that because Christen held or used the Cessna in the course of its retailing activity, the sale of the Cessna does not constitute an occasional sale.

The Sales and Use Tax Act was enacted by the legislature in 1955 and approved by the people of Nevada in a referendum vote in 1956. The act imposes an excise tax "on the storage, use or other consumption in this state of tangible personal property purchased from any retailer on or after July 1, 1955, for storage, use or other consumption in this state...." NRS 372.185. Exempted from this tax are the "gross receipts from occasional sales of tangible personal property and the storage, use or other consumption in this state of tangible personal property, the transfer of which to the purchaser is an occasional sale." NRS 372.320. NRS 372.035(1)(a) defines an "occasional sale" as including:

> A sale of property not held or used by a seller in the course of an activity for which he is required to hold a seller's permit, provided such sale is not one of a series of sales sufficient in number, scope and character to constitute an activity requiring the holding of a seller's permit.

At issue in this case is whether Christen can be considered to have held or used the Cessna in the course of an activity for which Christen was required to hold a seller's permit. The court below held that because Christen was not in the business of selling airplanes such as the Cessna, the sale was an occasional sale. The tax commission, on the other hand, acknowledges that Christen was not in the business of selling airplanes such as the Cessna, but asserts that because the Cessna was used in the course of Christen's activity in selling airplane kits, the "occasional sale" exemption is not available.

The issue presented is one of the proper construction of the definition of an "occasional sale." In construing a law approved by referrendum, the normal rules of statutory construction apply. *Pershing Co. v. Humboldt Co.,* 43 Nev. 78, 183 P. 314 (1919) (*opn. on rehrg.*). Where the meaning of a particular provision is doubtful, the courts will give consideration to the effect or consequences of proposed constructions. See *NL Industries v. Eisenman Chemical Co.,* 98 Nev. 253, 645 P.2d 976 (1982); *Alper v. State ex rel.* Dep't Hwys., 96 Nev. 925, 621 P.2d 492 (1980). If the language of the provision fairly permits, the courts will avoid construing it in a manner which will lead to an unreasonable result. *NL Industries v. Eisenman Chemical Co., supra*; *School Trustees v. Bray,* 60 Nev. 345, 109 P.2d 274 (1941). Additionally, in determining the meaning of a specific provision of an act, the act should be read as a whole. See *White v. Warden,* 96 Nev. 634, 614 P.2d 536 (1980); *Midwest Livestock v. Griswold,* 78 Nev. 358, 372 P.2d 689 (1962). Finally, where possible, a statute should be read to give meaning to all of its parts. See *Sheriff v. Morris,* 99 Nev. 109, 659 P.2d 852 (1983); *Nevada State Personnel Div. v. Haskins,* 90 Nev. 425, 529 P.2d 795 (1974).

In this case it is apparent from the act as a whole that transactions denominated occasional sales were intended to be exempt from sales and use taxes. The tax commission's interpretation of the definition of occasional sale would make the exemption available only in rare circumstances, and not to merchants at all. Such an interpretation would largely nullify the occasional sale provision and thereby violate the rule of statutory construction that effect should be given to all of a statute's parts. See *Big Three Industries, Inc. v. Keystone Industries, Inc.,* 472 S.W.2d 850 (Tex.Civ.App. 1971). Accordingly, we hold that the language of the exemption was intended to cover the circumstances of this case.

Affirmed.

As this case illustrates, a determination as to whether a sale is isolated or casual will require a close look at the seller's history with respect to aircraft transactions. Thankfully, the court recognized that the tax collector's interpretation of the exemption would lead to unreasonable results.

In some states, you may also find an exemption that applies to aircraft that are used in interstate commerce. Most often, this exemption will only apply to aircraft being used in commercial operations. In some states, it may apply to any aircraft being used for business purposes and flying regularly across state borders.

If you plan to lease your aircraft, there may be states that will exempt your lease transactions from taxation. Depending on the state, dry and/or wet aircraft leases may be exempt from taxation. There are some states where you may get an exemption because you intend to resell your aircraft through a leasing arrangement. This may require that you obtain a resale certificate. But with potentially thousands of tax dollars at stake, it may be worth the effort.

If you are leasing your aircraft through a flight school or FBO, you may have the interesting problem of determining whether there is a sales tax when your aircraft is leased. In some states, the law considers the leasing of the aircraft with a flight instructor to be a service that is not subject to sales tax. On the other hand, if you are leasing your aircraft to a licensed pilot, the lease is looked upon as a "sale" subject to sales tax that you may be responsible for collecting. Take a close look at your state's requirements and interpretations to determine how the law will treat your aircraft leasing activity.

One more set of commonplace exemptions worth note is exemptions for sale of aircraft repair parts and/or labor. With the cost of aircraft maintenance, this issue should not be ignored. States vary on this exemption. Some allow exemptions on aircraft repair parts and/or labor whereas others tax one or the other or both. You should familiarize yourself with the laws where you get your repair and maintenance work done. It is not unusual to see aircraft owners in nonexempt states taking expensive maintenance to other states if the overall cost works to their benefit.

Compliance issues

A very practical consideration for aircraft owners is compliance with sales and use tax laws. How do you notify the state of your aircraft purchase? Are you required to notify the state? How long do you have until you must make payment on sales and/or use taxes? Again, the answers to these questions will vary from state to state. However, there are enough common approaches to complying with sales and use tax laws that we can make some general observations.

The first and most important observation is that you are responsible for initiating compliance. You are presumed to have knowledge of the law and to properly comply with the law. Every state that has a sales and/or use tax has a compliance office and many even designate a particular person assigned to aircraft matters. A quick telephone call to the state department of revenue or taxation should get you started in the right direction. In today's digital age, you should even be able to download many necessary forms on the Internet.

It is also important to understand that, although aircraft transactions are generally no different from the sales of other assets, there is one important difference when it comes to sales and use tax compliance. Unlike the usual transaction where you are buying goods from a vendor, your seller will typically not be collecting sales and use tax. One exception to this general observation is the case where you are buying from a broker or dealer who may be obligated by state or local law to collect sales taxes.

In any case, if you fail to report your transaction, it is most likely that you will receive a friendly reminder from the tax collector after a few months or weeks. This reminder may come after the state receives FAA records documenting your purchase of the aircraft. Often your state will use the state registration of your aircraft as a trigger to inquire about the payment of sales and use taxes. Sometimes the payment of the sales and use tax is a prerequisite to obtaining necessary state registration for your aircraft.

What happens if you have based your aircraft in one state for a few years and you are now preparing to move the aircraft to a new state? The issue here is whether the state you move the aircraft to will require a use tax. The answer is most often yes. The more important question may be the valuation of the aircraft for purposes of assessing the use tax. Some states will actually depreciate the value of the aircraft by a certain percentage for each year you held the aircraft outside the state. Other states will simply use an updated valuation based on a chosen commercial valuation service. Under either scenario, you should still be able to claim a credit for taxes paid to the state where your aircraft was previously based.

Another sales/use tax compliance issue often arises if you decide to transfer your aircraft to a newly formed partnership, limited liability company, or corporation. Technically the transfer is a sale. Do you now have to pay sales tax on another transaction involving the same aircraft? Check the law carefully before you make such a transfer. Many states will exempt this type of transfer. In those states the transfer is merely looked upon as part of the initial capitalization of the new legal entity. However, in states without such an exemption, you may get tagged again for sales and use tax when the state gets FAA's records indicating an aircraft newly registered in the name of a partnership, limited liability company, or corporation.

Personal Property Taxes and Registration Fees

If you pay a sales and/or use tax, are you off the hook for taxes and fees at a state and local level? Maybe—but you should first check to see if the state where your aircraft is based has a personal property tax or registration fee. A few states collect both registration fees and personal property taxes, whereas some collect only one or the other. Still others collect neither. Tables 17-1 and 17-2 are lists of states that have personal property taxes and registration fees that apply to lighter, general aviation aircraft used for business or pleasure. Keep in mind that states regularly change their laws related to registration and personal property taxes for aircraft. In order to make informed decisions you should seek up-to-date information from the applicable state.

Aircraft registration fees

Registration fees seem to be more popular among states in current years. Essentially, states with registration fees require that you register your aircraft for state purposes in addition to your federal registration through the FAA.

Registration of your aircraft with a state is often a way for the state to locate you and your aircraft for purposes of levying a sales or use tax. Often, local FBOs or maintenance shops bear the burden of helping the state tax collector police state aircraft registrations.

The size and shape of aircraft registration fees vary widely. Sometimes they are imposed on the basis of a flat rate per aircraft. Other times, they are imposed on the basis of weight or seating capacity. There are even states where the registration fee is calculated on the basis of the fair market value of your airplane—this makes the fee look a lot more like a sales and/or use tax.

Personal property tax

As indicated in Table 17-2, many states permit the taxation of aircraft as personal property. More specifically, aircraft are looked upon as tangible personal property. Again, this simply means that your aircraft has a physical presence and is capable of movement. On the basis of these characteristics, your aircraft may be subject to a tax on tangible personal property.

Personal property tax rules may not only vary from state to state. They may also vary from county to county or locality to locality within a state. Rates of tax may also vary widely. For many reasons, aircraft are generally a worthwhile target for personal property taxes. Your aircraft is carefully regulated and registered—often at the federal and state level. Beyond that, your aircraft

TABLE 17-1   States Requiring Registration of Lighter, General Aviation Aircraft

Arizona	Indiana	Minnesota	North Dakota	South Dakota
Connecticut	Iowa	Mississippi*	Ohio*	Utah
Hawaii	Maine*	Montana	Oklahoma	Virginia
Idaho	Massachusetts	New Hampshire	Oregon	Washington
Illinois	Michigan	New Mexico	Rhode Island	Wisconsin

*If you register late in this state, the state's personal property tax will be applied to your aircraft.

TABLE 17-2   States with Personal Property Taxes on Aircraft

Alabama	Georgia	Missouri	South Carolina	Virginia
Alaska	Kansas	Nebraska	Tennessee*	Washington
Arkansas	Kentucky	Nevada	Texas*	West Virginia
California	Louisiana	North Carolina	Utah	

*In this state, taxes apply only to aircraft used for business.

is also a relatively high-value property that may yield significant taxes without a lot of hassle for the tax collector.

What triggers a personal property tax? Typically, the tax is triggered because your aircraft has established a permanent presence in a taxing jurisdiction. Therefore, it is important to note that the personal property tax is not necessarily related to the state and/or county where you live. Instead it is related to the state and/or county where your aircraft is typically located. As Case 17-3 indicates, if you want to avoid taxation in a particular county or district, you must be careful to establish that your aircraft has a permanent presence in another jurisdiction.

Case 17-3

Tom H. Davis, Petitioner v. City of Austin, Respondent
Supreme Court of Texas
632 S.W.2d 331

OPINION BY: Pope
OPINION: The City of Austin sued Tom Davis to recover delinquent taxes for 1975 on an aircraft owned by Davis. Davis answered that his aircraft did not have a tax situs within the city. After trial to the court, the judge rendered a take-nothing judgment against the city. In his findings of facts, the trial judge stated that the aircraft departed permanently from Austin in December 1974 and acquired, as of January 1, 1975, a permanent location outside the city and separate from Davis' domicile. The trial judge concluded from the facts that the aircraft did not have a tax situs within the city. The court of civil appeals disagreed and held that no evidence supported the trial court's conclusion. 615 S.W.2d 316, 318-19 (Tex.Civ.App.). We hold that Davis failed to prove that the plane had acquired a tax situs outside of Austin on January 1, 1975, and affirm the judgment of the court of civil appeals.

Tom Davis, a resident of Austin, Texas, owned an aircraft and hangared it with Ragsdale Aviation at the Austin Municipal Airport within the City of Austin. During the fall of 1974, Davis notified Ragsdale that he would not need its facilities after December 30, 1974, and arranged to hangar his aircraft, beginning January 1, 1975, with Aviation Training Center, Inc., at Tim's Airpark located outside Austin's city limits and within the Pflugerville Independent School District. On December 30, 1974, Davis returned to Austin Municipal Airport in his aircraft from a trip to South Dakota and told his pilot to fly the plane to Addison Airpark near Dallas that same day for its regular inspection and maintenance. The plane remained at Addison until January 9, 1975, when it was flown to Tim's Airpark. Except for business and pleasure trips, Davis hangared the plane at Tim's until 1977. The parties stipulated that the plane left Austin Municipal Airport on December 30, 1974; it was flown that day to Addison Airpark near Dallas and remained there until January 9, 1975; it arrived at Tim's Airpark on January 9, 1975; and Davis hangared his plane at Tim's Airpark until February, 1977. There is no evidence that the plane had ever been at Tim's prior to January 9, 1975.

In declaring that all property shall be assessed for valuation and taxes paid in the county "where situated," the Texas Constitution adopted the common law rule for determining the tax situs of personalty. *Great Southern Life Insurance Co. v. City of Austin,* 112 Tex. 1, 10, 243 S.W. 778, 780 (1922); Tex.Const. art. VIII, § 11; see also Tex.Rev.Civ.Stat.Ann. art. 7153. Under early common law, situs of personal property was inconsequential. Tax units taxed all personalty at its owner's domicile, and the maxim mobilia sequuntur personam, "movables follow the person," described the standard for determining the tax situs of personal property. See *Pullman's Car Co. v. Pennsylvania,* 141 U.S. 18, 22-23, 11 S. Ct. 876, 877-78, 35 L. Ed. 613, 616 (1891). Courts recognized, however, that the mobilia rule merely expressed a legal fiction that should yield if inconvenience or injustice resulted from its application. See 2 T. Cooley, The Law of Taxation § 440 (4th ed. 1924).

Our Texas courts, while acknowledging the principle expressed by the mobilia rule, have created exceptions to the rule. In Texas, personal property is taxable at the domicile of its owner unless (1) tangible personal property has acquired an actual situs of its own apart from its owner's domicile, *Llano Cattle Co. v. Faught,* 69 Tex. 402, 404-05, 5 S.W. 494, 495 (1897), or (2) a statute directs otherwise, *Great Southern Life Insurance Co. v. City of Austin,* 112 Tex. 1, 10, 243 S.W. 778, 785 (1922); accord, *Nacogdoches Independent School District v. McKinney,* 504 S.W.2d 832, 837-38 (Tex.), modified on other grounds, 513 S.W.2d 5 (Tex.1974); *Ferris v. Kemble,* 75 Tex. 476, 480, 12 S.W. 689, 690 (1889).

A tax authority establishing its prima facie case in a tax delinquency suit enjoys a rebuttable presumption of law that the personalty in question has a tax situs within the taxing unit's jurisdiction. *Whaley v. Nocona Independent School District,* 339 S.W.2d 265, 267 (Tex.Civ.App.-Fort Worth 1960, writ ref'd). This presumption imposes upon the defendant taxpayer the burden of producing evidence sufficient to justify a finding that the tax situs of the property was outside the tax authority's jurisdiction. *State v. Whittenburg* 153 Tex. 205, 209, 265 S.W.2d 569, 572 (1954); see generally *Farley v. MM Cattle Co.,* 529 S.W.2d 751, 756 (Tex.1975); R. Ray. & W. Young, Jr., 1 Texas Practice, Law of Evidence §§ 41-55 (3d ed. 1980). In light, therefore, of the mobilia rule and its exceptions, a taxpayer who believes the tax situs of the tangible personalty in question lies outside the tax authority's jurisdiction must present evidence that (1) the taxed personalty has no acquired situs and its owner is not domiciled within the tax authority's boundaries, (2) the taxed personalty has an acquired situs outside the tax authority's boundaries, or (3) a statute directs that the personalty be taxed elsewhere. See, generally, *Empire Gas and Fuel Co. v. Muegge,* 135 Tex. 520, 529, 143 S.W.2d 763, 768 (1940). Otherwise, the rebuttable presumption of law arising from the tax unit's prima facie case becomes conclusive. *Adams v. Royse City,* 61 S.W.2d 853, 855 (Tex.Civ.App.-Dallas 1933, writ ref'd); see also *Keystone Operating Co. v. Runge Independent School District,* 558 S.W.2d 82, 84 (Tex.Civ.App.-San Antonio 1977, writ ref'd n. r. e.).

Under these principles, the taxing authority established its prima facie case as to every material fact necessary to establish the cause of action when it introduced a copy of the delinquent tax record, certified by the proper taxing authority to be true and correct with the amount stated thereon to be unpaid. *Alamo Barge Lines, Inc. v.*

City of Houston, 453 S.W.2d 132, 133-34 (Tex.1970); *Whaley v. Nocona Independent School District,* 339 S.W.2d 265, 267 (Tex.Civ.App.-Fort Worth 1960, writ ref'd); *Stone v. City of Dallas,* 244 S.W.2d 937, 943-44 (Tex.Civ.App.-Waco 1952, writ dism'd); Tex.Rev.Civ.Stat.Ann. arts. 7326, 7328.1. Davis, moreover, testified on cross-examination that he had not paid the 1975 tax.

In order, at that point, to avoid the force of the common law mobilia rule that would tax the plane at his residence, Davis had to prove the tax situs of the plane was governed by the acquired situs exception to the mobilia rule—the plane was situated outside the Austin taxing district. The force of the mobilia rule and its exceptions requires all taxable property to be somewhere on January 1. If the property has not acquired a situs where it is situated, then it is taxed at the owner's residence, which in this instance would be in Austin. In his defense, Davis testified that after receiving and paying 1974 personal property taxes on his plane, he resolved to remove it permanently from the city before January 1, 1975. Davis also testified that he intended to relocate the plane permanently at Tim's Airpark before the 1975 tax year, that he visited Tim's to look at their facilities and found them in good condition, that he told Mr. John Baker at Tim's the plane would be moved there as of January 1, and that Baker replied, "We will be looking for you." Davis produced bills, issued by Aviation Training Center at Tim's Airpark and dated January 27 and 31, 1975, that charged him for hangar rent and aviation fuel during January and February 1975.

Davis' intentions and preparatory acts are legally insufficient evidence to establish that his plane had acquired an actual physical situs at Tim's Airpark on January 1, 1975. In order for tangible personal property to acquire an actual situs outside the taxpayer's domicile, the property must be situated at a location with a "degree of permanency" that distinguishes it from personalty having a purely temporary or transitory basis within that jurisdiction. *Greyhound Lines, Inc. v. Board of Equalization,* 419 S.W.2d 345, 349 (Tex.1967). Courts addressing the acquired situs issue usually concentrate on determining whether the personalty in question enjoys that "degree of permanency" necessary to establish actual situs. See *Nacogdoches Independent School District v. McKinney,* 504 S.W.2d 832 (Tex.), modified on other grounds, 513 S.W.2d 5 (Tex.1974); *Greyhound Lines, Inc. v. Board of Equalization,* 419 S.W.2d 345 (Tex.1967); *A&M Consolidated Independent School District v. Fickey,* 542 S.W.2d 735 (Tex.Civ.App.-Waco 1976, writ ref'd n. r. e.); *Brown v. City of Dallas,* 508 S.W.2d 134 (Tex.Civ.App.-Dallas 1974, no writ); *City of Bryan v. Texas Services, Inc.,* 499 S.W.2d 750 (Tex.Civ.App.-Waco 1973, writ ref'd n. r. e.); *Lawson v. City of Groves,* 487 S.W.2d 439 (Tex.Civ.App.-Beaumont 1972, no writ); *Dennis v. City of Waco,* 445 S.W.2d 56 (Tex.Civ.App.-Waco 1969, no writ); *City of Dallas v. Overton,* 363 S.W.2d 821 (Tex.Civ.App.-Dallas 1962, writ ref'd n. r. e.); *State v. Crown Central Petroleum Corp.,* 242 S.W.2d 457 (Tex.Civ.App.-San Antonio 1951, writ ref'd). Without fail, however, the courts' inquiries assume that an owner must have physically brought the tangible personalty within the jurisdiction before it can acquire an actual situs there. See *State v. Crown Central Petroleum Corp.,* 242 S.W.2d 457, 461 (Tex.Civ.App.-San Antonio 1951, writ ref'd). This physical presence requirement stems from the dual policies that support the mobilia rule's actual situs exception: permitting an authority to tax non-domiciled owners' tangible personal property enjoying that authority's services, and protecting taxpayers by insuring that the property taxed by an acquired situs jurisdiction benefits from that jurisdiction's services. See *Nacogdoches Independent School District v. McKinney,* 504 S.W.2d 832, 837-38 (Tex.), modified on other grounds, 513 S.W.2d 5 (Tex.1974).

If, as in the present cause, an owner of property does not bring it within a tax authority's boundaries by January 1, and only its owner's intentions and preparatory acts link the personalty to that jurisdiction, we cannot say the tangible personal property will enjoy that authority's benefits or that its owner, consequently, will incur an obligation to pay the authority taxes. The actual situs exception to the mobilia rule, therefore, requires showings that the property in question was located within the situs jurisdiction on or before tax day and that the property was "permanently" located there. See *City of Dallas v. Overton*, 363 S.W.2d 821 (Tex.Civ.App.-Dallas 1962, writ ref'd n. r. e.). Here, Davis' evidence fails to address the issue of actual presence. It is only after the proponent of an acquired situs has shown the personalty actually has been within the jurisdiction that intent and preparatory acts become important considerations in determining whether personalty is "permanently" situated within that jurisdiction.

We hold, therefore, that Davis presented no evidence justifying the trial court's finding that the aircraft in question had acquired an actual situs at Tim's Airpark on January 1, 1975. Consequently, Davis failed to rebut the city's presumption of law that his plane enjoyed a tax situs in Austin. We accordingly agree with the court of civil appeals that the plane had a tax situs in the City of Austin and affirm that portion of its judgment.

* * *

As indicated in this case, you must show that your aircraft had a permanent presence elsewhere at the time the tax is assessed. This issue comes into play in many places where the state or local government has an annual "tax day" where all aircraft and other personal property in place on that day are subject to personal property tax. Some aircraft owners think they can beat the tax by simply flying to another base on tax day. That probably won't be an effective strategy. Just as illustrated in Case 17-3, you've got to show more than the fact that your aircraft was not in the taxing jurisdiction on tax day. You've got to demonstrate that it had a permanent home elsewhere as of that tax day. This can also work to your advantage. If you happen to land somewhere on tax day to get fuel or stop for weather, you should not be subject to a personal property tax on your aircraft.

Selected Bibliography

CCH Editorial Staff. *Aviation Law Reporter.* 4 vols. Chicago: CCH Incorporated, November 2001.

Holmes, Eric Mills. *Holmes's Appleman on Insurance,* Vol. 1, 2d ed. Minneapolis: West Publishing, 1996.

Krause, Charles F., and Krause, Kent C. *Aviation Tort and Regulatory Law,* 2d ed. Eagan, Minn.: West Group, 2002.

Kuchta, Joseph D., *Federal Aviation Decisions.* 5 vols. New York: Clark Boardman Callaghan, 1999.

Nelson, Deborah L., and Howicz, Jennifer L. *Williston on Sales,* Vol. 1, 5th ed. New York: Clark Boardman Callaghan, 2000.

Pike & Fischer, Inc. *Uniform Commercial Code Case Digest.* New York: Clark Boardman Callaghan, 1995.

Rollo, Vera Foster. *Aviation Insurance.* Lanham, Md.: Maryland Historical Press, 1986.

Rollo, Vera Foster. *Aviation Law: An Introduction.* Lanham, Md.: Maryland Historical Press, 1985.

Speiser, Stuart M., and Krause, Charles F. *Aviation Tort Law.* Eagan, Minn.: West Group, 2000.

U.S. Department of Transportation, Federal Aviation Administration. *Advisory Circular: Maintenance Records* (AC No. 43-9C), June 1998.

U.S. Department of Transportation, Federal Aviation Administration. *Advisory Circular: Airworthiness Directives* (AC No. 39-7C), November 1995.

U.S. Department of Transportation, Federal Aviation Administration. *Advisory Circular: Identification and Registration Marking* (AC No. 45-2A), April 1992.

U.S. Department of Transportation, Federal Aviation Administration. *Advisory Circular: Preventive Maintenance* (AC No. 43-12A), October 1983.

Yodice, John S. *Aviation Lawyer's Manual: Representing the Pilot in FAA Enforcement Actions.* Lanham, Md.: Maryland Historical Press, 1986.

Index

Pages shown in **boldface** have illustrations on them.

ABOUT THE AUTHOR

Raymond C. Speciale is a lawyer and certified public accountant in Frederick, Maryland. He has provided legal counsel to hundreds of aircraft owners and pilots since he began his work as an aviation attorney over 15 years ago.

Mr. Speciale is an active pilot and flight instructor (CFII). He has written several booklets and articles for the Aircraft Owners and Pilots Association (AOPA) related to aircraft ownership and taxation issues.

He teaches courses in law and accounting at Mount Saint Mary's College in Emmitsburg, Maryland, where he serves on the full-time faculty as an assistant professor. He has been associated with the Law Offices of Yodice Associates (AOPA Counsel) since 1988. Mr. Speciale is an AOPA Legal Services Plan panel attorney.

CD-ROM WARRANTY

This software is protected by both United States copyright law and international copyright treaty provision. You must treat this software just like a book. By saying "just like a book," McGraw-Hill means, for example, that this software may be used by any number of people and may be freely moved from one computer location to another, so long as there is no possibility of its being used at one location or on one computer while it also is being used at another. Just as a book cannot be read by two different people in two different places at the same time, neither can the software be used by two different people in two different places at the same time (unless, of course, McGraw-Hill's copyright is being violated).

LIMITED WARRANTY

Customers who have problems installing or running a McGraw-Hill CD should consult our online technical support site at http://books.mcgraw-hill.com/techsupport. McGraw-Hill takes great care to provide you with top-quality software, thoroughly checked to prevent virus infections. McGraw-Hill warrants the physical CD-ROM contained herein to be free of defects in materials and workmanship for a period of sixty days from the purchase date. If McGraw-Hill receives written notification within the warranty period of defects in materials or workmanship, and such notification is determined by McGraw-Hill to be correct, McGraw-Hill will replace the defective CD-ROM. Send requests to:

> McGraw-Hill
> Customer Services
> P.O. Box 545
> Blacklick, OH 43004-0545

The entire and exclusive liability and remedy for breach of this Limited Warranty shall be limited to replacement of a defective CD-ROM and shall not include or extend to any claim for or right to cover any other damages, including, but not limited to, loss of profit, data, or use of the software, or special, incidental, or consequential damages or other similar claims, even if McGraw-Hill has been specifically advised of the possibility of such damages. In no event will McGraw-Hill's liability for any damages to you or any other person ever exceed the lower of suggested list price or actual price paid for the license to use the software, regardless of any form of the claim.

McGRAW-HILL SPECIFICALLY DISCLAIMS ALL OTHER WARRANTIES, EXPRESS OR IMPLIED, INCLUDING, BUT NOT LIMITED TO, ANY IMPLIED WARRANTY OF MERCHANTABILITY OR FITNESS FOR A PARTICULAR PURPOSE.

Specifically, McGraw-Hill makes no representation or warranty that the software is fit for any particular purpose and any implied warranty of merchantability is limited to the sixty-day duration of the Limited Warranty covering the physical CD-ROM only (and not the software) and is otherwise expressly and specifically disclaimed.

This limited warranty gives you specific legal rights; you may have others which may vary from state to state. Some states do not allow the exclusion of incidental or consequential damages, or the limitation on how long an implied warranty lasts, so some of the above may not apply to you.